Graduate Texts in Mathematics **210**

Springer
New York
Berlin
Heidelberg
Barcelona
Hong Kong
London
Milan
Paris
Singapore
Tokyo

Graduate Texts in Mathematics

(continued after index)

Michael Rosen

Number Theory in
Function Fields

Springer

Michael Rosen
Department of Mathematics
Brown University
Providence, RI 02912-1917
USA
michael_rosen@brown.edu

Mathematics Subject Classification (2000): 11R29, 11R58, 14H05

Library of Congress Cataloging-in-Publication Data
Rosen, Michael I. (Michael Ira), 1938–
        Number theory in function fields / Michael Rosen.
          p. cm. — (Graduate texts in mathematics ; 210)
        Includes bibliographical references and index.
        ISBN 0-387-95335-3 (alk. paper)
          1. Number theory.   2. Finite fields (Algebra).   I. Title.   II. Series.
        QA241 .R675 2001
        512.7—dc21   2001042962

Printed on acid-free paper.

Production managed by Allan Abrams; manufacturing supervised by Jacqui Ashri.
Typeset by TeXniques, Inc., Boston, MA.
Printed and bound by R.R. Donnelley and Sons, Harrisonburg, VA.
Printed in the United States of America.

9 8 7 6 5 4 3 2 1

ISBN 0-387-95335-3   SPIN 10844406

Springer-Verlag   New York Berlin Heidelberg
A member of BertelsmannSpringer Science+Business Media GmbH

This book is dedicated to the memory
of my parents, Fred and Lee Rosen

Preface

Elementary number theory is concerned with the arithmetic properties of the ring of integers, $\mathbb{Z}$, and its field of fractions, the rational numbers, $\mathbb{Q}$. Early on in the development of the subject it was noticed that $\mathbb{Z}$ has many properties in common with $A = \mathbb{F}[T]$, the ring of polynomials over a finite field. Both rings are principal ideal domains, both have the property that the residue class ring of any non-zero ideal is finite, both rings have infinitely many prime elements, and both rings have finitely many units. Thus, one is led to suspect that many results which hold for $\mathbb{Z}$ have analogues of the ring A. This is indeed the case. The first four chapters of this book are devoted to illustrating this by presenting, for example, analogues of the little theorems of Fermat and Euler, Wilson's theorem, quadratic (and higher) reciprocity, the prime number theorem, and Dirichlet's theorem on primes in an arithmetic progression. All these results have been known for a long time, but it is hard to locate any exposition of them outside of the original papers.

Algebraic number theory arises from elementary number theory by considering finite algebraic extensions K of $\mathbb{Q}$, which are called algebraic number fields, and investigating properties of the ring of algebraic integers $O_K \subset K$, defined as the integral closure of $\mathbb{Z}$ in K. Similarly, we can consider $k = \mathbb{F}(T)$, the quotient field of A and finite algebraic extensions L of k. Fields of this type are called algebraic function fields. More precisely, an algebraic function fields with a finite constant field is called a global function field. A global function field is the true analogue of algebraic number field and much of this book will be concerned with investigating properties of global function fields. In Chapters 5 and 6, we will discuss function fields over arbitrary constant fields and review (sometimes in detail) the basic theory up to and including the fundamental theorem of Riemann-Roch and its corollaries. This will serve as the basis for many of the later developments.

It is important to point out that the theory of algebraic function fields is but another guise for the theory of algebraic curves. The point of view of this book will be very arithmetic. At every turn the emphasis will be on the analogy of algebaic function fields with algebraic number fields. Curves will be mentioned only in passing. However, the algebraic-geometric point of view is very powerful and we will freely borrow theorems about algebraic curves (and their Jacobian varieties) which, up to now, have no purely arithmetic proof. In some cases we will not give the proof, but will be content to state the result accurately and to draw from it the needed arithmetic consequences.

This book is aimed primarily at graduate students who have had a good introductory course in abstract algebra covering, in addition to Galois theory, commutative algebra as presented, for example, in the classic text of Atiyah and MacDonald. In the interest of presenting some advanced results in a relatively elementary text, we do not aspire to prove everything. However, we do prove most of the results that we present and hope to inspire the reader to search out the proofs of those important results whose proof we omit. In addition to graduate students, we hope that this material will be of interest to many others who know some algebraic number theory and/or algebraic geometry and are curious about what number theory in function field is all about. Although the presentation is not primarily directed toward people with an interest in algebraic coding theory, much of what is discussed can serve as useful background for those wishing to pursue the arithmetic side of this topic.

Now for a brief tour through the later chapters of the book.

Chapter 7 covers the background leading up to the statement and proof of the Riemann-Hurwitz theorem. As an application we discuss and prove the analogue of the ABC conjecture in the function field context. This important result has many consequences and we present a few applications to diophantine problems over function fields.

Chapter 8 gives the theory of constant field extensions, mostly under the assumption that the constant field is perfect. This is basic material which will be put to use repeatedly in later chapters.

Chapter 9 is primarily devoted to the theory of finite Galois extensions and the theory of Artin and Hecke L-functions. Two versions of the very important Tchebatorov density theorem are presented: one using Dirichlet density and the other using natural density. Toward the end of the chapter there is a sketch of global class field theory which enables one, in the abelian case, to identify Artin L-series with Hecke L-series.

Chapter 10 is devoted to the proof of a theorem of Bilharz (a studentof Hasse) which is the function field version of Artin's famous conjecture on primitive roots. This material, interesting in itself, illustrates the use of many of the results developed in the preceding chapters.

Chapter 11 discusses the behavior of the class group under constant field extensions. It is this circle of ideas which led Iwasawa to develop "Iwasawa theory," one of the most powerful tools of modern number theory.

Chapters 12 and 13 provide an introduction to the theory of Drinfeld modules. Chapter 12 presents the theory of the Carlitz module, which was developed by L. Carlitz in the 1930s. Drinfeld's papers, published in the 1970s, contain a vast generalization of Carlitz's work. Drinfeld's work was directed toward a proof of the Langlands' conjectures in function fields. Another consequence of the theory, worked out separately by Drinfeld and Hayes, is an explicit class field theory for global function fields. These chapters present the basic definitions and concepts, as well as the beginnings of the general theory.

Chapter 14 presents preliminary material on S-units, S-class groups, and the corresponding L-functions. This leads up to the statement and proof of a special case of the Brumer-Stark conjecture in the function field context. This is the content of Chapter 15. The Brumer-Stark conjecture in function fields is now known in full generality. There are two proofs — one due to Tate and Deligne, another due to Hayes. It is the author's hope that anyone who has read Chapters 14 and 15 will be inspired to go on to master one or both of the proofs of the general result.

Chapter 16 presents function field analogues of the famous class number formulas of Kummer for cyclotomic number fields together with variations on this theme. Once again, most of this material has been generalized considerably and the material in this chapter, which has its own interest, can also serve as the background for further study.

Finally, in Chapter 17 we discuss average value theorems in global fields. The material presented here generalizes work of Carlitz over the ring $A = \mathbb{F}[T]$. A novel feature is a function field analogue of the Wiener-Ikehara Tauberian theorem. The beginning of the chapter discusses average values of elementary number-theoretic functions. The last part of the chapter deals with average values for class numbers of hyperelliptic function fields.

In the effort to keep this book reasonably short, many topics which could have been included were left out. For example, chapters had been contemplated on automorphisms and the inverse Galois problem, the number of rational points with applications to algebraic coding theory, and the theory of character sums. Thought had been given to a more extensive discussion of Drinfeld modules and the subject of explicit class field theory in global fields. Also omitted is any discussion of the fascinating subject of transcendental numbers in the function field context (for an excellent survey see J. Yu [1]). Clearly, number theory in function fields is a vast subject. It is of interest for its own sake and because it has so often served as a stimulous to research in algebraic number theory and arithmetic geometry. We hope this book will arouse in the reader a desire to learn more and explore further.

I would like to thank my friends David Goss and David Hayes for their encouragement over the years and for their work which has been a constant source of delight and inspiration.

I also want to thank Allison Pacelli and Michael Reid who read several chapters and made valuable suggestions. I especially want to thank Amir Jafari and Hua-Chieh Li who read most of the book and did a thorough job spotting misprints and inaccuracies. For those that remain I accept full responsibility.

This book had its origins in a set of seven lectures I delivered at KAIST (Korean Advanced Institute of Science and Technology) in the summer of 1994. They were published in: "Lecture Notes of the Ninth KAIST Mathematics Workshop, Volume 1, 1994, Taejon, Korea." For this wonderful opportunity to bring my thoughts together on these topics I wish to thank both the Institute and my hosts, Professors S.H. Bae and J. Koo.

Years ago my friend Ken Ireland suggested the idea of writing a book together on the subject of arithmetic in function fields. His premature death in 1991 prevented this collaboration from ever taking place. This book would have been much better had we been able to do it together. His spirit and great love of mathematics still exert a deep influence over me. I hope something of this shows through on the pages that follow.

Finally, my thanks to Polly for being there when I became discouraged and for cheering me on.

December 30, 2000   Michael Rosen
                                                     Brown University

Contents

1
Polynomials over Finite Fields

In all that follows $\mathbb{F}$ will denote a finite field with q elements. The model for such a field is $\mathbb{Z}/p\mathbb{Z}$, where p is a prime number. This field has p elements. In general the number of elements in a finite field is a power of a prime, $q = p^f$. Of course, p is the characteristic of $\mathbb{F}$.

Let $A = \mathbb{F}[T]$, the polynomial ring over $\mathbb{F}$. A has many properties in common with the ring of integers $\mathbb{Z}$. Both are principal ideal domains, both have a finite unit group, and both have the property that every residue class ring modulo a non-zero ideal has finitely many elements. We will verify all this shortly. The result is that many of the number theoretic questions we ask about $\mathbb{Z}$ have their analogues for A. We will explore these in some detail.

Every element in A has the form $f(T) = \alpha_0 T^n + \alpha_1 T^{n-1} + \cdots + \alpha_n$. If $\alpha_0 \neq 0$ we say that f has degree n, notationally $\deg(f) = n$. In this case we set $\mathrm{sgn}(f) = \alpha_0$ and call this element of $\mathbb{F}^*$ the sign of f. Note the following very important properties of these functions. If f and g are non-zero polynomials we have

$$\deg(fg) = \deg(f) + \deg(g) \quad \text{and} \quad \mathrm{sgn}(fg) = \mathrm{sgn}(f)\mathrm{sgn}(g).$$

$$\deg(f + g) \leq \max(\deg(f), \deg(g)).$$

In the second line, equality holds if $\deg(f) \neq \deg(g)$.

If $\mathrm{sgn}(f) = 1$ we say that f is a monic polynomial. Monic polynomials play the role of positive integers. It is sometimes useful to define the sign of the zero polynomial to be 0 and its degree to be $-\infty$. The above properties of degree then remain true without restriction.

Proposition 1.1. *Let $f, g \in A$ with $g \neq 0$. Then there exist elements $q, r \in A$ such that $f = qg + r$ and r is either 0 or $\deg(r) < \deg(g)$. Moreover, q and r are uniquely determined by these conditions.*

Proof. Let $n = \deg(f)$, $m = \deg(g)$, $\alpha = \text{sgn}(f)$, $\beta = \text{sgn}(g)$. We give the proof by induction on $n = \deg(f)$. If $n < m$, set $q = 0$ and $r = f$. If $n \geq m$, we note that $f_1 = f - \alpha\beta^{-1}T^{n-m}g$ has smaller degree than f. By induction, there exist $q_1, r_1 \in A$ such that $f_1 = q_1 g + r_1$ with r_1 being either 0 or with degree less than $\deg(g)$. In this case, set $q = \alpha\beta^{-1}T^{n-m} + q_1$ and $r = r_1$ and we are done.

If $f = qg + r = q'g + r'$, then g divides $r - r'$ and by degree considerations we see $r = r'$. In this case, $qg = q'g$ so $q = q'$ and the uniqueness is established.

This proposition shows that A is a Euclidean domain and thus a principal ideal domain and a unique factorization domain. It also allows a quick proof of the finiteness of the residue class rings.

Proposition 1.2. *Suppose $g \in A$ and $g \neq 0$. Then A/gA is a finite ring with $q^{\deg(g)}$ elements.*

Proof. Let $m = \deg(g)$. By Proposition 1.1 one easily verifies that $\{r \in A \mid \deg(r) < m\}$ is a complete set of representatives for A/gA. Such elements look like

$$r = \alpha_0 T^{m-1} + \alpha_1 T^{m-2} + \cdots + \alpha_{m-1} \quad \text{with } \alpha_i \in \mathbb{F}.$$

Since the α_i vary independently through $\mathbb{F}$ there are q^m such polynomials and the result follows.

Definition. Let $g \in A$. If $g \neq 0$, set $|g| = q^{\deg(g)}$. If $g = 0$, set $|g| = 0$.

$|g|$ is a measure of the size of g. Note that if n is an ordinary integer, then its usual absolute value, $|n|$, is the number of elements in $\mathbb{Z}/n\mathbb{Z}$. Similarly, $|g|$ is the number of elements in A/gA. It is immediate that $|fg| = |f| |g|$ and $|f + g| \leq \max(|f|, |g|)$ with equality holding if $|f| \neq |g|$.

It is a simple matter to determine the group of units in A, A^*. If g is a unit, then there is an f such that $fg = 1$. Thus, $0 = \deg(1) = \deg(f) + \deg(g)$ and so $\deg(f) = \deg(g) = 0$. The only units are the non-zero constants and each such constant is a unit.

Proposition 1.3. *The group of units in A is $\mathbb{F}^*$. In particular, it is a finite cyclic group with $q - 1$ elements.*

Proof. The only thing left to prove is the cyclicity of $\mathbb{F}^*$. This follows from the very general fact that a finite subgroup of the multiplicative group of a field is cyclic.

In what follows we will see that the number $q - 1$ often occurs where the number 2 occurs in ordinary number theory. This stems from the fact that the order of $\mathbb{Z}^*$ is 2.

By definition, a non-constant polynomial $f \in A$ is irreducible if it cannot be written as a product of two polynomials, each of positive degree. Since every ideal in A is principal, we see that a polynomial is irreducible if and only if it is prime (for the definitions of divisibility, prime, irreducible, etc., see Ireland and Rosen [1]). These words will be used interchangeably. Every non-zero polynomial can be written uniquely as a non-zero constant times a monic polynomial. Thus, every ideal in A has a unique monic generator. This should be compared with the statement that evey non-zero ideal in $\mathbb{Z}$ has a unique positive generator. Finally, the unique factorization property in A can be sharpened to the following statement. Every $f \in A$, $f \neq 0$, can be written uniquely in the form

$$f = \alpha P_1^{e_1} P_2^{e_2} \ldots P_t^{e_t},$$

where $\alpha \in \mathbb{F}^*$, each P_i is a monic irreducible, $P_i \neq P_j$ for $i \neq j$, and each e_i is a non-negative integer.

The letter P will often be used for a monic irreducible polynomial in A. We use P instead of p since the latter letter is reserved for the characteristic of $\mathbb{F}$. This is a bit awkward, but it is compensated for by being less likely to lead to confusion.

The next order of business will be to investigate the structure of the rings A/fA and the unit groups $(A/fA)^*$. A valuable tool is the Chinese Remainder Theorem.

Proposition 1.4. *Let* $m_1, m_2, \ldots, m_t$ *be elements of A which are pairwise relatively prime. Let* $m = m_1 m_2 \ldots m_t$ *and* ϕ_i *be the natural homomorphism from A/mA to A/m_iA. Then, the map* $\phi : A/mA \to A/m_1A \oplus A/m_2A \oplus \cdots \oplus A/m_tA$ *given by*

$$\phi(a) = (\phi_1(a), \phi_2(a), \ldots, \phi_t(a))$$

is a ring isomorphism.

Proof. This is a standard result which holds in any principal ideal domain (properly formulated it holds in much greater generality).

Corollary. *The same map* ϕ *restricted to the units of A, A^*, gives rise to a group isomorphism*

$$(A/mA)^* \simeq (A/m_1A)^* \times (A/m_2A)^* \times \cdots \times (A/m_tA)^*.$$

Proof. This is a standard exercise. See Ireland and Rosen [1], Proposition 3.4.1.

Now, let $f \in A$ be non-zero and not a unit and suppose that $f = \alpha P_1^{e_1} P_2^{e_2} \ldots P_t^{e_t}$ is its prime decomposition. From the above considerations we have

$$(A/fA)^* \simeq (A/P_1^{e_1}A)^* \times (A/P_2^{e_2}A)^* \times \cdots \times (A/P_t^{e_t}A)^*.$$

This isomorphism reduces our task to that of determining the structure of the groups $(A/P^e A)^*$ where P is an irreducible polynomial and e is a positive integer. When $e = 1$ the situation is very similar to that is $\mathbb{Z}$.

Proposition 1.5. *Let $P \in A$ be an irreducible polynomial. Then, $(A/PA)^*$ is a cyclic group with $|P| - 1$ elements.*

Proof. Since A is a principal ideal domain, PA is a maximal ideal and so A/PA is a field. A finite subgroup of the multiplicative group of a field is cyclic. Thus $(A/PA)^*$ is cyclic. That the order of this group is $|P| - 1$ is immediate.

We now consider the situation when $e > 1$. Here we encounter something which is quite different in A from the situation in $\mathbb{Z}$. If p is an odd prime number in $\mathbb{Z}$ then it is a standard result that $(\mathbb{Z}/p^e\mathbb{Z})^*$ is cyclic for all positive integers e. If $p = 2$ and $e \geq 3$ then $(\mathbb{Z}/2^e\mathbb{Z})^*$ is the direct product of a cyclic group of order 2 and a cyclic group of order 2^{e-2}. The situation is very different in A.

Proposition 1.6. *Let $P \in A$ be an irreducible polynomial and e a positive integer. The order of $(A/P^e A)^*$ is $|P|^{e-1}(|P| - 1)$. Let $(A/P^e A)^{(1)}$ be the kernel of the natural map from $(A/P^e A)^*$ to $(A/PA)^*$. It is a p-group of order $|P|^{e-1}$. As e tends to infinity, the minimal number of generators of $(A/P^e A)^{(1)}$ tends to infinity.*

Proof. The ring $A/P^e A$ has only one maximal ideal $PA/P^e A$ which has $|P|^{e-1}$ elements. Thus, $(A/P^e A)^* = A/P^e A - PA/P^e A$ has $|P|^e - |P|^{e-1} = |P|^{e-1}(|P| - 1)$ number of elements. Since $(A/P^e A)^* \to (A/PA)^*$ is onto, and the latter group has $|P| - 1$ elements the assertion about the size of $(A/P^e A)^{(1)}$ follows. It remains to prove the assertion about the minimal number of generators.

It is instructive to first consider the case $e = 2$. Every element in $(A/P^2 A)^{(1)}$ can be represented by a polynomial of the form $a = 1 + bP$. Since we are working in characteristic p we have $a^p = 1 + b^p P^p \equiv 1 \pmod{P^2}$. So, we have a group of order $|P|$ with exponent p. If $q = p^f$ it follows that $(A/P^2 A)^{(1)}$ is a direct sum of $f \deg(P)$ number of copies of $\mathbb{Z}/p\mathbb{Z}$. This is a cyclic group only under the very restrictive conditions that $q = p$ and $\deg(P) = 1$.

To deal with the general case, suppose that s is the smallest integer such that $p^s \geq e$. Since $(1 + bP)^{p^s} = 1 + (bP)^{p^s} \equiv 1 \pmod{P^e}$ we have that raising to the p^s-power annihilates $G = (A/P^e A)^{(1)}$. Let d be the minimal number of generators of this group. It follows that there is an onto map from $(\mathbb{Z}/p^s\mathbb{Z})^d$ onto G. Thus, $p^{ds} \geq p^{f \deg(P)(e-1)}$ and so

$$d \geq \frac{f \deg(P)(e - 1)}{s}.$$

Since s is the smallest integer bigger than or equal to $\log_p(e)$ it is clear that $d \to \infty$ as $e \to \infty$.

It is possible to do a much closer analysis of the structure of these groups, but it is not necessary to do so now. The fact that these groups get very complicated does cause problems in the more advanced parts of the theory.

We have developed more than enough material to enable us to give interesting analogues of the Euler ϕ-function and the little theorems of Euler and Fermat.

To begin with, let $f \in A$ be a non-zero polynomial. Define $\Phi(f)$ to be the number of elements in the group $(A/fA)^*$. We can give another characterization of this number which makes the relation to the Euler ϕ-function even more evident. We have seen that $\{r \in A \mid \deg(r) < \deg(f)\}$ is a set of representatives for A/fA. Such an r represents a unit in A/fA if and only if it is relatively prime to f. Thus $\Phi(f)$ is the number of non-zero polynomials of degree less than $\deg(f)$ and relatively prime to f.

Proposition 1.7.

$$\Phi(f) = |f| \prod_{P|f} (1 - \frac{1}{|P|}).$$

Proof. Let $f = \alpha P_1^{e_1} P_2^{e_2} \ldots P_t^{e_t}$ be the prime decomposition of f. By the corollary to Propositions 1.4 and by Proposition 1.6, we see that

$$\Phi(f) = \prod_{i=1}^{t} \Phi(P_i^{e_i}) = \prod_{i=1}^{t} (|P_i|^{e_i} - |P_i|^{e_i - 1}),$$

from which the result follows immediately.

The similarity of the formula in this proposition to the classical formula for $\phi(n)$ is striking.

Proposition 1.8. *If* $f \in A$, $f \neq 0$, *and* $a \in A$ *is relatively prime to* f, *i.e.*, $(a, f) = 1$, *then*

$$a^{\Phi(f)} \equiv 1 \pmod{f}.$$

Proof. The group $(A/fA)^*$ has $\Phi(f)$ elements. The coset of a modulo f, $\bar{a}$, lies in this group. Thus, $\bar{a}^{\Phi(f)} = \bar{1}$ and this is equivalent to the congruence in the proposition.

Corollary. *Let* $P \in A$ *be irreducible and* $a \in A$ *be a polynomial not divisible by* P. *Then,*

$$a^{|P|-1} \equiv 1 \pmod{P}.$$

Proof. Since P is irreducible, it is relatively prime to a if and only if it does not divide a. The corollary follows from the proposition and the fact that for an irreducible P, $\Phi(P) = |P| - 1$ (Proposition 1.5).

It is clear that Proposition 1.8 and its corollary are direct analogues of Euler's little theorem and Fermat's little theorem. They play the same very important role in this context as they do in elementary number theory. By way of illustration we proceed to the analogue of Wilson's theorem. Recall that this states that $(p-1)! \equiv -1 \pmod{p}$ where p is a prime number.

Proposition 1.9. *Let $P \in A$ be irreducible of degree d. Suppose X is an indeterminate. Then,*

$$X^{|P|-1} - 1 \equiv \prod_{0 \leq \deg(f) < d} (X - f) \pmod{P}.$$

Proof. Recall that $\{f \in A \mid \deg(f) < d\}$ is a set of representatives for the cosets of A/PA. If we throw out $f = 0$ we get a set of representatives for $(A/PA)^*$. We find

$$X^{|P|-1} - \bar{1} = \prod_{0 \leq \deg(f) < d} (X - \bar{f}),$$

where the bars denote cosets modulo P. This follows from the corollary to Proposition 1.8 since both sides of the equation are monic polynomials in X with the same set of roots in the field A/PA. Since there are $|P| - 1$ roots and the difference of the two sides has degree less than $|P| - 1$, the difference of the two sides must be 0. The congruence in the Proposition is equivalent to this assertion.

Corollary 1. *Let d divide $|P| - 1$. The congruence $X^d \equiv 1 \pmod{P}$ has exactly d solutions. Equivalently, the equation $X^d = \bar{1}$ has exactly d solutions in $(A/PA)^*$.*

Proof. We prove the second assertion. Since $d \mid |P| - 1$ it follows that $X^d - 1$ divides $X^{|P|-1} - 1$. By the proposition, the latter polynomial splits as a product of distinct linear factors. Thus so does the former polynomial. This establishes the result.

Corollary 2. *With the same notation,*

$$\prod_{0 \leq \deg(f) < \deg P} f \equiv -1 \pmod{P}.$$

Proof. Just set $X = 0$ in the proposition. If the characteristic of $\mathbb{F}$ is odd $|P| - 1$ is even and the result follows. If the characteristic is 2 then the result also follows since in characteristic 2 we have $-1 = 1$.

The above corollary is the polynomial version of Wilson's theorem. It's interesting to note that the left-hand side of the congruence only depends on the degree of P and not on P itself.

As a final topic in this chapter we give some of the theory of d-th power residues. This will be of importance in Chapter 3 when we discuss quadratic reciprocity and more general reciprocity laws for A.

If $f \in A$ is of positive degree and $a \in A$ is relatively prime to f, we say that a is a d-th power residue modulo f if the equation $x^d \equiv a \pmod{f}$ is solvable in A. Equivalently, $\bar{a}$ is a d-th power in $(A/fA)^*$.

Suppose $f = \alpha P_1^{e_1} P_2^{e_2} \ldots P_t^{e_t}$ is the prime decomposition of f. Then it is easy to check that a is a d-th power residue modulo f if and only if a is a d-th power residue modulo $P_i^{e_i}$ for all i between 1 and t. This reduces the problem to the case where the modulus is a prime power.

Proposition 1.10. *Let P be irreducible and $a \in A$ not divisible by P. Assume d divides $|P| - 1$. The congruence $X^d \equiv a \pmod{P^e}$ is solvable if and only if*

$$a^{\frac{|P|-1}{d}} \equiv 1 \pmod{P}.$$

There are $\frac{\Phi(P^e)}{d}$ d-th power residues modulo P^e.

Proof. Assume to begin with that $e = 1$.

If $b^d \equiv a \pmod{P}$, then $a^{\frac{|P|-1}{d}} \equiv b^{|P|-1} \equiv 1 \pmod{P}$ by the corollary to Proposition 1.8. This shows the condition is necessary. To show it is sufficient recall that by Corollary 1 to Proposition 1.9 all the d-th roots of unity are in the field A/PA. Consider the homomorphism from $(A/PA)^*$ to itself given by raising to the d-th power. It's kernel has order d and its image is the d-th powers. Thus, there are precisely $\frac{|P|-1}{d}$ d-th powers in $(A/PA)^*$. We have seen that they all satisfy $X^{\frac{|P|-1}{d}} - 1 = 0$. Thus, they are precisely the roots of this equation. This proves all assertions in the case $e = 1$.

To deal with the remaining cases, we employ a little group theory. The natural map (i.e., reduction modulo P) is a homomorphism from $(A/P^e A)^*$ onto $(A/PA)^*$ and the kernel is a p-group as follows from Proposition 1.6. Since the order of $(A/PA)^*$ is $|P| - 1$ which is prime to p it follows that $(A/P^e A)^*$ is the direct product of a p-group and a copy of $(A/PA)^*$. Since $(d, p) = 1$, raising to the d-th power in an abelian p-group is an automorphism. Thus, $a \in A$ is a d-th power modulo P^e if and only if it is a d-th power modulo P. The latter has been shown to hold if and only if $a^{\frac{|P|-1}{d}} \equiv 1 \pmod{P}$. Now consider the homomorphism from $(A/P^e A)^*$ to itself given by raising to the d-th power. It easily follows from what has been said that the kernel has d elements and the image is the subgroup of d-th powers. It follows that the latter group has order $\frac{\Phi(P^e)}{d}$. This concludes the proof.

Exercises

1. If $m \in A = \mathbb{F}[T]$, and $\deg(m) > 0$, show that $q - 1 \mid \Phi(m)$.

2. If $q = p$ is a prime number and $P \in A$ is an irreducible, show $(\mathbb{F}[T]/P^2 A)^*$ is cyclic if and only if $\deg P = 1$.

3. Suppose $m \in A$ is monic and that $m = m_1 m_2$ is a factorization into two monics which are relatively prime and of positive degree. Show $(A/mA)^*$ is not cyclic except possibly in the case $q = 2$ and m_1 and m_2 have relatively prime degrees.

4. Assume $q \neq 2$. Determine all m for which $(A/mA)^*$ is cyclic (see the proof of Proposition 1.6).

5. Suppose $d \mid q - 1$. Show $x^d \equiv -1 \pmod{P}$ is solvable if and only if $(-1)^{\frac{q-1}{d} \deg P} = 1$.

6. Show $\prod_{\alpha \in \mathbb{F}^*} \alpha = -1$.

7. Let $P \in A$ be a monic irreducible. Show

$$\prod_{\substack{\deg f < d \\ f \text{ monic}}} f \equiv \pm 1 \pmod{P},$$

where $d = \deg P$. Determine the sign on the right-hand side of this congruence.

8. For an integer $m \geq 1$ define $[m] = T^{q^m} - T$. Show that $[m]$ is the product of all monic irreducible polynomials $P(T)$ such that $\deg P(T)$ divides m.

9. Working in the polynomial ring $\mathbb{F}[u_0, u_1, \ldots, u_n]$, define $D(u_0, u_1, \ldots, u_n) = \det |u_i^{q^j}|$, where $i, j = 0, 1, \ldots, n$. This is called the Moore determinant. Show

$$D(u_0, u_1, \ldots, u_n) = \prod_{i=0}^{n} \prod_{c_{i-1} \in \mathbb{F}} \cdots \prod_{c_0 \in \mathbb{F}} (u_i + c_{i-1} u_{i-1} + \cdots + c_0 u_0).$$

Hint: Show each factor on the right divides the determinant and then count degrees.

10. Define $F_j = \prod_{i=0}^{j-1} (T^{q^j} - T^{q^i}) = \prod_{i=0}^{j-1} [j - i]^{q^i}$. Show that

$$D(1, T, T^2, \ldots, T^n) = \prod_{j=0}^{n} F_j.$$

Hint: Use the fact that $D(1, T, T^2, \ldots, T^n)$ can be viewed as a Vandermonde determinant.

11. Show that F_j is the product of all monic polynomials in A of degee j.

12. Define $L_j = \prod_{i=1}^{j} (T^{q^i} - T) = \prod_{i=1}^{j} [i]$. Use Exercise 8 to prove that L_j is the least common multiple of all monics of degree j.

13. Show
$$\prod_{\deg f < d} (u + f) = \frac{D(1, T, T^2, \ldots, T^{d-1}, u)}{D(1, T, T^2, \ldots, T^{d-1})} .$$

14. Deduce from Exercise 13 that
$$\prod_{\deg f < d} (u + f) = \sum_{j=0}^{d} (-1)^{d-j} \frac{F_d}{F_j L_{d-j}^{q^j}} u^{q^j} .$$

15. Show that the product of all the non-zero polynomials of degree less than d is equal to $(-1)^d F_d / L_d$.

16. Prove that
$$u \prod_{\deg f < d}' \left(1 - \frac{u}{f}\right) = \sum_{j=0}^{d} (-1)^j \frac{L_d}{F_j L_{d-j}^{q^j}} u^{q^j} .$$

In the product the term corresponding to $f = 0$ is omitted.

2

Primes, Arithmetic Functions, and the Zeta Function

In this chapter we will discuss properties of primes and prime decomposition in the ring $A = \mathbb{F}[T]$. Much of this discussion will be facilitated by the use of the zeta function associated to A. This zeta function is an analogue of the classical zeta function which was first introduced by L. Euler and whose study was immeasurably enriched by the contributions of B. Riemann. In the case of polynomial rings the zeta function is a much simpler object and its use rapidly leads to a sharp version of the prime number theorem for polynomials without the need for any complicated analytic investigations. Later we will see that this situation is a bit deceptive. When we investigate arithmetic in more general function fields than $\mathbb{F}(T)$, the corresponding zeta function will turn out to be a much more subtle invariant.

Definition. The zeta function of A, denoted $\zeta_A(s)$, is defined by the infinite series

$$\zeta_A(s) = \sum_{\substack{f \in A \\ f \text{ monic}}} \frac{1}{|f|^s}.$$

There are exactly q^d monic polynomials of degree d in A, so one has

$$\sum_{\deg(f) \leq d} |f|^{-s} = 1 + \frac{q}{q^s} + \frac{q^2}{q^{2s}} + \cdots + \frac{q^d}{q^{ds}},$$

and consequently

$$\zeta_A(s) = \frac{1}{1 - q^{1-s}} \tag{1}$$

for all complex numbers s with $\Re(s) > 1$. In the classical case of the Riemann zeta function, $\zeta(s) = \sum_{n=1}^{\infty} n^{-s}$, it is easy to show the defining series converges for $\Re(s) > 1$, but it is more difficult to provide an analytic continuation. Riemann showed that it can be analytically continued to a meromorphic function on the whole complex plane with the only pole being a simple pole of residue 1 at $s = 1$. Moreover, if $\Gamma(s)$ is the classical gamma function and $\xi(s) = \pi^{-\frac{s}{2}}\Gamma(\frac{s}{2})\zeta(s)$, Riemann showed the functional equation $\xi(1 - s) = \xi(s)$. What can be said about $\zeta_A(s)$?

By Equation 1 above, we see clearly that $\zeta_A(s)$, which is initially defined for $\Re(s) > 1$, can be continued to a meromorphic function on the whole complex plane with a simple pole at $s = 1$. A simple computation shows that the residue at $s = 1$ is $\frac{1}{\log(q)}$. Now define $\xi_A(s) = q^{-s}(1 - q^{-s})^{-1}\zeta_A(s)$. It is easy to check that $\xi_A(1-s) = \xi_A(s)$ so that a functional equation holds in this situation as well. As opposed to case of the classical zeta-function, the proofs are very easy for $\zeta_A(s)$. Later we will consider generalizations of $\zeta_A(s)$ in the context of function fields over finite fields. Similar statements will hold, but the proofs will be more difficult and will be based on the Riemann-Roch theorem for algebraic curves.

Euler noted that the unique decomposition of integers into products of primes leads to the following identity for the Riemann zeta-function:

$$\zeta(s) = \prod_{\substack{p \text{ prime} \\ p>0}} \left(1 - \frac{1}{p^s}\right)^{-1}.$$

This is valid for $\Re(s) > 1$. The exact same reasoning (which we won't repeat here) leads to the following identity:

$$\zeta_A(s) = \prod_{\substack{P \text{ irreducible} \\ P \text{ monic}}} \left(1 - \frac{1}{|P|^s}\right)^{-1}. \tag{2}$$

This is also valid for all $\Re(s) > 1$.

One can immediately put Equation 2 to use. Suppose there were only finitely many irreducible polynomials in A. The right-hand side of the equation would then be defined at $s = 1$ and even have a non-zero value there. On the other hand, the left hand side has a pole at $s = 1$. This shows there are infinitely many irreducibles in A. One doesn't need the zeta-function to show this. Euclid's proof that there are infinitely many prime integers works equally well in polynomial rings. Suppose S is a finite set of irreducibles. Multiply the elements of S together and add one. The result is a polynomial of positive degree not divisible by any element of S. Thus, S cannot contain all irreducible polynomials. It follows, once more, that there are infinitely many irreducibles.

Let x be a real number and $\pi(x)$ be the number of positive prime numbers less than or equal to x. The classical prime number theorem states that

$\pi(x)$ is asymptotic to $x/\log(x)$. Let d be a positive integer and $x = q^d$. We will show that the number of monic irreducibles P such that $|P| = x$ is asymptotic to $x/\log_q(x)$ which is clearly in the spirit of the classical result.

Define a_d to be the number of monic irreducibles of degree d. Then, from Equation 2 we find

$$\zeta_A(s) = \prod_{d=1}^{\infty}(1 - q^{-ds})^{-a_d}.$$

If we recall that $\zeta_A(s) = 1/(1 - q^{1-s})$ and substitute $u = q^{-s}$ (note that $|u| < 1$ if and only if $\Re(s) > 1$) we obtain the identity

$$\frac{1}{1 - qu} = \prod_{d=1}^{\infty}(1 - u^d)^{-a_d}.$$

Taking the logarithmic derivative of both sides and multiplying the result by u yields

$$\frac{qu}{1 - qu} = \sum_{d=1}^{\infty}\frac{da_d\, u^d}{1 - u^d}.$$

Finally, expand both sides into power series using the geometric series and compare coefficients of u^n. The result is the beautiful formula,

Proposition 2.1.

$$\sum_{d|n} da_d = q^n.$$

This formula is often attributed to Richard Dedekind. It is interesting to note that it appears, with essentially the above proof, in a manuscript of C.F. Gauss (unpublished in his lifetime), "Die Lehre von den Resten." See Gauss [1], pages 608–611.

Corollary

$$a_n = \frac{1}{n}\sum_{d|n}\mu(d)q^{\frac{n}{d}}. \tag{3}$$

Proof. This formula follows by applying the Möbius inversion formula to the formula given in the proposition.

The formula in the above proposition can also be proven by means of the algebraic theory of finite fields. In fact, most books on abstract algebra contain the formula and the purely algebraic proof. The zeta-function approach has the advantage that the same method can be used to prove many other things as we shall see in this and later chapters.

The next task is to write a_n in a way which makes it easy to see how big it is. In Equation 3 the highest power of q that occurs is q^n and the next highest power that may occur is $q^{\frac{n}{2}}$ (this occurs if and only if $2|n$. All the other terms have the form $\pm q^m$ where $m \le \frac{n}{3}$. The total number of terms is

$\sum_{d|n} |\mu(d)|$, which is easily seen to be 2^t, where t is the number of distinct prime divisors of n. Let $p_1, p_2, \ldots, p_t$ be the distinct primes dividing n. Then, $2^t \le p_1 p_2 \ldots p_t \le n$. Thus, we have the following estimate:

$$\left| a_n - \frac{q^n}{n} \right| \le \frac{q^{\frac{n}{2}}}{n} + q^{\frac{n}{3}}.$$

Using the standard big O notation, we have proved the following theorem.

Theorem 2.2. (The prime number theorem for polynomials) *Let a_n denote the number of monic irreducible polynomials in $A = \mathbb{F}[T]$ of degree n. Then,*

$$a_n = \frac{q^n}{n} + O\left(\frac{q^{\frac{n}{2}}}{n} \right).$$

Note that if we set $x = q^n$ the right-hand side of this equation is $x/\log_q(x) + O(\sqrt{x}/\log_q(x))$ which looks like the conjectured precise form of the classical prime number theorem. This is still not proven. It depends on the truth of the Riemann hypothesis (which will be discussed later).

We now show how to use the zeta function for other counting problems. What is the number of square-free monics of degree n? Let this number be b_n. Consider the product

$$\prod_P (1 + \frac{1}{|P|^s}) = \sum \frac{\delta(f)}{|f|^s}. \tag{4}$$

As usual, the product is over all monic irreducibles P and the sum is over all monics f. We will maintain this convention unless otherwise stated. The function $\delta(f)$ is 1 when f is square-free, and 0 otherwise. This is an easy consequence of unique factorization in A and the definition of square-free. Making the substitution $u = q^{-s}$ once again, the right-hand side of Equation 4 becomes $\sum_{n=0}^{\infty} b_n u^n$. Consider the identity $1 + w = (1 - w^2)/(1 - w)$. If we substitute $w = |P|^{-s}$ and then take the product over all monic irreducibles P, we see that the left-hand side of Equation 4 is equal to $\zeta_A(s)/\zeta_A(2s) = (1 - q^{1-2s})/(1 - q^{1-s})$. Putting everything in terms of u leads to the identity

$$\frac{1 - qu^2}{1 - qu} = \sum_{n=0}^{\infty} b_n u^n.$$

Finally, expand the left-hand side in a geometric series and compare the coefficients of u^n on both sides. We have proven—

Proposition 2.3. *Let b_n be the number of square-free monics in A of degree n. Then $b_1 = q$ and for $n > 1$, $b_n = q^n(1 - q^{-1})$.*

It is amusing to compare this result with what is known to be true in $\mathbb{Z}$. If B_n is the number of positive square-free integers less than or equal to n, then $\lim_{n \to \infty} B_n/n = 6/\pi^2$. In less precise language, the probability that a positive integer is square-free is $6/\pi^2$. The probablity that a monic polynomial of degree n is square-free is b_n/q^n, and this equals $(1 - q^{-1})$ for $n > 1$. Thus the probabilty that a monic polynomial in A is square-free is $(1 - q^{-1})$. Now, $6/\pi^2 = 1/\zeta(2)$, so it is interesting to note that $(1 - q^{-1}) = 1/\zeta_A(2)$. This is, of course, no accident and one can give good heuristic reasons why this should occur. The interested reader may want to find these reasons and to investigate the probablity that a polynomial be cube-free, fourth-power-free, etc.

Our next goal is to introduce analogues of some well-known number-theoretic functions and to discuss their properties. We have already introduced $\Phi(f)$. Let $\mu(f)$ be 0 if f is not square-free, and $(-1)^t$ if f is a constant times a product of t distinct monic irreducibles. This is the polynomial version of the Möbius function. Let $d(f)$ be the number of monic divisors of f and $\sigma(f) = \sum_{g|f} |g|$ where the sum is over all monic divisors of f.

These functions, like their classical counterparts, have the property of being multiplicative. More precisely, a complex valued function λ on $A - \{0\}$ is called multiplicative if $\lambda(fg) = \lambda(f)\lambda(g)$ whenever f and g are relatively prime. We assume λ is 1 on $\mathbb{F}^*$. Let

$$f = \alpha P_1^{e_1} P_2^{e_2} \ldots P_t^{e_t}$$

be the prime decomposition of f. If λ is multiplicative,

$$\lambda(f) = \lambda(P_1^{e_1})\lambda(P_2^{e_2}) \ldots \lambda(P_t^{e_t}).$$

Thus, a multiplicative function is completely determined by its values on prime powers. Using multiplicativity, one can derive the following formulas for these functions.

Proposition 2.4. *Let the prime decomposition of f be given as above. Then,*

$$\Phi(f) = |f| \prod_{P|f}(1 - |P|^{-1}),$$
$$d(f) = (e_1 + 1)(e_2 + 1) \ldots (e_t + 1).$$
$$\sigma(f) = \frac{|P_1|^{e_1+1} - 1}{|P_1| - 1} \cdot \frac{|P_2|^{e_2+1} - 1}{|P_2| - 1} \ldots \frac{|P_t|^{e_t+1} - 1}{|P_t| - 1}.$$

Proof. The formula for $\Phi(n)$ has already been given in Proposition 1.7.

If P is a monic irreducible, the only monic divisors of P^e are $1, P$, $P^2, \ldots, P^e$ so $d(P^e) = e + 1$ and the second formula follows.

By the above paragraph, $\sigma(P^e) = 1 + |P| + |P|^2 + \ldots |P|^e = (|P|^{e+1} - 1/(|P| - 1)$, and the formula for $\sigma(f)$ also follows.

As a final topic in this chapter we shall introduce the notion of the average values in the context of polynomials. Suppose $h(x)$ is a complex-valued function on $\mathbb{N}$, the set of positive integers. Suppose the following limit exists

$$\lim_{n \to \infty} \frac{1}{n} \sum_{k=1}^{n} h(n) = \alpha.$$

We then define α to be the average value of the function h. For example, suppose $h(n) = 1$ if n is square-free and 0 otherwise. Then, as noted above, the average value of h is known to be $6/\pi^2$. The sum $\sum_{k=1}^{n} h(k)$ sometimes grows too fast for the average value to exist. Often though, one can show the growth is dominated by a simple function of n. An example of this is the Euler ϕ-function. One can show

$$\sum_{k=1}^{n} \phi(k) = \frac{3}{\pi^2} n^2 + O(n \log(n)).$$

For this and other results of a similar nature, see Chapter VIII of the classic book by G.H. Hardy and E.M. Wright, Hardy and Wright [1]. Another good reference for this material is Chapter 3 of Apostol [1].

In the ring A the analogue of the positive integers is the set of monic polynomials. Let $h(x)$ be a function on the set of monic polynomials. For $n > 0$ we define

$$\text{Ave}_n(h) = \frac{1}{q^n} \sum_{\substack{f \text{ monic} \\ \deg(f) = n}} h(f).$$

This is clearly the average value of h on the set of monic polynomials of degree n. We define the average value of h to be $\lim_{n \to \infty} \text{Ave}_n(h)$ provided this limit exists. This is the natural way in which average values arise in the context of polynomials. It is an exercise to show that if the average value exists in the sense just given, then it is also equal to the following limit:

$$\lim_{n \to \infty} \frac{1}{1 + q + q^2 \cdots + q^n} \sum_{\substack{f \text{ monic} \\ \deg(f) \leq n}} h(f).$$

As we pointed out above, this limit does not always exist. However, even when it doesn't exist, one can speak of the average rate of growth of $h(f)$. Define $H(n)$ to equal the sum of $h(f)$ over all monic polynomials of degree n. As we will see, the function $H(n)$ sometimes behaves in a quite regular manner even though the values $h(f)$ vary erratically.

Instead of approaching these problems directly we use the method of Carlitz which uses Dirichlet series. Given a function h as above, we define the associated Dirichlet series to be

$$D_h(s) = \sum_{f \text{ monic}} \frac{h(f)}{|f|^s} = \sum_{n=0}^{\infty} \frac{H(n)}{q^{ns}}. \tag{5}$$

In what follows, we will work in a formal manner with these series. If one wants to worry about convergence, it is useful to remark that if $|h(f)| =$

$O(|f|^\beta)$, then $D_h(s)$ converges for $\Re(s) > 1 + \beta$. The proof just uses the comparison test and the fact that $\zeta_A(s)$ converges for $\Re(s) > 1$.

The right-hand side of 5 is simply $\sum_{n=0}^{\infty} H(n)u^n$, so the Dirichlet series in s becomes a power series in u whose coefficients are the averages $H(n)$. To see how this is useful, recall the function $d(f)$ which is the number of monic divisors of f. Let $D(n)$ be the sum of $d(f)$ over all monics of degree n (hopefully, this notation will not cause too much confusion). Then,

Proposition 2.5. $D_d(s) = \zeta_A(s)^2 = (1 - qu)^{-2}$. Consequently, $D(n) = (n+1)q^n$.

Proof.

$$\zeta_A(s)^2 = \Big(\sum_h \frac{1}{|h|^s}\Big)\Big(\sum_g \frac{1}{|g|^s}\Big) =$$

$$\sum_f \Big(\sum_{\substack{h,g \\ hg=f}} 1\Big)\frac{1}{|f|^s} = \sum_f \frac{d(f)}{|f|^s} = D_d(s) \ .$$

This proves the first assertion. To prove the second assertion, notice

$$D_d(s) = \sum_{n=0}^{\infty} D(n)u^n = (1 - qu)^{-2} \ .$$

It is easily seen that $(1 - qu)^{-2} = \sum_{n=0}^{\infty}(n+1)q^n u^n$. Thus, the second assertion follows by comparing the coefficients of u^n on both sides of this identity.

A few remarks are in order. Notice that $\mathrm{Ave}_n(d) = n+1$ so the average value of $d(f)$ in the way we have defined it doesn't exist. On average, the number of divisors of f grows with the degree. If we set $x = q^n$ then our result reads $D(n) = x \log_q(x) + x$ which resembles closely the analogous result for the integers $\sum_{k=1}^n d(k) = x \log(x) + (2\gamma - 1)x + O(\sqrt{x})$ (here $\gamma \approx .577216$ is Euler's constant). This formula is due to Dirichlet. It is a famous problem in elementary number theory to find the best possible error term. In the polynomial case, there is no error term! This is because of the very simple nature of the zeta function $\zeta_A(s)$. Similar sums in the general function field context lead to more difficult problems. We shall have more to say in this direction in Chapter 17.

It is an interesting fact that many multiplicative functions have corresponding Dirichlet series which can be simply expressed in terms of the zeta function. We have just seen this for $d(f)$. More generally, let $h(f)$ be multiplicative. The multiplicativity of $h(f)$ leads to the identity

$$D_h(s) = \prod_P \Big(\sum_{k=0}^{\infty} \frac{h(P^k)}{|P|^{ks}}\Big) \ .$$

As an example, consider the function $\mu(f)$. Since $\sum_{k=0}^{\infty} \frac{\mu(P^k)}{|P|^{ks}} = 1 - |P|^{-s}$, we find $D_\mu(s) = \zeta_A(s)^{-1}$. The same method would enable us to determine the Dirichlet series for $\Phi(f)$ and $\sigma(f)$. However, we will follow a slightly different path to this goal.

Let λ and ρ be two complex valued functions on the monic polynomials. We define their Dirichlet product by the following formula (all polynomials involved are assumed to be monic)

$$(\lambda * \rho)(f) = \sum_{\substack{h,g \\ hg=f}} \lambda(h)\rho(g) \ .$$

This definition is exactly similar to the corresponding notion in elementary number theory. As is the case there, the Dirichlet product is closely related to multiplication of Dirichlet series.

Proposition 2.6.

$$D_\lambda(s)D_\rho(s) = D_{\lambda*\rho}(s) \ .$$

Proof. The calculation is just like that of Proposition 2.5.

$$D_\lambda(s)D_\rho(s) = \left(\sum_h \frac{\lambda(h)}{|h|^s}\right)\left(\sum_g \frac{\rho(g)}{|g|^s}\right) =$$

$$\sum_f \left(\sum_{\substack{h,g \\ hg=f}} \lambda(h)\rho(g)\right)\frac{1}{|f|^s} = D_{\lambda*\rho}(s) \ .$$

We now proceed to calculate the average value of $\Phi(f)$. We have seen that

$$\Phi(f) = |f| \prod_{P|f}(1 - |P|^{-1}) \ .$$

Define $\lambda(f) = |f|$. A moment's reflection shows that the right hand side of the above equation can be rewritten as $\sum_{g|f} \mu(g)|f/g| = (\mu * \lambda)(f)$. Thus, by Proposition 2.6 we find

$$D_\Phi(s) = D_{\mu*\lambda}(s) = D_\mu(s)D_\lambda(s) = \zeta_A(s)^{-1}\zeta_A(s-1) \ . \tag{6}$$

Proposition 2.7.

$$\sum_{\substack{\deg f=n \\ f \text{ monic}}} \Phi(f) = q^{2n}(1 - q^{-1}) \ .$$

Proof. Let $A(n)$ be the left-hand side of the above equation. Then, with the usual transformation $u = q^{-s}$, Equation 6 becomes

$$\sum_{n=0}^{\infty} A(n)u^n = \frac{1 - qu}{1 - q^2u} \ .$$

Now, expand $(1 - q^2 u)^{-1}$ into a power series using the geometric series, multiply out, and equate the coefficients of u^n on both sides. One finds $A(n) = q^{2n} - q^{2n-1}$. The result follows.

Finally, we want to do a similar analysis for the function $\sigma(f)$. Let $\mathbf{1}(f)$ denote the function which is identically equal to 1 on all monics f. For any complex valued function λ on monics, we see immediately that $(\mathbf{1} * \lambda)(f) = \sum_{g|f} \lambda(g)$. In particular, if $\lambda(f) = |f|$, then $(\mathbf{1} * \lambda)(f) = \sigma(f)$. Thus,

$$D_\sigma(s) = D_{\mathbf{1}*\lambda}(s) = D_{\mathbf{1}}(s) D_\lambda(s) = \zeta_A(s) \zeta_A(s-1) \ . \tag{7}$$

Proposition 2.8.

$$\sum_{\substack{\deg(f)=n \\ f \text{ monic}}} \sigma(f) = q^{2n} \cdot \frac{1 - q^{-n-1}}{1 - q^{-1}} \ .$$

Proof. Define $S(n)$ to be the sum on the left hand side of the above equation. Then, making the substitution $u = q^{-s}$ in Equation 7 we find

$$\sum_{n=0}^{\infty} S(n) u^n = (1 - qu)^{-1} (1 - q^2 u)^{-1} \ .$$

Expanding the two terms on the right using the geometric series, multiplying out, and collecting terms, we deduce

$$S(n) = \sum_{k+l=n} q^k q^{2l} \ .$$

The result follows after applying a little algebra.

The method of obtaining average value results via the zeta function has now been amply demonstrated. The reader who wants to pursue this further can consult the original article of Carlitz [1]. Alternatively, it is an interesting exercise to look at Chapter VII of Hardy and Wright [1] or Chapter 3 of Apostol [1] , formulate the results given there for $\mathbb{Z}$ in the context of the polynomial ring $A = \mathbb{F}[T]$, and prove them by the methods developed above.

In Chapter 17, we will return to the subject of average value results, but in the broader context of global function fields.

Exercises

1. Let $f \in A$ be a polynomial of degree at least $m \geq 1$. For each $N \geq m$ show that the number of polynomials of degree N divisible by f divided by the number of polynomials of degree N is just $|f|^{-1}$. Thus, it makes sense to say that the probability that an arbitrary polynomial is divisible by f is $|f|^{-1}$.

2. Let $P_1, P_2, \ldots, P_t \in A$ be distinct monic irreducibles. Give a probabilistic argument that the probability that a polynomial not be divisible by any P_i^2 for $1 = 1, 2, \ldots, t$ is give by $\prod_{i=1}^{t}(1 - |P_i|^{-2})$.

3. Based on Exercise 2, give a heuristic argument to show that the probability that a polynomial in A is square-free is given by $\zeta_A(2)^{-1}$.

4. Generalize Exercise 3 to give a heuristic argument to show that the probability that a polynomial in A be k-th power free is given by $\zeta_A(k)^{-1}$.

5. Show $\sum |m|^{-1}$ diverges, where the sum is over all monic polynomials $m \in A$.

6. Use the fact that every monic m can be written uniquely in the form $m = m_0 m_1^2$ where m_0 and m_1 are monic and m_0 is square-free to show $\sum |m_0|^{-1}$ diverges where the sum is over all square-free monics m_0.

7. Use Exercise 6 to show

$$\prod_{\substack{P \text{ irreducible} \\ \deg P \leq d}} (1 + |P|^{-1}) \to \infty \quad \text{as} \quad d \to \infty .$$

8. Use the obvious inequality $1 + x \leq e^x$ and Exercise 7 to show $\sum |P|^{-1}$ diverges where the sum is over all monic irreducibles $P \in A$.

9. Use Theorem 2.2 to give another proof that $\sum |P|^{-1}$ diverges.

10. Suppose there were only finitely many monic irreducibles in A. Denote them by $\{P_1, P_2, \ldots, P_n\}$. Let $m = P_1 P_2 \ldots P_n$ be their product. Show $\Phi(m) = 1$ and derive a contradiction.

11. Suppose h is a complex valued function on monics in A and that the limit as n tends to infinity of $\text{Ave}_n(h)$ is equal to α. Show

$$\lim_{n \to \infty} (1 + q + \cdots + q^n)^{-1} \sum_{\substack{f \text{ monic} \\ \deg f \leq n}} h(f) = \alpha .$$

12. Let $\mu(m)$ be the Möbius function on monic polynomials which we introduced in the text. Consider the sum $\sum_{\deg m = n} \mu(m)$ over monic polynomials of degree n. Show the value of this sum is 1 if $n = 0$, $-q$ if $n = 1$, and 0 if $n > 1$.

13. For each integer $k \geq 1$ define $\sigma_k(m) = \sum_{f | m} |f|^k$. Calculate $\text{Ave}_n(\sigma_k)$.

14. Define $\Lambda(m)$ to be $\log|P|$ if $m = P^t$, a prime power, and zero otherwise. Show

$$\sum_{f|m} \Lambda(f) = \log|m| .$$

15. Show that

$$D_\Lambda(s) = -\zeta'_A(s)/\zeta_A(s).$$

Use this to evaluate $\sum_{\deg m=n} \Lambda(m)$.

16. Recall that $d(m)$ is the number of monic divisors of m. Show

$$\sum_{m \text{ monic}} \frac{d(m)^2}{|m|^s} = \frac{\zeta_A(s)^4}{\zeta_A(2s)} .$$

Use this to evaluate $\sum_{\deg m=n} d(m)^2$.

3

The Reciprocity Law

Gauss called the quadratic reciprocity law "the golden theorem." He was the first to give a valid proof of this theorem. In fact, he found nine different proofs. After this he worked on biquadratic reciprocity, obtaining the correct statement, but not finding a proof. The first to do so were Eisenstein and Jacobi. The history of the general reciprocity law is long and complicated involving the creation of a good portion of algebraic number theory and class field theory. By contrast, it is possible to formulate and prove a very general reciprocity law for $A = \mathbb{F}[T]$ without introducing much machinery. Dedekind proved an analogue of the quadratic reciprocity law for A in the last century. Carlitz thought he was the first to prove the general reciprocity law for $\mathbb{F}[T]$. However O. Ore pointed out to him that F.K. Schmidt had already published the result, albeit in a somewhat obscure place (Erlanger Sitzungsberichte, Vol. 58–59, 1928). See Carlitz [2] for this remark and also for a number of references in which Carlitz gives different proofs the reciprocity law. We will present a particularly simple and elegant proof due to Carlitz. The only tools necessary will be a few results from the theory of finite fields.

Let $P \in A$ be an irreducible polynomial and d a divisor of $q - 1$ (recall that q is the cardinality of $\mathbb{F}$). If $a \in A$ and P does not divide a, then, by Proposition 1.10, we know $x^d \equiv a \pmod{P}$ is solvable if and only if

$$a^{\frac{|P|-1}{d}} \equiv 1 \pmod{P}.$$

The left-hand side of this congruence is, in any case, an element of order dividing d in $(A/PA)^*$. Since $\mathbb{F}^* \to (A/PA)^*$ is one to one, there is a unique $\alpha \in \mathbb{F}^*$ such that

$$a^{\frac{|P|-1}{d}} \equiv \alpha \pmod{P}.$$

Definition. If P does not divide a, let $(a/P)_d$ be the unique element of $\mathbb{F}^*$ such that

$$a^{\frac{|P|-1}{d}} \equiv \left(\frac{a}{P}\right)_d \pmod{P}.$$

If $P|a$ define $(a/P)_d = 0$. The symbol $(a/P)_d$ is called the d-th power residue symbol.

When $d = 2$, this symbol is just like the Legendre symbol of elementary number theory. The situation is a bit more flexible in A since $A^* = \mathbb{F}^*$ is cyclic of order $q - 1$, whereas $\mathbb{Z}^*$ is just $\{\pm 1\}$. Notice that the value of the residue symbol is in the finite field $\mathbb{F}$ and not in the complex numbers.

Proposition 3.1. *The d-th power residue symbol has the following properties:*

1) $\left(\frac{a}{P}\right)_d = \left(\frac{b}{P}\right)_d$ *if $a \equiv b \pmod{P}$.*

2) $\left(\frac{ab}{P}\right)_d = \left(\frac{a}{P}\right)_d \left(\frac{b}{P}\right)_d.$

3) $\left(\frac{a}{P}\right)_d = 1$ *iff $x^d \equiv a \pmod{P}$ is solvable.*

4) *Let $\zeta \in \mathbb{F}^*$ be an element of order dividing d. There exists an $a \in A$ such that $\left(\frac{a}{P}\right)_d = \zeta$.*

Proof. The first assertion follows immediately from the definition. The second follows from the definition and the fact that if two constants are congruent modulo P then they are equal. The third assertion follows from the definition and Proposition 1.10. Finally, note that the map from $(A/PA)^* \to \mathbb{F}^*$ given by $a \to (a/P)_d$ is a homomorphism whose kernel is the d-th powers in $(A/PA)^*$ by part 3. Since $(A/PA)^*$ is a cyclic group of order $|P| - 1$, the order of the kernel is $(|P| - 1)/d$. Consequently, the image has order d and part 4 follows from this.

It is an easy matter to evaluate the residue symbol on a constant.

Proposition 3.2. *Let $\alpha \in \mathbb{F}$. Then,*

$$\left(\frac{\alpha}{P}\right)_d = \alpha^{\frac{q-1}{d} \deg(P)}.$$

Proof. Let $\delta = \deg(P)$. Then,

$$\frac{|P|-1}{d} = \frac{q^\delta - 1}{d} = (1 + q + \cdots + q^{\delta-1})\frac{q-1}{d}.$$

The result now follows from the definition and the fact that for all $\alpha \in \mathbb{F}$ we have $\alpha^q = \alpha$.

Notice that if $d \mid \deg(P)$ every constant is automatically a d'th power residue modulo P.

We are now in a position to state the reciprocity law.

Theorem 3.3. (The d-th power reciprocity law) *Let P and Q be monic irreducible polynomials of degrees δ and ν respectively. Then,*

$$\left(\frac{Q}{P}\right)_d = (-1)^{\frac{q-1}{d}\delta\nu}\left(\frac{P}{Q}\right)_d.$$

Proof. Let's define $(a/P) = (a/P)_{q-1}$. Then $(a/P)_d = (a/P)^{\frac{q-1}{d}}$. The theorem would follow in full generality if we could show

$$\left(\frac{Q}{P}\right) = (-1)^{\delta\nu}\left(\frac{P}{Q}\right),$$

since the general result would follow by raising both sides to the $(q-1)/d$ power.

Let α be a root of P and β a root of Q. Let $\mathbb{F}'$ be a finite field which contains $\mathbb{F}$, α, and β. Using the theory of finite fields we find

$$P(T) = (T - \alpha)(T - \alpha^q)\cdots(T - \alpha^{q^{\delta-1}}) \quad \text{and}$$

$$Q(T) = (T - \beta)(T - \beta^q)\cdots(T - \beta^{q^{\nu-1}}). \tag{1}$$

We now take congruences in the ring $A' = \mathbb{F}'[T]$. Note that if $f(T) \in A'$ we have $f(T) \equiv f(\alpha) \pmod{(T - \alpha)}$. Also note that if $g(T) \in A$ then $g(T)^q = g(T^q)$ which follows readily from the fact that the coefficients of $g(T)$ are in $\mathbb{F}$. From this remark, and the definition, we compute that (Q/P) is congruent to

$$Q(T)^{1+q+\cdots+q^{\delta-1}} \equiv Q(T)Q(T^q)\cdots Q(T^{q^{\delta-1}})$$
$$\equiv Q(\alpha)Q(\alpha^q)\cdots Q(\alpha^{q^{\delta-1}}) \pmod{(T - \alpha)}.$$

By symmetry this congruence holds modulo $(T - \alpha^{q^i})$ for all i and it follows that it holds modulo P. Combining this result with Equation 1 yields the following congruence:

$$\left(\frac{Q}{P}\right) \equiv \prod_{i=0}^{\delta-1}\prod_{j=0}^{\nu-1}(\alpha^{q^i} - \beta^{q^j}) \pmod{P}.$$

Both sides of this congruence are in $\mathbb{F}'$ so they must be equal. Thus,

$$\left(\frac{Q}{P}\right) = \prod_{i=0}^{\delta-1}\prod_{j=0}^{\nu-1}(\alpha^{q^i} - \beta^{q^j}) = (-1)^{\delta\nu}\prod_{j=0}^{\nu-1}\prod_{i=0}^{\delta-1}(\beta^{q^j} - \alpha^{q^i}) = (-1)^{\delta\nu}\left(\frac{P}{Q}\right).$$

This concludes the proof.

This beautiful proof is due to Carlitz. It is contained in a set of lecture notes for a course on polynomials over finite fields which he gave at Duke in the 1950s. We will outline another proof, also due to Carlitz, in the exercises to Chapter 12.

As in the classical theory, it is convenient to extend the definition of the d-th power reciprocity symbol to the case where the prime P is replaced with an arbitrary non-zero element $b \in A$.

Definition. Let $b \in A$, $b \neq 0$, and $b = \beta Q_1^{f_1} Q_2^{f_2} \ldots Q_s^{f_s}$ be the prime decomposition of b. If $a \in A$, define

$$\left(\frac{a}{b}\right)_d = \prod_{j=1}^{s} \left(\frac{a}{Q_j}\right)_d^{f_j}. \tag{2}$$

Notice that this definition ignores $\beta = \text{sgn}(b)$ and so the symbol only depends on the principal ideal bA generated by b. The basic properties of this extended symbol are easily derived from those of the d-th power residue symbol.

Proposition 3.4. *The symbol $(a/b)_d$ has the following properties.*

1) *If $a_1 \equiv a_2 \pmod{b}$ then $(a_1/b)_d = (a_2/b)_d$.*

2) *$(a_1 a_2/b)_d = (a_1/b)_d (a_2/b)_d$.*

3) *$(a/b_1 b_2)_d = (a/b_1)_d (a/b_2)_d$.*

4) *$(a/b)_d \neq 0$ iff $(a, b) = 1$ (a is relatively prime to b).*

5) *If $x^d \equiv a \pmod{b}$ is solvable, then $(a/b)_d = 1$, provided that $(a, b) = 1$.*

Proof. Properties $1 - 4$ follow from the definition and the properties of the symbol $(a/P)_d$.

To show property 5, suppose $c^d \equiv a \pmod{b}$. Then, by properties 1 and 2, $(a/b)_d = (c^d/b)_d = (c/b)_d^d = 1$.

The converse of assertion 5 in Proposition 3.4 is not true in general. For example, suppose Q is a monic irreducible not dividing a and $b = Q^d$. Then, by property 3 above we have $(a/b)_d = (a/Q^d)_d = (a/Q)_d^d = 1$. However, not every element of $(A/bA)^* = (A/Q^d A)^*$ is a d-th power. In fact, the group of d-th powers has index d as we saw in Proposition 1.10.

The same example shows that property 4 of Proposition 3.1 doesn't hold for the generalized symbol. As a mapping from $(A/Q^d A)^* \to \mathbb{F}^*$ the symbol $(a/Q^d)_d$ only takes on the value 1 and no other element of order dividing d.

It is useful to have a form of the reciprocity law which works for arbitrary (i.e., not necessarily monic or irreducible) elements of A. For $f \in A$, $f \neq 0$, define $\text{sgn}_d(f)$ to be the leading coefficient of f raised to the $\frac{q-1}{d}$ power.

Theorem 3.5. (The general reciprocity law). *Let $a, b \in A$ be relatively prime, non-zero elements. Then,*

$$\left(\frac{a}{b}\right)_d \left(\frac{b}{a}\right)_d^{-1} = (-1)^{\frac{q-1}{d} \deg(a) \deg(b)} \operatorname{sgn}_d(a)^{\deg(b)} \operatorname{sgn}_d(b)^{-\deg(a)}.$$

Proof. When a and b are monic irreducibles this reduces to Theorem 3.3. In general, the proof proceeds by appealing to Proposition 3.2, Theorem 3.3, the definitions, and the fact that the degree of a product of two polynomials is equal to the sum of their degrees. We omit the details.

The reciprocity law can be thought of as a pretty formula, but its importance lies in the fact that it relates two natural questions in an intrinsic way. Given a polynomial m of positive degree, what are the d-th powers modulo m? Since $(A/mA)^*$ is finite, one can answer this question in principle by just writing down the elements of $(A/mA)^*$, raising them to the d-th power, and making a list of the results. The answer will be a list of cosets or residue classes modulo m. In practice this may be hard because of the amount of calculation involved. One can appeal to the structure of $(A/mA)^*$ to find shortcuts. Parenthetically, it is an interesting question to determine the number of d-th powers modulo m. Recall that we are assuming $d|(q-1)$. Under this assumption, the answer is $\Phi(m)/d^{\lambda(m)}$, where $\lambda(m)$ is the number of distinct monic prime divisors of m. This follows from Proposition 1.10 and the Chinese Remainder Theorem.

Now, let's turn things around somewhat. Given m, find all primes P such that m is a d-th power modulo P. It turns out that there are infinitely many such primes, so that it is not possible to answer the question by making a list. One has to characterize the primes with this property in some natural way. This is what the reciprocity law allows us to do.

For simplicity, we will assume that m is monic. It is no loss of generality to assume that all the primes we deal with are monic as well. Let $\{a_1, a_2, \ldots, a_t\}$ be coset representatives for the classes in $(A/mA)^*$ which have the property $(a/m)_d = 1$. If there is a $b \in A$ such that $(b/m)_d = -1$ let $\{b_1, b_2, \ldots, b_t\}$ be coset representatives for all classes with this property.

Proposition 3.6. *With the above assumptions we have*

1)   If $\deg(m)$ is even, $(q-1)/d$ is even, or $p = \operatorname{char}(F) = 2$, m is a d-th power modulo P iff $P \equiv a_i \pmod{m}$ for some $i = 1, 2, \ldots, t$.

2)   If $\deg(m)$ is odd, $(q-1)/d$ is odd, and $p = \operatorname{char}(F)$ is odd, then m is a d-th power modulo P iff either $\deg(P)$ is even and $P \equiv a_i \pmod{m}$ for some $i = 1, 2, \ldots, t$ or $\deg(P)$ is odd and $P \equiv b_i \pmod{m}$ for some $i = 1, 2, \ldots, t$.

Proof. By Theorem 3.5, we have

$$\left(\frac{m}{P}\right)_d = (-1)^{\frac{q-1}{d} \deg(m) \deg(P)} \left(\frac{P}{m}\right)_d.$$

If any of the conditions in Part 1 hold, we have $(m/P)_d = (P/m)_d$ and this gives the result by Part 3 of Proposition 3.1 and the fact that $(P/m)_d$ only depends on the residue class of P modulo m.

If the conditions of Part 2 hold, then $(m/P)_d = (-1)^{\deg(P)}(P/m)_d$. Thus, if $\deg(P)$ is even, $(m/P)_d = 1$ iff $P \equiv a_i \pmod{m}$ for some i, and if $\deg(P)$ is odd, $(m/P)_d = 1$ iff $P \equiv b_i \pmod{m}$ for some i. That there is a $b \in A$ with $(b/m)_d = -1$ under the conditions of Part 2 follows from the fact that

$$\left(\frac{-1}{m}\right)_d = (-1)^{\frac{q-1}{d}\deg(m)} = -1.$$

A number of interesting number-theoretic questions are of the following form: if a certain property holds modulo all but finitely many primes, does it hold in A? One such property is that of being a d-th power. In this case the question has a positive answer. The key to the proof, as we shall see, is the reciprocity law.

Theorem 3.7. *Let $m \in A$ be a polynomial of positive degree. Let d be an integer dividing $q - 1$. If $x^d \equiv m \pmod{P}$ is solvable for all but finitely many primes P, then $m = m_o^d$ for some $m_o \in A$.*

Proof. Let $m = \mu Q_1^{e_1} Q_2^{e_2} \ldots Q_t^{e_t}$ be the prime decomposition of m. We begin by showing that if some e_i is not divisible by d, then there are infinitely many primes L such that $(m/L)_d \neq 1$. This will contradict the hypothesis and we can conclude that the hypothesis implies $m = \mu m_o'^d$ for some $m_o' \in A$.

We may as well assume that e_1 is not divisible by d. Let $\{L_1, L_2 \ldots, L_s\}$ be a set of primes not dividing m such that $(m/L_j)_d \neq 1$ for $j = 1, 2, \ldots, s$. For any $a \in A$ we have

$$\left(\frac{a}{m}\right)_d = \prod_{i=1}^{t}\left(\frac{a}{Q_i}\right)_d^{e_i}. \tag{3}$$

By Part 4 of Proposition 3.1, there exists an element $c \in A$ such that $(c/Q_1)_d = \zeta_d$, a primitive dth root of 1. By the Chinese Remainder Theorem, we can find an $a \in A$ such that $a \equiv c \pmod{Q_1}$ and $a \equiv 1 \pmod{Q_i}$ for $i \geq 2$, and $a \equiv 1 \pmod{L_j}$ for all j. Once such an a is chosen we can add to it any A-multiple of $Q_1 Q_2 \ldots Q_t L_1 L_2 \ldots L_s$ and it will satisfy the same congruences as a. Thus we may assume, by choosing a suitable such multiple of large degree, that a is monic and of degree divisible by $2d$. Assuming that a has these properties, we substitute it into Equation 3 and derive

$$\left(\frac{a}{m}\right)_d = \zeta_d^{e_1} \neq 1.$$

By the reciprocity law,

$$\left(\frac{m}{a}\right)_d = \left(\frac{a}{m}\right)_d \neq 1.$$

It follows that there must be a prime $L|a$ such that $(m/L)_d \neq 1$. Since $a \equiv 1 \pmod{L_j}$ for every j we must have $L \neq L_j$ for all j. This shows there must be infinitely many primes L such that $(m/L)_d \neq 1$ if e_1 is not divisible by d. The same assertion holds for each e_i.

We have shown that under the hypothesis of the theorem $m = \mu m_o'^d$, where $\mu \in \mathbb{F}^*$. It remains to show that μ must be a d-th power. Consider

$$\left(\frac{m}{P}\right)_d = \left(\frac{\mu}{P}\right)_d = \mu^{\frac{q-1}{d} \deg(P)}. \tag{4}$$

By Theorem 2.2, there are infinitely many irreducibles of degree relatively prime to d. In fact, there are irreducibles of every degree. Thus there is an irreducible P of degree prime to d and such that $(m/P)_d = 1$. It then follows from equation number (4) that $\mu^{\frac{q-1}{d} \deg P} = 1$ and so, $\mu^{\frac{q-1}{d}} = 1$. This shows that μ is a d-th power, $\mu = \mu_o^d$, in F. Set $m_o = \mu_o m_o'$ and we have $m = m_o^d$, as asserted.

In the statement and proof of Theorem 3.7 we have been assuming that d divides $q-1$. Is this necessary? The statement of the theorem is not true for all d. For example, consider $p = \text{char}(\mathbb{F})$. For every prime P and any $a \in A$ we have that a is a p-th power modulo P. This follows from the fact that raising to the p-th power is an automorphism of the finite field A/PA. Thus, the theorem fails if $d = p$ or indeed if d is a power of p. However,

Fact. The assertion of Theorem 3.7 remains true if p does not divide d. In other words, if d is not divisible by p it is not necessary to assume that $d|\,q-1$.

We will sketch a proof. We rely on Theorem 3.7 together with some elementary facts about finite fields.

Since p does not divide d, q and d are relatively prime. Thus, there is a positive integer n such that $q^n \equiv 1 \pmod{d}$. Let $\mathbb{F}'$ be a field extension of $\mathbb{F}$ of degree n. $\mathbb{F}'^*$ has $q^n - 1$ elements and so must contain a primitive d-th root of unity. Set $A' = \mathbb{F}'[T]$.

Now, suppose that $m \in A$ and that m is a d-th power for all but finitely many primes P of A. If P' is a prime of A' it is easy to check that $P'A' \cap A = PA$ where P is a prime of A. It follows that m is a d-th power modulo all but finitely many primes of A'. Invoking Theorem 3.7, we see that $m = m'^d$ is a d-th power in A'. We need to show that m' can be chosen to be in A.

Let P be a prime of A and consider it as an element of A'. It factors as a product of primes in A'; $P = P_1' P_2' \cdots P_s'$ where the P_i' are all distinct (over a finite field, every irreducible polynomial has no repeated roots in any algebraic extension). For a prime P of A, let e be the highest power to which P divides m. If P' is a prime of A' dividing P, then e is also the highest power of P' dividing m. Since $m = m'^d$, unique factorization in A' implies $d|e$. This being true for all primes P of A, it follows that $m = \mu m_o^d$ with $m_o \in A$ and $\mu \in \mathbb{F}$. It remains to show that μ is a d-th power in $\mathbb{F}^*$.

By the hypothesis on m and the equation $m = \mu m_o^d$ we see that μ is a d-th power for all but finitely many primes P. Let $d' = (d, q - 1)$. μ is a d'-th power for all but finitely many primes P (since $d'|d$). Moreover, $d'|(q - 1)$. Using Theorem 3.7 once again, we see that μ is a d'-th power. Since $\mathbb{F}^*$ is cyclic of order $q - 1$ it is easy to see that $\mathbb{F}^{*d} = \mathbb{F}^{*d'}$. Thus, μ is a d-th power, and we are done.

Exercises

1. Fill in the details of the proof of Proposition 3.4.

2. Fill in the details of the proof of Theorem 3.5.

3. Suppose $d \mid q - 1$ and that $m \in A$ is a polynomial of positive degree. Show that the number of d-th powers in $(A/mA)^*$ is given by $\Phi(m)/d^{\lambda(m)}$, where $\lambda(m)$ is the number of distinct monic prime divisors of m.

4. Let $P \in A$ be a prime and consider the congruence $X^2 \equiv -1$ (mod P). Show this congruence is solvable except in the case where $q \equiv 3$ (mod 4) and $\deg P$ is odd.

5. Suppose $d' \mid q - 1$ and $\alpha \in \mathbb{F}^*$ is an element of order d'. Let $P \in A$ be a prime of positive degree and suppose that d is a divisor of $|P| - 1$. Show that $X^d \equiv \alpha$ (mod P) is solvable if and only if dd' divides $|P| - 1$. Show how Exercise 4 is a special case of this result.

6. Suppose that d is a positive integer and that $q \equiv 1$ (mod $4d$). Let $P \in A$ be a monic prime. Show that $X^d \equiv T$ (mod P) if and only if the constant term of P, i.e. $P(0)$, is a d-th power in $\mathbb{F}$.

7. Suppose d divides $q - 1$ and that $P \in A$ is a prime. Show that the number of solutions to $X^d \equiv a$ (mod P) is given by

$$1 + \left(\frac{a}{P}\right)_d + \left(\frac{a}{P}\right)_d^2 + \cdots + \left(\frac{a}{P}\right)_d^{d-1}.$$

8. Let $b \in A$ and suppose $b = \beta P_1^{e_1} P_2^{e_2} \cdots P_t^{e_t}$ is the prime decomposition of b. Here, $\beta \in \mathbb{F}^*$ and the P_i are distinct monic primes. Consider $(a/b)_d$ as a homomorphism from $(A/bA)^*$ to the cyclic group $< \zeta_d >$ generated by an element $\zeta_d \in \mathbb{F}^*$ of order d. Show that this map is onto if and only if the greatest common divisor of the set $\{e_1, e_2, \ldots, e_t\}$ is relatively prime to d.

9. Suppose $d \mid q - 1$ and $a, b_1, b_2 \in A$. Show that $(a/b_1)_d = (a/b_2)_d$ if the following conditions hold: $b_1 \equiv b_2$ (mod a), $\deg b_1 \equiv \deg b_2$ (mod d), and $\text{sgn}_d(b_1) = \text{sgn}_d(b_2)$.

10. In this exercise we give an analogue of the classical Gauss criterion
    for the Legendre symbol. Let $P \in A$ be a prime. Show that every non-
    zero residue class modulo P has a unique representative of the form
    μm where $\mu \in \mathbb{F}^*$ and m is a monic polynomial of degree less than
    $\deg P$. Let $\mathcal{M}$ denote the set of monics of degree less than $\deg P$.
    Suppose $a \in A$ with $P \nmid a$. For each $m \in \mathcal{M}$ write $am \equiv \mu_m m'$
    (mod P) where $\mu_m \in \mathbb{F}^*$ and $m' \in \mathcal{M}$. Show

    $$\left(\frac{a}{P}\right)_{q-1} = \prod_{m \in \mathcal{M}} \mu_m \, .$$

In the exercises to Chapter 12, we will use this criterion to outline another
proof of the Reciprocity Law (also due to Carlitz).

10. In this exercise we glance in the margin of the physical characterization

$$\left(\frac{m}{n}\right) = \prod \cdots$$

4

Dirichlet L-Series and Primes in an Arithmetic Progression

Our principal goal in this chapter will be to prove the analogue of Dirichlet's famous theorem about primes in arithmetic progressions. This was first proved by H. Kornblum in his PhD thesis written, just before the onset of World War I, under the direction of Edmund Landau. After completing the work on his thesis, but before writing it up, Kornblum enlisted in the army. He died in the fighting on the Eastern Front. After the war, Landau completed the sad duty of writing up and publishing his student's results, see Kornblum [1].

The proof of the theorem uses the theory of Dirichlet series. After giving the definitions and proving the elementary properties of these series, we outline the connection with primes in arithmetic progressions and isolate the main difficulty which is the proof that $L(1, \chi) \neq 0$ for non-trivial characters χ. We then give a proof of this fact which differs from the Kornblum-Landau approach. It is an adaptation to polynomial rings of a proof of the corresponding number-theoretic fact due to de la Vallee Poussin. Finally, to complete the chapter, we give a refinement of Dirichet's theorem, which shows that given an arithmetic progression $\{a+mx \mid a, m \in A, (a, m) = 1\}$, then, for all sufficiently large integers N, there is a prime P of degree N which lies in this arithmetic progression.

Before beginning we discuss the notion of the Dirichlet density of a set of primes in A. This will give a quantitative measure of how big such a set is. Let $f(s)$ and $g(s)$ be two complex valued functions of a real variable s both defined on some open interval $(1, b)$. We define $f \approx g$ to mean that $f - g$ remains bounded as $s \to 1$ inside $(1, b)$.

Proposition 4.1. *We have*

$$\log \zeta_A(s) \approx \log \left(\frac{1}{s-1} \right) \approx \sum_P |P|^{-s} \, ,$$

where the sum is over all irreducible monic polynomials P.

Proof. Since $\zeta_A(s) = (1 - q^{1-s})^{-1}$ we see that $\lim_{s \to 1}(s - 1)\zeta_A(s) = 1/\log(q)$. Thus, $\log \zeta_A(s) - \log(s-1)^{-1}$ is bounded as $s \to 1$, which establishes the first relation. As for the second relation we see, using the Euler product for $\zeta_A(s)$

$$\log \zeta_A(s) = -\sum_P \log(1-|P|^{-s}) = \sum_{P,k} |P|^{-ks}/k = \sum_P |P|^{-s} + \sum_{P,k \geq 2} |P|^{-ks}/k \, .$$

Now, $\sum_{k \geq 2} |P|^{-ks}/k < \sum_{k \geq 2} |P|^{-ks} = |P|^{-2s}(1 - |P|^{-s})^{-1} < 2|P|^{-2s}$. Thus the last sum in the above equation is bounded by $2\zeta_A(2)$. This shows that $\log \zeta_A(s) \approx \sum_P |P|^{-s}$ which completes the proof.

Definition. Henceforth the word "prime" will denote a monic irreducible in A. Let S be a set of primes in A. The Dirichlet density of S, $\delta(S)$ is defined to be

$$\delta(S) = \lim_{s \to 1} \frac{\sum_{P \in S} |P|^{-s}}{\sum_P |P|^{-s}} \, ,$$

provided that the limit exists. The limit is assumed to be taken over the values of s lying in a real interval $(1, b)$.

Several remarks are in order. First note that $0 \leq \delta(S) \leq 1$ and if $S = S_1 \cup S_2$, then $\delta(S) = \delta(S_1) + \delta(S_2)$ provided S_1 and S_2 both have densities and are disjoint. Thus, Dirichlet density is something like a probability measure. One must not carry this too far, however. Dirichlet density is not countably additive.

It is obvious that the Dirichlet density of a finite set is zero. Thus, if the Dirichlet density of a set exists and is positive, we are assured that the set is infinite. One of the two main results of this chapter asserts that if a and m are relatively prime polynomials, then the Dirichlet density of the set $S = \{P \in A \mid P \text{ prime}, P \equiv a \pmod{m}\}$ exists and is equal to $1/\Phi(m)$. It is in this refined form that we prove Dirichlet's famous theorem in the context of the polynomial ring A.

The next step is to introduce the main tools necessary to the proof, Dirichlet characters and Dirichlet L-series.

Let m be an element of A of positive degree. A Dirichlet character modulo m is a function from $A \to \mathbb{C}$ such that

(a) $\chi(a + bm) = \chi(a)$ for all $a, b \in A$.

(b) $\chi(a)\chi(b) = \chi(ab)$ for all $a, b \in A$

(c) $\chi(a) \neq 0$ if and only if $(a, m) = 1$.

A Dirichlet character modulo m induces a homomorphism from $(A/mA)^* \to \mathbb{C}^*$ and conversely, given such a homomorphism there is a uniquely corresponding Dirichlet character. The trivial Dirichlet character χ_o is defined by the property that $\chi_o(a) = 1$ if $(a, m) = 1$ and $\chi_o(a) = 0$ if $(a, m) \neq 1$.

It can be shown that there are exactly $\Phi(m)$ Dirichlet characters modulo m which is the same cardinality as that of the group $(A/mA)^*$. Let X_m be the set of Dirichlet characters modulo m. If $\chi, \psi \in X_m$ define their product, $\chi\psi$, by the formula $\chi\psi(a) = \chi(a)\psi(a)$. This makes X_m into a group. The identity of this group is the trivial character χ_o. The inverse of a character is given by $\chi^{-1}(a) = \chi(a)^{-1}$ if $(a, m) = 1$, and $\chi^{-1}(a) = 0$ if $(a, m) \neq 1$. It can be shown, but we will not do so here, that X_m is isomorphic to $(A/mA)^*$, which is a much better result than the bare statement that they have the same number of elements. This is a special case of a general result which asserts that a finite abelian group G is isomorphic to its character group $\hat{G}$, see Lang [4], Chapter 1, Section 9.

Another definition is useful. If $\chi \in X_m$ let $\bar{\chi}$ be defined by $\bar{\chi}(a) = \overline{\chi(a)}$ = complex conjugate of $\chi(a)$. Since the value of a character is either zero or a root of unity, it is easy to see that $\bar{\chi} = \chi^{-1}$. Moreover, we have the following very important proposition, the orthogonality relations.

Proposition 4.2. *Let χ and ψ be two Dirichlet characters modulo m and a and b two elements of A relatively prime to m. Then*

(1) $\sum_a \chi(a)\overline{\psi(a)} = \Phi(m)\delta(\chi, \psi)$.

(2) $\sum_\chi \chi(a)\overline{\chi(b)} = \Phi(m)\delta(a, b)$.

The first sum is over any set of representative for A/mA and the second sum is over all Dirichlet characters modulo m. By definition, $\delta(\chi, \psi) = 0$ if $\chi \neq \psi$ and 1 if $\chi = \psi$. Similarly, $\delta(a, b) = 0$ if $a \neq b$ and 1 if $a = b$.

The proofs of all these facts are standard. For the corresponding facts over the integers, $\mathbb{Z}$, the reader can consult, for example, Ireland-Rosen [1], Chapter 16, Section 3. The relations given in the above proposition are called the orthogonality relations.

Definition. Let χ be a Dirichlet character modulo m. The Dirichlet L-series corresponding to χ is defined by

$$L(s, \chi) = \sum_{f \text{ monic}} \frac{\chi(f)}{|f|^s}.$$

From the definition and by comparison with the zeta function $\zeta_A(s)$ one sees immediately that the series for $L(s, \chi)$ converges absolutely for $\Re(s) > 1$. Also, since characters are multiplicative we can deduce that the following product decomposition is valid in the same region.

$$L(s,\chi) = \prod_P \left(1 - \frac{\chi(P)}{|P|^s}\right)^{-1}.$$

An immediate consequence of this product decomposition is the fact that the L-series corresponding to the trivial character is almost the same as $\zeta_A(s)$. More precisely,

$$L(s,\chi_o) = \prod_{P|m} \left(1 - \frac{1}{|P|^s}\right) \zeta_A(s).$$

This shows that $L(s,\chi_o)$ can be analytically continued to all of $\mathbb{C}$ and has a simple pole at $s = 1$ since the same is true of $\zeta_A(s)$. On the other hand,

Proposition 4.3. *Let χ be a non-trivial Dirichlet character modulo m. Then, $L(s,\chi)$ is a polynomial in q^{-s} of degree at most $\deg(m) - 1$.*

Proof. Define

$$A(n,\chi) = \sum_{\substack{\deg(f) = n \\ f \text{ monic}}} \chi(f).$$

It is clear from the definition of $L(s,\chi)$ that

$$L(s,\chi) = \sum_{n=0}^{\infty} A(n,\chi) q^{-ns}.$$

Thus, the result will follow if we can show that $A(n,\chi) = 0$ for all $n \geq \deg(m)$.

Let's assume that $n \geq \deg(m)$. If $\deg(f) = n$, we can write $f = hm + r$, where r is a polynomial of degree less than $\deg(m)$ or $r = 0$. Here, h is a polynomial of degree $n - \deg(m) \geq 0$, whose leading coefficient is $\text{sgn}(m)^{-1}$ (since f is monic). Conversely, all monic polynomials of degree $n \geq \deg(m)$ can be uniquely written in this fashion. Since χ is periodic modulo m and since h can be chosen in $q^{n-\deg(m)}$ ways, we have

$$A(n,\chi) = q^{n-\deg(m)} \sum_r \chi(r) = 0,$$

by the first orthogonality relation (Proposition 4.2, part (1)) since $\chi \neq \chi_o$, and the sum is over all r with $\deg(r) < \deg(m)$, which is a set of representatives for A/mA.

Proposition 4.3 shows that if χ is non-trivial, then $L(s,\chi)$ which was initially defined for $\Re(s) > 1$ can be analytically continued to an entire function on all of $\mathbb{C}$. We have already seen that $L(s,\chi_o)$ can be analytically continued to all of $\mathbb{C}$ with a simple pole at $s = 1$. These facts are much harder to establish when working over $\mathbb{Z}$ rather than A.

In the proof of Dirichlet's theorem on primes in arithmetic progressions the most difficult part is the proof that $L(1, \chi) \neq 0$ if χ is non-trivial. This turns out to be substantially easier in function fields because the L-series are essentially polynomials. We begin with a lemma.

Lemma 4.4. *Let χ vary over all Dirichlet characters modulo m. Then, for each prime P not dividing m, there exist positive integers f_P and g_P such that $f_P g_P = \Phi(m)$ and*

$$\prod_\chi L(s, \chi) = \prod_{P \nmid m} (1 - |P|^{-f_P s})^{-g_P}.$$

Proof. For a fixed prime P not dividing m, the map $\chi \to \chi(P)$ is a homomorphism from the group $X_m \to \mathbb{C}^*$. The image must be a cyclic group of order f_P, say, generated by ζ_{f_P}. If g_P is the order of the kernel, clearly $f_P g_P = \Phi(m)$.

With these preliminaries, we calculate for fixed P.

$$\prod_\chi (1 - \chi(P)|P|^{-s}) = \prod_{i=0}^{f_P - 1} (1 - \zeta_{f_P}^i |P|^{-s})^{g_P} = (1 - |P|^{-f_P s})^{g_P}.$$

Now take the inverse of both sides, multiply over all P, and the lemma follows.

Lemma 4.5. *Suppose χ is a complex Dirichlet character modulo m , i.e. $\bar{\chi} \neq \chi$. Then, $L(1, \chi) \neq 0$.*

Proof. The right-hand side of the equation in the statement of Lemma 4.4 is equal to a Dirichlet series with positive coefficients and constant term 1. Consequently, its value at real numbers s such that $s > 1$ is a real number greater than 1. Suppose χ is a complex Dirichlet character and that $L(1, \chi) = 0$. Then, by complex conjugation we see $L(1, \bar{\chi}) = 0$ as well. In the product $\Pi_\chi L(s, \chi)$ the term corresponding to the trivial character has a simple pole at $s = 1$. All the other terms are regular there and two of them have zeros. Thus, the product is zero at $s = 1$. This contradicts the fact, established above, that for all $s > 1$ the value of the product is greater than 1. Thus, $L(1, \chi) \neq 0$, as asserted.

The next step is to deal with real-valued characters. It is not hard to see that these coincide with characters of order 2. The proof for such characters will be a modification of a proof of the classical case due to de la Vallée Poussin.

Assume now that χ has order 2 and consider the function

$$G(s) = \frac{L(s, \chi_0) L(s, \chi)}{L(2s, \chi_0)} .$$

This can be written as a product over all monic irreducibles not dividing m. Let P be such a prime. Then $\chi(P) = \pm 1$. The factor of the above series corresponding to P is

$$\frac{(1 - |P|^{-s})^{-1}(1 - \chi(P)|P|^{-s})^{-1}}{(1 - |P|^{-2s})^{-1}}.$$

If $\chi(P) = -1$ this whole factor reduces to 1. If $\chi(P) = 1$ it simplifies to

$$\frac{(1 + |P|^{-s})}{(1 - |P|^{-s})} = 1 + 2\sum_{k=1}^{\infty} |P|^{-ks}.$$

It follows from these remarks that $G(s)$ is a Dirichlet series with non-negative coefficients. This will shortly play a crucial role.

First, we look more carefully at $L(s, \chi_0)/L(2s, \chi_0)$. As we have already seen,

$$L(s, \chi_0) = \prod_{P|m}(1 - |P|^{-s})\zeta_A(s) = \prod_{P|m}(1 - |P|^{-s}) \frac{1}{1 - q^{1-s}}.$$

A short calculation shows

$$\frac{L(s, \chi_0)}{L(2s, \chi_0)} = \prod_{P|m}(1 + |P|^{-s})^{-1} \frac{1 - q^{1-2s}}{1 - q^{1-s}}.$$

From this identity and what we have already proven about $G(s)$ we deduce that

$$\frac{(1 - q^{1-2s})L(s, \chi)}{(1 - q^{1-s})} = \sum_{n} \frac{a(n)}{|n|^s},$$

a Dirichlet series with non-negative coefficients.

It is now convenient to switch to a new variable, $u = q^{-s}$. The above equation becomes

$$\frac{(1 - qu^2)L^*(u, \chi)}{1 - qu} = \sum_{d} A(d)u^d,$$

where $L^*(u, \chi)$ is a polynomial in u by Proposition 4.3, and

$$A(d) = \sum_{n, \deg(n)=d} a(n)$$

is non-negative for all $d \geq 0$ and $A(0) = 1$. The Dirichlet series converges for $Re(s) > 1$ which implies the power series in u converges for $|u| < q^{-1}$. Finally, notice that $s = 1$ corresponds to q^{-1} so what we are trying to prove is that $L^*(q^{-1}, \chi) \neq 0$. We now have developed everything we need to give a quick proof of this.

We argue by contradiction. Suppose $L^*(q^{-1}, \chi) = 0$. Then $(1-qu)$ divides $L^*(u, \chi)$ and the left-hand side of the above equation is a polynomial in u. It follows that the right-hand side is a polynomial in u with non-negative coefficients and constant term 1. It therefore cannot have a positive root. However, the left-hand side vanishes when $u = 1/\sqrt{q}$. This is a contradiction, so $L^*(q^{-1}, \chi) \neq 0$ and thus, $L(1, \chi) \neq 0$. We have proven the following key result.

Proposition 4.6. *Let χ be a non-trivial Dirichlet character modulo m. Then, $L(1, \chi) \neq 0$.*

From Proposition 4.6 and previous remarks we see that as $s \to 1$ with s real and greater than 1 we have

$$\lim_{s \to 1} \log L(s, \chi_o) = \infty \quad \text{and} \quad \lim_{s \to 1} \log L(s, \chi) \quad \text{exists, for } \chi \neq \chi_o.$$

Here, and in what follows we take for $\log(z)$ the principal branch of the logarithm.

Theorem 4.7. *Let $a, m \in A$ be two relatively prime polynomials with m of positive degree. Consider the set of primes, $S = \{P \in A \mid P \equiv a \;(\mathrm{mod}\; m)\}$. Then, $\delta(S) = 1/\Phi(m)$. In particular, S is an infinite set.*

Proof. Using the product formula for $L(s, \chi)$ and the same technique used in the proof of Proposition 4.1, one finds

$$\log L(s, \chi) = \sum_p \frac{\chi(P)}{|P|^s} + R(s, \chi),$$

where the function $R(s, \chi)$ is bounded as s tends to 1 from above. Multipy both sides by $\bar{\chi}(a)$ and sum over all χ. Using the orthogonality relation for Dirichlet characters, Proposition 4.2, part (2), we obtain

$$\sum_\chi \chi(a) \log L(s, \chi) = \Phi(m) \sum_{P \equiv a \;(\mathrm{mod}\; m)} \frac{1}{|P|^s} + R(s),$$

where $R(s)$ is a function which remains bounded as $s \to 1$.

Divide each summand on the left-hand side of the above equation by $\sum_P |P|^{-s}$ and let s tend to 1 from above. By Proposition 4.1 and the remarks preceding the theorem, the summand corresponding to the trivial character tends to 1, while each summand corresponding to a non-trivial character tends to zero. If we divide the right-hand side by $\sum_P |P|^{-s}$ and let s tend to 1 from above, we get $\Phi(m)\delta(S)$. The result follows.

Theorem 4.7 is the original form of Dirichlet's theorem. It is possible, with more work, to prove a much stronger form of the theorem. Suppose $a, m \in A$ are relatively prime and that m has positive degree. Consider the set of primes

$$S_N(a, m) = \{P \in A \mid P \equiv a \;(\mathrm{mod}\; m), \; \deg(P) = N\}.$$

We claim that for all large integers N this set is not empty. The following theorem proves this and more.

Theorem 4.8.

$$\#S_N(a,m) = \frac{1}{\Phi(m)} \frac{q^N}{N} + O\left(\frac{q^{\frac{N}{2}}}{N}\right) .$$

It will take us several steps to prove this result, but first, a remark. Let S_N be the set of primes of degree N. We have seen (Theorem 2.2) that

$$\#S_N = \frac{q^N}{N} + O\left(\frac{q^{\frac{N}{2}}}{N}\right) .$$

Putting this together with the statement of the theorem we find

$$\lim_{N \to \infty} \frac{\#S_N(a,m)}{\#S_N} = \frac{1}{\Phi(m)} .$$

This is a natural density analogue to the Dirichlet density form of the main theorem.

Proof of Theorem 4.8. The idea of the proof is to realize that the L-series $L(s,\chi)$ can be expressed as a product in two ways. One way, which we have already considered, is as an Euler product. The other is as a product over its complex zeros. This is made easier by rewriting, as we have done before, everything in terms of the variable $u = q^{-s}$. If χ is not trivial, then by Proposition 4.3, $L(s,\chi)$ is a polynomial in q^{-s} of degree at most $M-1$ where $M = \deg(m)$. We have

$$L^*(u,\chi) = \sum_{k=0}^{M-1} a_k(\chi)u^k = \prod_{i=1}^{M-1} (1 - \alpha_i(\chi)u) . \tag{1}$$

The second expression for $L^*(s,\chi)$ comes from rewriting the Euler product for $L(s,\chi)$ in terms of u. We first regroup the terms in the Euler product.

$$L(s,\chi) = \prod_{P \nmid m} (1 - \chi(P)|P|^{-s})^{-1} = \prod_{d=1}^{\infty} \prod_{\substack{P \nmid m \\ \deg(P)=d}} (1 - \chi(P)q^{-ds})^{-1} .$$

Now, make the substitution $u = q^{-s}$. We obtain the expression

$$L^*(u,\chi) = \prod_{d=1}^{\infty} \prod_{\substack{P \nmid m \\ \deg(P)=d}} (1 - \chi(P)u^d)^{-1} . \tag{2}$$

Our intention is to take the logarithmic derivative of both expressions, write the results as power series in u and compare coefficients. Afterwards we apply the orthogonality relations to isolate the primes congruent to a modulo m. However, in addition to the algebra involved, we will have to do a number of estimates. One of these estimates will involve invoking a deep result of A. Weil. The others are more elementary.

We begin by writing down an identity which will be used repeatedly. Namely,

$$u\frac{d}{du}\left(\log(1-\alpha u)^{-1}\right) = \sum_{k=1}^{\infty} \alpha^k u^k . \tag{3}$$

Here α is a complex number. The sum converges for all u such that $|u| < |\alpha|^{-1}$. The proof of this identity is a simple exercise using the geometric series.

For each character χ modulo m define the numbers $c_N(\chi)$ by

$$u\frac{d}{du}\log(L^*(u,\chi) = \sum_{N=1}^{\infty} c_N(\chi)u^N .$$

We claim that

$$c_N(\chi_o) = q^N + O(1) \quad \text{and that} \quad c_N(\chi) = O(q^{\frac{N}{2}}) \text{ if } \chi \neq \chi_o . \tag{4}$$

The easy case is when $\chi = \chi_o$. Recall that

$$L(s,\chi_o) = \prod_{P|m}(1-|P|^{-s})\,\zeta_A(s) .$$

Thus,

$$L^*(u,\chi_o) = \prod_{P|m}(1-u^{\deg P})\,\frac{1}{1-qu} .$$

It now follows immediately, using Equation 3 and the additivity of the logarithmic derivative, that $c_N(\chi_o) = q^N + O(1)$. For $\chi \neq \chi_o$, by combining Equation 1 with Equation 3 we find

$$c_N(\chi) = -\sum_{k=1}^{M-1} \alpha_k(\chi)^N .$$

It follows from the analogue of the Riemann hypothesis for function fields over a finite field that each of the roots $\alpha_k(\chi)$ has absolute value either 1 or $\sqrt{q}$. This is the deepest part of the proof and is due to A. Weil (see Weil [1]). We will discuss it in some detail in the next chapter. In the Appendix to this book we will present an "elementary" proof, due to E. Bombieri, of this important result . Assuming it for now, we see immediately from the last equation that $c_N(\chi) = O(q^{N/2})$. Thus, we have verified both assertions of (4) above.

It should be remarked that one can prove much more easily, a weaker result than the Riemann hypothesis which has the effect of replacing the error term in the theorem with $O(q^{\theta N})$ where θ is some real number less than 1. This still gives the corollary that the set $S_N(a, m)$ is non-empty for all large N. We will indicate how to prove this in the next chapter.

We now continue with the proof of the theorem. Consider the Euler product expansion of $L^*(s, \chi)$ given by Equation 2. Take the logarithmic derivative of both sides and multiply both sides of the resulting equation by u. Again using Equation 3 we find

$$c_N(\chi) = \sum_{\substack{k,P \\ k \deg P = N}} \deg P \; \chi(P)^k \; .$$

In the sum on the right-hand side separate out the terms corresponding to $k = 1$. The result is $N \sum_{\deg P = N} \chi(P)$. The rest of the terms can be written as follows:

$$\sum_{\substack{d \mid N \\ d \leq N/2}} d \sum_{\deg P = d} \chi(P)^{N/d} \; .$$

The inner sum in absolute value is less than or equal to $\#\{P \in A \mid \deg P = d\} = q^d/d + O(q^{d/2}/d)$ by Theorem 2.2. Thus the double sum is bounded by

$$1 + q + q^2 + \cdots + q^{[N/2]} + O(1 + q + q^2 + \cdots + q^{[N/4]}) = O(q^{\frac{N}{2}}) \; .$$

We have proven

$$c_N(\chi) = N \sum_{\deg P = N} \chi(P) + O(q^{\frac{N}{2}}) \; . \tag{5}$$

Finally we compute the expression $\sum_{\chi} \bar{\chi}(a)c_N(\chi)$ in two ways. First we use Equation 5 and then we use Equation 4.

From the orthogonality relations and Equation 5 we find

$$\frac{1}{\Phi(m)} \sum_{\chi} \bar{\chi}(a)c_N(\chi) = N \#S_N(a, m) + O(q^{\frac{N}{2}}) \; .$$

Next, from Equation 4 we see

$$\sum_{\chi} \bar{\chi}(a)c_N(\chi) = q^N + O(q^{\frac{N}{2}}) \; .$$

So, we finally arrive at the main result:

$$\#S_N(a, m) = \frac{1}{\Phi(m)} \frac{q^N}{N} + O\left(\frac{q^{\frac{N}{2}}}{N} \right) \; .$$

Exercises

1. Let $S = \{P_1, P_2, \dots\}$ be the set of monic primes in A. Let $S_i = \{P_i\}$ be the set consisting of P_i alone. Then, $S = \cup_{i=1}^{\infty} S_i$. Show that this implies that Dirichlet density is not countably additive.

2. Let $P(T) \in A$ and define $N(U^2 = P(T))$ to be the number of pairs $(\alpha, \beta) \in \mathbb{F} \times \mathbb{F}$ such that $\beta^2 = P(\alpha)$. Show that

$$N(U^2 = P(T)) = \sum_{\alpha \in \mathbb{F}} (1 + P(\alpha)^{\frac{q-1}{2}}) \ .$$

3. Suppose q is odd and let $P \in A$ is a monic irreducible of degree two and that $\chi(a) = (a/P)_2$ for all $a \in A$. Show that $L_A(s, \chi) = 1 \pm q^{-s}$. (Hint: Use the Reciprocity Law and Exercise 2).

4. In general, suppose $P \in A$ is a monic irreducible of positive degree and set $\chi(a) = (a/P)_2$. Show that

$$\sum_{\substack{a \text{ monic} \\ \deg a = 1}} \chi(a) = \pm(q - N(U^2 = P(T))) \ .$$

5. With the same notation as in Exercise 4, consider the coefficient of q^{-s} in $L(s, \chi)$. Use Exercise 4 and the Riemann Hypothesis for function fields to prove

$$|N(U^2 = P(T)) - q| \le (\deg P - 1)\sqrt{q} \ .$$

6. Let $h(T) \in A$ be a polynomial of degree m with a non-zero constant term. Show that there are infinitely many primes in A whose first $m+1$ terms coincide with $h(T)$. What is the Dirichlet density of this collection of primes?

7. Let $\{\alpha_1, \alpha_2, \dots, \alpha_q\}$ be the elements of $\mathbb{F}$ labeled in some order and choose elements $\beta_i \in \mathbb{F}^*$ for $i = 1, \dots, q$, where repetition is allowed. Prove that thee are infinitely many primes, $P(T)$, such that $P(\alpha_i) = \beta_i$ for $i = 1, \dots, q$. What is the Dirichlet density of this set of primes?

5
Algebraic Function Fields and Global Function Fields

So far we have been working with the polynomial ring A inside the rational function field $k = F(T)$. In this section we extend our considerations to more general function fields of transcendence degree one over a general constant field. This process is somewhat like passing from elementary number theory to algebraic number theory. The Riemann-Roch theorem is the fundamental result needed to accomplish this generalization. We will give a proof of this fundamental result in Chapter 6. In this chapter we give the basic definitions, state the theorem, and derive a number of important corollaries. After this is accomplished, attention will be shifted to function fields over a finite constant field. Such fields are called global function fields. The other class of global fields are algebraic number fields. All global fields share a great number of common features. We introduce the zeta function of a global function field and explore its properties. The Riemann hypothesis for such zeta functions will be explained in some detail, and we will derive several very important consequences, among others an analogue for the prime number theorem for arbitrary global function fields. A proof of the Riemann hypothesis will be given in the appendix. In this chapter we will prove a weak version. This is enough to yield the analogue of the prime number theorem , albeit with a poor error term. In later chapters we will also explore L-functions associated to global function fields - both Hecke L-functions (generalizations of Dirichlet L-functions) and Artin L-functions.

One final comment before we begin. Our treatment of this subject is very arithmetic. The geometric underpinnings will not be much in evidence. The whole subject can be dealt with under the aspect of curves over finite fields. We have chosen the arithmetic approach because our guiding theme in this book will be the exploration of the rich analogies that exist between algebraic number fields and global function fields.

To begin with it is not necessary to restrict the constant field F to be finite. In fact, in this first part of the chapter we make no restrictions on F whatsoever. A function field in one variable over F is a field K, containing F and at least one element x, transcendental over F, such that $K/F(x)$ is a finite algebraic extension. Such a field is said to have transcendence degree one over F. It is not hard to show that the algebraic closure of F in K is finite over F. One way to see this is to note that if E is a subfield of K, which is algebraic over F, then $[E : F] = [E(x) : F(x)] \leq [K : F(x)]$. So, replacing F with its algebraic closure in K, if necessary, we assume that F is algebraically closed in K. In that case, F is called the constant field of K. Note the following simple consequence of this definition. If F is the constant field of K and $y \in K$ is not in F, then y is transcendental over F. It is also true that $K/F(y)$ is a finite extension. To see this, note that y is algebraic over $F(x)$ which shows there is a non-zero polynomial in two variables $g(X, Y) \in F[X, Y]$ such that $g(x, y) = 0$. Since y is transcendental over F we must have that $g(X, Y) \notin F[Y]$. It follows that x is algebraic over $F(y)$. Since K is finite over $F(x, y)$ and $F(x, y)$ is finite over $F(y)$, it follows that K is finite over $F(y)$.

A prime in K is, by definition, a discrete valuation ring R with maximal ideal P such that $F \subset R$ and the quotient field of R equal to K. As a shorthand such a prime is often referred to as P, the maximal ideal of R. The ord function associated with R is denoted $\mathrm{ord}_P(*)$. The degree of P, $\deg P$, is defined to be the dimension of R/P over F which can be shown to be finite. We sketch the proof. Choose an element $y \in P$ which is not in F. By the deductions of the last paragraph, $K/F(y)$ is finite . We claim that $[R/P : F] \leq [K : F(y)]$. To see this let $u_1, u_2, \ldots, u_m \in R$ be such that the residue classes modulo P, $\bar{u}_1, \bar{u}_2, \ldots, \bar{u}_m$, are linearly independent over F. We claim that $u_1, u_2, \ldots, u_m$ are linearly independent over $F(y)$. Suppose not. Then we could find polynomials in y, $\{f_1(y), f_2(y), \ldots, f_m(y)\}$, such that

$$f_1(y)u_1 + f_2(y)u_2 + \cdots + f_m(y)u_m = 0 \ .$$

It is no loss of generality to assume that not all the polynomials $f_i(y)$ are divisible by y. Now, reducing this relation modulo P gives a non-trivial linear relation for the elements $\bar{u}_i$ over F, a contradiction. Thus, $\{u_1, u_2, \ldots, u_m\}$ is a set linearly independent over $F(y)$ and it follows that $m \leq [K : F(y)]$ which proves the assertion.

To illustrate these definitions, consider the case of the rational function field $F(x)$. Let $A = F[x]$. Every non-zero prime ideal in A is generated by a unique monic irreducible P. The localization of A at P, A_P, is a discrete valuation ring. We continue to use the letter P to denote the unique maximal ideal of A_P. It is clear that P is a prime of $F(x)$ in the above sense. This collection of primes can be shown to almost exhaust the set of primes of $F(x)$. In fact, there is just one more. Consider the ring $A' = F[x^{-1}]$ and the prime ideal P' generated by x^{-1} in A'. The localization of A' at P' is a discrete valuation ring which defines a prime of $F(x)$ called the prime at infinity. This is usually denoted by P_∞ or, more simply, by "∞" alone. The corresponding ord-function, ord_∞, attaches the value $-\deg(f)$ to any polynomial $f \in A$ and thus the value $\deg(g) - \deg(f)$ to any rational function f/g where $f, g \in A$. The reader may wish to supply the proof that the only primes of $F(x)$ are the ones attached to the monic irreducibles, called the finite primes, together with the prime at infinity. The degree of any finite prime is equal to the degree of the monic irreducible to which it corresponds, and the degree of the prime at infinity is 1.

Returning to the general case, the group of divisors of K, $\mathcal{D}_K$, is by definition the free abelian group generated by the primes. We write these additively so that a typical divisor looks like $D = \sum_P a(P)P$. The coefficients, $a(P)$, are uniquely determined by D and we will sometimes denote them as $\text{ord}_P(D)$. The degree of such a divisor is defined as $\deg(D) = \sum_P a(P) \deg P$. This gives a homomorphism from $\mathcal{D}_K$ to $\mathbb{Z}$ whose kernel is denoted by $\mathcal{D}_K^o$, the group of divisors of degree zero.

Let $a \in K^*$. The divisor of a, (a), is defined to be $\sum_P \text{ord}_P(a)P$. It is not hard to see that (a) is actually a divisor, i.e., that $\text{ord}_P(a)$ is zero for all but finitely many P. The idea of the proof will be included in the proof of Proposition 5.1 (given below). The map $a \to (a)$ is a homomorphism from K^* to $\mathcal{D}_K$. The image of this map is denoted by $\mathcal{P}_K$ and is called the group of principal divisors.

If P is a prime such that $\text{ord}_P(a) = m > 0$, we say that P is a zero of a of order m. If $\text{ord}_P(a) = -n < 0$ we say that P is a pole of a of order n. Let

$$(a)_o = \sum_{\substack{P \\ \text{ord}_P(a) > 0}} \text{ord}_P(a)\, P \quad \text{and} \quad (a)_\infty = -\sum_{\substack{P \\ \text{ord}_P(a) < 0}} \text{ord}_P(a)\, P .$$

The divisor $(a)_o$ is called the divisor of zeros of a and the divisor $(a)_\infty$ is called the divisor of poles of a. Note that $(a) = (a)_o - (a)_\infty$.

Proposition 5.1. *Let $a \in K^*$. Then, $\text{ord}_P(a) = 0$ for all but finitely many primes P. Secondly, $(a) = 0$, the zero divisor, if and only if $a \in F^*$, i.e., a is a non-zero constant. Finally, $\deg(a)_o = \deg(a)_\infty = [K : F(a)]$. It follows that $\deg(a) = 0$, i.e., the degree of a principal divisor is zero.*

Proof. (Sketch) If $a \in F^*$, it is easy to see from the definitions that $(a) = 0$. So, suppose $a \in K^* - F^*$. Then, as we have seen, K is finite over $F(a)$. Let R be the integral closure of $F[a]$ in K. R is a Dedekind domain (see Samuel and Zariski [1], Chapter V, Theorem 19). Let $Ra = \mathfrak{P}_1^{e_1} \mathfrak{P}_2^{e_2} \cdots \mathfrak{P}_g^{e_g}$ be the prime decomposition of the principal ideal Ra in R. The localizations of R at the prime ideals $\mathfrak{P}_i$ are primes of the field K. If we denote by P_i the maximal ideals of these discrete valuation rings we find that $\text{ord}_{P_i}(a) = e_i$.

It is now not hard to show that the finite set $\{P_1, P_2, \ldots, P_g\}$ is the set of zeros of a. Applying the same reasoning to a^{-1} we see that the set of poles of a are is also finite. This proves the first assertion. It also proves the second assertion since if a is not in F^* we see that the set of P such that $\text{ord}_P(a) > 0$ is not empty.

To show $[K : F(a)] = \deg(a)_o = \deg(a)_\infty$ we can apply Theorem 7.6 of this book if we assume that F is a perfect field. For the general case, see Deuring [1], Chevalley [1], or Stichtenoth [1].

For emphasis we point out that implicit in the above sketch is the fact that every non-constant element of K has at least one zero and at least one pole.

Two divisors, D_1 and D_2, are said to be linearly equivalent, $D_1 \sim D_2$ if their difference is principal, i.e., $D_1 - D_2 = (a)$ for some $a \in K^*$. Define $Cl_K = \mathcal{D}_K / \mathcal{P}_K$, the group of divisor classes. Since the degree of a principal divisor is zero, the degree function gives rise to a homomorphism from Cl_K to $\mathbb{Z}$. The kernel of this map is denoted Cl_K^o, the group of divisor classes of degree zero.

We are almost ready to state the Riemann-Roch theorem. Just two more definitions are needed. A divisor, $D = \sum_P a(P)P$, is said to be an effective divisor if for all P, $a(P) \geq 0$. We denote this by $D \geq 0$.

Definition. Let D be a divisor. Define $L(D) = \{x \in K^* \,|\, (x) + D \geq 0\} \cup \{0\}$. It is easy to see that $L(D)$ has the structure of a vector space over F and it can be proved that it is finite dimensional over F (see Exercises 17 and 18). The dimension of $L(D)$ over F is denoted by $l(D)$. The number $l(D)$ is sometimes referred to as the dimension of D.

Lemma 5.2. *If A and B are linearly equivalent divisors, then $L(A)$ and $L(B)$ are isomorphic. In particular, $l(A) = l(B)$.*

Proof. Suppose $A = B + (h)$. Then a short calculation shows that $x \to xh$ is an isomorphism from $L(A)$ with $L(B)$.

Lemma 5.3. *If $\deg(A) \leq 0$ then $l(A) = 0$ unless $A \sim 0$ in which case $l(A) = 1$.*

Proof. If $\deg(A) < 0$ and $x \in L(A)$, then $\deg((x) + A)$ is both < 0 and ≥ 0 which is a contradiction. If $\deg(A) = 0$ and $L(A)$ is not empty, let $x \in L(A)$. Then $(x) + A \geq 0$ and has degree zero, so it must be the zero divisor. Thus, $A \sim 0$. Conversely, if $A \sim 0$, then $l(A) = l(0) = 1$ since $L(0) = F$ because $x \in L(0)$ implies x has no poles and so $x \in F$.

Before stating the Riemann-Roch theorem it is worth pointing out that Lemma 5.2 shows $l(A)$ depends only on the class of A. Similarly, $\deg(A)$ depends only on the class of A. Thus we could define $l(\bar{A})$ and $\deg(\bar{A})$ and state Riemann-Roch in terms of divisor classes. However, we prefer to state it in terms of divisors which is more customary.

Theorem 5.4. (Riemann-Roch) *There is an integer $g \geq 0$ and a divisor class C such that for $C \in \mathcal{C}$ and $A \in \mathcal{D}_K$ we have*

$$l(A) = \deg(A) - g + 1 + l(C - A).$$

The proof will be given in the next chapter. For other treatments see Chevalley [1], Deuring [1], Eichler [1], Moreno [1], or Stichtenoth [1]. The integer g is uniquely determined by K, as we shall see, and is called the genus of K. The genus of a function field is a key invariant. The divisor class C is also uniquely determined and is called the canonical class. It is related to differentials of K. In the next chapter we will define the notion of a Weil differential. To each Weil differential will be associated a divisor. It turns out that all such divisors are equivalent and that C, the canonical class, is the equivalence class of divisors of Weil differentials.

We now give a series of corollaries to this important theorem.

Corollary 1. (Riemann's inequality) *For all divisors A, we have $l(A) \geq \deg(A) - g + 1$.*

Corollary 2. *For $C \in \mathcal{C}$ we have $l(C) = g$.*

Proof. Set $A = 0$ in the theorem.

Corollary 3. *For $C \in \mathcal{C}$ we have $\deg(C) = 2g - 2$.*

Proof. Set $A = C$ in the theorem, and use Corollary 2.

Corollary 4. *If $\deg(A) \geq 2g - 2$, then $l(A) = \deg(A) - g + 1$ except in the case $\deg(A) = 2g - 2$ and $A \in \mathcal{C}$.*

Proof. If $\deg(A) \geq 2g - 2$, then $\deg(C - A) \leq 0$. Now use Lemma 5.3.

Corollary 5. *Suppose that g' and C' have the same properties as those of g and C stated in the theorem. Then, $g = g'$ and $C \sim C'$.*

Proof. Find a divisor A whose degree is larger than $\max(2g - 2, 2g' - 2)$ (a large positive multiple of a prime will do). By Corollary 4, $l(A) = \deg(A) - g + 1 = \deg(A) - g' + 1$. Thus, $g = g'$. Now set $A = C'$ in the statement of the theorem. Using Corollaries 2 and 3, applied to C', we see that $l(C - C') = 1$. There is an $x \in K^*$ such that $(x) + C - C' \geq 0$. On the other hand, $(x) + C - C'$ has degree zero by Corollary 3. Thus, it is the zero divisor, and $C \sim C'$.

As an example of these results, consider the rational function field $F(x)$. Let (R_∞, P_∞) be the prime which is, as we have seen, the localization of the ring $F[1/x]$ at the prime ideal generated by $1/x$. The corresponding ord function is $\mathrm{ord}_\infty(f) = -\deg(f)$. By Corollary 4, for n large and positive we must have $l(nP_\infty) = n - g + 1$. On the other hand, one can prove that $f \in L(nP_\infty)$ if and only if f is a polynomial in T of degree $\leq n$. Thus, $l(nP_\infty) = n + 1$. It follows that $g = 0$. From this and Corollary 3 one sees that C has degree -2. It can be shown that $Cl_K^o = (1)$ so there is only one class of degree -2 and we can choose any divisor of degree -2 for C. A conventional choice is $C = -2P_\infty$.

We can characterize the rational function field intrinsically as follows: K/F is a rational function field if and only if there exists a prime P of K degree 1 and the genus of K is 0. We have seen that rational function fields have this property. Now, assume these conditions and consider $l(P)$. Since $g = 0$ we have $l(D) = \deg D - g + 1 = \deg D + 1$ for $\deg D > 2g - 2 = -2$. Thus, $l(P) = 2$ and we can find a non-constant function x such that $(x) + P \geq 0$. Since $\deg ((x) + P) = 1$, it follows that $(x) + P = Q$, a prime of degree 1. Thus, $(x) = Q - P$ and it follows that $[K : F(x)] = 1$. Thus, $K = F(x)$ as asserted.

In the same way one can investigate fields of genus 1. Assume K is a function field of genus 1 and that there is a prime P of degree 1. Such a field is called an elliptic function field. By the above results we have , for any divisor D, $l(D) = \deg(D)$ if $\deg(D) > 2g - 2 = 0$. Thus, $l(nP) = n$ for positive integers n. Taking $n = 2$ and $n = 3$ we see there exist functions x and y with polar divisors $2P$ and $3P$, respectively. It follows that $[K : F(x)] = 2$ and $[K : F(y)] = 3$ so that $K = F(x, y)$. We see that y must satisfy a quadratic equation over $F(x)$. One can prove much more. If the characteristic of F is different from 2 one can show that by a small change of variables y can be chosen so that $y^2 = f(x)$ where $f(x)$ is a cubic polynomial of degree 3 without repeated roots. See Silverman [3] for more details.

For the rest of this section we assume that $F = \mathbb{F}$ is a finite field with q elements. A function field in one variable over a finite constant field is called a global function field. Our next goal is to define the zeta function of a global function field $K/\mathbb{F}$ and to investigate its properties.

It was proven by F.K. Schmidt, Schmidt [1], that a function field over a finite field always has divisors of degree 1. We will assume this, although it is possible to give a proof without introducing any new concepts. Using Schmidt's theorem, we have an exact sequence

$$(0) \to Cl_K^o \to Cl_K \to \mathbb{Z} \to (0).$$

We will prove shortly that the group Cl_K^o is finite. Denote its order by h_K. The number h_K is called the class number of the field K. This number is an important invariant of K and has been the object of much study. The above exact sequence shows that for any integer n there are exactly h_K classes of degree n.

Lemma 5.5. *For any integer $n \geq 0$ the number of effective divisors of degree n is finite.*

Proof. (Sketch) Choose an $x \in K$ such that x is transcendental over $\mathbb{F}$. $K/\mathbb{F}(x)$ is finite. The primes of $\mathbb{F}(x)$ are in one to one correspondence with the monic irreducible polynomials in $\mathbb{F}[x]$ with the one exception of the prime at infinity. Thus, there are only finitely many primes of $\mathbb{F}(x)$ of any fixed degree. By standard theorems on extensions of primes (see Chapter 7) one sees that there are only finitely many primes of K of fixed degree. If $\sum_P a(P)P$ is an effective divisor of degree n then each prime that occurs with positive coefficient must have degree $\leq n$. There are only finitely many such primes. Moreover the coefficients must be $\leq n$, so there are at most finitely many such effective divisors.

We define a_n to be the number of primes of degree n and b_n to be the number of effective divisors of degree n. Both these numbers are of considerable interest.

Lemma 5.6. *The number of divisor classes of degree zero, h_K, is finite.*

Proof. Let D be a divisor of degree 1. If A is any divisor of degree 0, then $\deg(gD+A) = g$ and so by Riemann's inequality, $l(gD+A) \geq g-g+1 = 1$. Let $f \in L(gD + A)$. Then, $B = (f) + gD + A \geq 0$ and so $A \sim B - gD$ where B is an effective divisor of degree g. It follows that the number of divisor classes of degree zero is bounded above by the number of effective divisors of degree g which is finite by Lemma 5.5. More precisely, what we have shown is that $h_k \leq b_g$.

We have now proved that the class number $h_K = |Cl_K^o|$ is finite. Later we will give estimates for the size of h_K derived from the Riemann hypothesis for function fields (see Proposition 5.11).

Lemma 5.7. *For any divisor A, the number of effective divisors in $\bar{A}$ is $\frac{q^{l(A)}-1}{q-1}$.*

Proof. We begin by showing that $\bar{A}$ contains effective divisors if and only if $l(A) > 0$.

Suppose $B \in \bar{A}$ and is effective. There is an $f \in K^*$ such that $(f) + A = B \geq 0$, so $f \in L(A)$ and $l(A) > 0$. The converse is obtained by just running this proof backwards.

Suppose $l(A) > 0$. The map from $L(A) - \{0\}$ to effective divisors in $\bar{A}$ given by $f \to (f)+A$ is onto. Two functions f and f' have the same image iff $(f) + A = (f') + A$ iff $(f) = (f')$ iff $(f'f^{-1}) = 0$. The last condition happens iff $f'f^{-1}$ is in $\mathbb{F}^*$ by Proposition 5.1. Since $L(A) - \{0\}$ has $q^{l(A)} - 1$ elements and the fibers of our map have $q - 1$ elements, the result follows.

Finally, if $l(A) = 0$, then $q^{l(A)} - 1 = 0$ and the result holds in this case as well.

For $A \in \mathcal{D}_K$ define the norm of A, NA, to be $q^{\deg(A)}$. Note that NA is a positive integer and that for any two divisors A and B we have $N(A+B) = NANB$.

Definition. The zeta function of K, $\zeta_K(s)$, is defined by

$$\zeta_K(s) = \sum_{A \geq 0} NA^{-s}.$$

Over the rational function field $k = \mathbb{F}(T)$ we did not discuss the zeta function of k but rather the zeta function associated to the ring $A = \mathbb{F}[T]$. These are closely related. In fact, it is not hard to prove that $\zeta_A(s) = \zeta_k(s)(1 - q^{-s})$, so $\zeta_k(s) = (1 - q^{1-s})^{-1}(1 - q^{-s})^{-1}$. Also in the general case it is sometimes useful to associate zeta functions with appropriate subrings of the field. However, for the purposes of the following discussion we will concentrate on the zeta function of the field

The term NA^{-s} in the definition of the zeta function is equal to q^{-ns} where n is the degree of A. Thus the zeta function can be rewritten in the form

$$\zeta_K(s) = \sum_{n=1}^{\infty} \frac{b_n}{q^{ns}} \ .$$

Another key fact is that we have an Euler product for $\zeta_K(s)$. Using the multiplicativity of the norm and the fact that $\mathcal{D}_K$ is a free abelian group on the set of primes we see, at least formally, that

$$\zeta_K(s) = \prod_{P} \left(1 - \frac{1}{NP^s}\right)^{-1} \ .$$

Recalling that a_n is the number of primes of degree n, we observe that this expression can be rewritten as follows:

$$\zeta_K(s) = \prod_{n=1}^{\infty} \left(1 - \frac{1}{q^{ns}}\right)^{-a_n} \ .$$

We shall soon see that all these expressions converge absolutely for $\Re(s) > 1$ and define analytic functions in this region.

Lemma 5.8. *Let $h = h_K$. For every integer n, there are h divisor classes of degree n. Suppose $n \geq 0$ and that $\{\bar{A}_1, \bar{A}_2, \ldots, \bar{A}_h\}$ are the divisor classes of degree n. Then the number of effective divisors of degree n, b_n, is given by $\sum_{i=1}^{h} \frac{q^{l(A_i)} - 1}{q - 1}$.*

Proof. The first assertion follows directly from Lemma 5.6 and the remarks preceding Lemma 5.5. The second follows just as directly from Lemmas 5.6 and 5.7.

By Lemma 5.7 and Corollary 4 to Theorem 5.4, we see that if $n > 2g - 2$, then $b_n = h_k \frac{q^{n-g+1} - 1}{q - 1}$. It follows that $b_n = O(q^n)$. From this fact, and the expression $\zeta_K(s) = \sum_{n=0}^{\infty} b_n q^{-ns}$, it follows that $\zeta_K(s)$ converges absolutely for all s with $\Re(s) > 1$.

In the same way we can prove the product expression for $\zeta_K(s)$ converges absolutely for $\Re(s) > 1$. To do this it suffices, by the theory of infinite products, to show that $\sum_{n=1}^{\infty} a_n |q^{-ns}|$ converges in this region. This follows immediately since $a_n \leq b_n = O(q^n)$.

The next thing to do is to investigate whether $\zeta_K(s)$ can be analytically continued to all of $\mathbb{C}$ and whether it satisfies a functional equation, etc. The next theorem shows that the answer to both these questions is yes, and that a lot more is true as well.

Theorem 5.9. *Let K be a global function field in one variable with a finite constant field $\mathbb{F}$ with q elements. Suppose that the genus of K is g. Then there is a polynomial $L_K(u) \in \mathbb{Z}[u]$ of degree $2g$ such that*

$$\zeta_K(s) = \frac{L_K(q^{-s})}{(1 - q^{-s})(1 - q^{1-s})} \ .$$

This holds for all s such that $\Re(s) > 1$ and the right-hand side provides an analytic continuation of $\zeta_K(s)$ to all of $\mathbb{C}$. $\zeta_k(s)$ has simple poles at $s = 0$ and $s = 1$. One has $L_K(0) = 1$, $L_K'(0) = a_1 - 1 - q$, and $L_K(1) = h_K$. Finally, set $\xi_K(s) = q^{(g-1)s}\zeta_K(s)$. Then for all s one has $\xi_K(1-s) = \xi_K(s)$ (this relationship is referred to as the functional equation for $\zeta_K(s)$).

Proof. It is convenient to work with the variable $u = q^{-s}$. Then

$$\zeta_K(s) \stackrel{def}{=} Z_K(u) = \sum_{n=0}^{\infty} b_n u^n \ .$$

We noted earlier that for $n > 2g - 2$ we have $b_n = h_K \frac{q^{n-g+1}-1}{q-1}$. Substituting this into the above formula and summing the resulting geometric series, yields

$$Z_K(u) = \sum_{n=0}^{2g-2} b_n u^n + \frac{h_K}{q-1}\left(\frac{q^g}{1-qu} - \frac{1}{1-u}\right)u^{2g-1}. \tag{1}$$

From this, simple algebraic manipulation shows

$$Z_K(u) = \frac{L_K(u)}{(1-u)(1-qu)} \quad \text{with } L_K(u) \in \mathbb{Z}[u]. \tag{2}$$

From Equation 2, we see the expression for $\zeta_k(s)$ given in the theorem is correct. We will show that $L_K(1)$ and $L_K(q^{-1})$ are both non-zero. Thus, $\zeta_K(s)$ has a pole at 0 and 1. The fact that $\deg L_K(u) \leq 2g$ also follows from this calculation. Substituting $u = 0$ yields $L_K(0) = 1$. Comparing the coefficients of u on both sides yields $b_1 = L_K'(0) + 1 + q$. It is easy to see that $b_1 = a_1 = $ the number of primes of K of degree one.

From Equation 1 above, we see that $\lim_{u \to 1}(u-1)Z_K(u) = h_K/(q-1)$. From Equation 2 we see

$$\lim_{u \to 1}(u-1)Z_K(u) = \frac{-L_K(1)}{1-q} \ .$$

Thus, $L_K(1) = h_K$, as asserted.

As for the functional equation, recall that $b_n = \sum_{\deg \bar{A}=n} (q^{l(\bar{A})} - 1)/(q-1)$. Then,

$$(q-1)Z_K(u) = \sum_{n=0}^{\infty} \left(\sum_{\deg \bar{A}=n} q^{l(\bar{A})} - 1 \right) u^n = \sum_{\deg \bar{A} \geq 0} q^{l(\bar{A})} u^{\deg \bar{A}} - h_K \frac{1}{1-u}$$

$$= \sum_{0 \leq \deg \bar{A} \leq 2g-2} q^{l(\bar{A})} u^{\deg \bar{A}} - h_K \frac{1}{1-u} + \sum_{2g-2 < \deg \bar{A} < \infty} q^{l(\bar{A})} u^{\deg \bar{A}}$$

$$= \sum_{0 \leq \deg \bar{A} \leq 2g-2} q^{l(\bar{A})} u^{\deg \bar{A}} - h_K \frac{1}{1-u} + h_K \frac{q^g u^{2g-1}}{1 - qu} \ .$$

Multiplying both sides by u^{1-g} we have $(q-1)u^{1-g} Z_K(u) = R(u) + S(u)$ where

$$R(u) = \sum_{0 \leq \deg \bar{A} \leq 2g-2} q^{l(\bar{A})} u^{\deg \bar{A} - g + 1} \quad \text{and} \quad S(u) = -h_K \frac{u^{1-g}}{1-u} + h_K \frac{q^g u^g}{1 - qu}.$$

A direct calculation shows that $S(u)$ is invariant under $u \to q^{-1} u^{-1}$. $R(u)$ is also invariant under this transformation. To see this, first note that

$$R(q^{-1}u^{-1}) = \sum_{\deg \bar{A} \leq 2g-2} q^{l(\bar{A})+g-1-\deg \bar{A}} u^{-\deg \bar{A}+g-1} \ .$$

From the Riemann-Roch Theorem, Theorem 5.4, and Corollary 3, we see

$$l(\mathcal{C} - \bar{A}) = \deg(\mathcal{C} - A) - g + 1 + l(\bar{A}) = g - 1 - \deg \bar{A} + l(\bar{A}) \ .$$

Substituting this expression into the formula for $R(q^{-1}u^{-1})$ yields

$$R(q^{-1}u^{-1}) = \sum_{\deg \bar{A} \leq 2g-2} q^{l(\mathcal{C}-\bar{A})} u^{\deg(\mathcal{C}-\bar{A})-g+1} \ .$$

Since $\bar{A} \to \mathcal{C} - \bar{A}$ is a permutation of the divisor classes of degree d with $0 \leq d \leq 2g - 2$ it follows that $R(q^{-1}u^{-1}) = R(u)$ as asserted. We have now completed the proof that $u^{1-g} Z_K(u)$ is invariant under the transformation $u \to q^{-1} u^{-1}$.

Since $u^{1-g} Z_K(u)$ is invariant under $u \to q^{-1} u^{-1}$, it follows easily that $q^{-g} u^{-2g} L_K(u) = L_K(q^{-1}u^{-1})$. Letting $u \to \infty$ we see that $\deg L_K(u) = 2g$ and that the highest degree term is $q^g u^{2g}$.

Finally, recalling that $u = q^{-s}$, we see that $u^{1-g} = q^{(g-1)s}$ and the transformation $u \to q^{-1} u^{-1}$ is the same as the transformation $s \to 1 - s$. So passing from the u language to the s language we see we have shown $\xi_K(s)$ is invariant under $s \to 1 - s$, as asserted.

The polynomial $L_K(u)$ defined in the theorem carries a lot of information. Since the coefficients are in $\mathbb{Z}$ we can factor this polynomial over the complex numbers,

$$L_K(u) = \prod_{i=1}^{2g}(1 - \pi_i u) \quad .$$

It is worth pointing out that the relation $L_K(q^{-1}u^{-1}) = q^{-g}u^{-2g}L_K(u)$ implies that the set $\{\pi_1, \pi_2, \ldots, \pi_{2g}\}$ is permuted by the transformation $\pi \to q/\pi$. This is easily seen to be equivalent to the functional equation for $\zeta_K(s)$.

Since $\zeta_K(s)$ has a convergent Euler product whose factors have no zeros in the region $Re(s) > 1$, it follows that $\zeta_K(s)$ has no zeros there. Consequently, $L_K(u)$ has no zeros in the region $\{u \in \mathbb{C} \mid |u| < q^{-1}\}$. For the inverse roots, π_i, the consequence is that $|\pi_i| \le q$. We will prove later, Proposition 5.13, that $|\pi_i| < q$ for all i and this will have a number of important applications. However, much more is true. The classical generalized Riemann hypothesis states that the zeros of $\zeta_K(s)$, the Dedekind zeta function of a number field K, has all its non-trivial zeros on the line $\Re(s) = 1/2$. Riemann conjectured this for $\zeta(s)$, the Riemann zeta function. Neither Riemann's conjecture nor its generalizations are known to be true. In fact, these are among the most important unsolved problems in all of mathematics. However, the analogous statement over global function fields was proved by A. Weil in the 1940s.

Theorem 5.10. (The Riemann Hypothesis for Function Fields) *Let K be a global function field whose constant field $\mathbb{F}$ has q elements. All the roots of $\zeta_K(s)$ lie on the line $\Re(s) = 1/2$. Equivalently, the inverse roots of $L_K(u)$ all have absolute value $\sqrt{q}$.*

Theorem 5.10 was first conjectured for hyper-elliptic function fields by E. Artin in his thesis, Artin [1]. The important special case when $g = 1$ was proven by H. Hasse. The first proof of the general result was published by Weil in 1948. Weil gave two, rather difficult, proofs of this theorem. The first used the geometry of algebraic surfaces and the theory of correspondences. The second used the theory of abelian varieties. See Weil [1] and Weil [2]. The whole project required revisions in the foundations of algebraic geometry since he needed these theories to be valid over arbitrary fields not just algebraically closed fields in characteristic zero. In the early seventies, a more elementary proof appeared due, in a special case to Stepanov, and in the general case to Bombieri [1]. We will give an exposition of Bombieri's proof in the appendix to this book.

Here are two simple but important consequences of the Riemann Hypothesis.

Proposition 5.11. *The number of prime divisors of degree 1 of K, a_1, satisfies the inequality $|a_1 - q - 1| \le 2g\sqrt{q}$. Also, $(\sqrt{q} - 1)^{2g} \le h_K \le (\sqrt{q} + 1)^{2g}$.*

Proof. By Theorem 5.9, $L'_K(0) = a_1 - q - 1$. From the above factorization of $L_K(u)$ we see $-L'_K(0) = \pi_1 + \pi_2 + \cdots + \pi_{2g}$. The first assertion is immediate from this and Theorem 5.10.

As for the second assertion, we have $h_K = L_K(1) = \prod_{i=1}^{2g}(1 - \pi_i)$, by Theorem 5.9. Now use Theorem 5.10 once again.

Here are several qualitative consequences of this proposition. If q is big compared to the genus, then there must exist primes of degree one. Indeed, $a_1/q \to 1$ if we fix g and let q grow. Secondly, if $q > 4$ we must have $h_K > 1$. Also, if we fix g and let q tend to infinity then $h_K/q^g \to 1$ (here, K is varying over global fields of fixed genus g with varying constant fields). Moreover, if we fix $q > 4$ and let g grow, then $h_K \to \infty$.

We can now present a generalization of Proposition 2.3, which, as we pointed out, is an analogue of the prime number theorem.

Theorem 5.12.

$$a_N = \#\{P \mid \deg(P) = N\} = \frac{q^N}{N} + O\left(\frac{q^{\frac{N}{2}}}{N}\right) .$$

Proof. Using the Euler product decomposition and Theorem 5.9, we see

$$Z_K(u) = \frac{\prod_{i=1}^{2g}(1 - \pi_i u)}{(1 - u)(1 - qu)} = \prod_{d=1}^{\infty}(1 - u^d)^{-a_d} .$$

Take the logarithmic derivative of both sides, multiply the result by u, and equate the coefficients of u^N on both sides. We find

$$q^N + 1 - \sum_{i=1}^{2g} \pi_i^N = \sum_{d|N} d a_d .$$

Using the Möbius inversion formula, yields

$$N a_N = \sum_{d|N} \mu(d) q^{\frac{N}{d}} + 0 + \sum_{d|N} \mu(d) \left(\sum_{i=1}^{2g} \pi_i^{\frac{N}{d}}\right) .$$

Let $e(N)$ be -1 if N is even and 0 if N is odd. Then, as we saw in the proof of Proposition 2.3,

$$\sum_{d|N} \mu(d) q^{\frac{N}{d}} = q^N - e(N) q^{\frac{N}{2}} + O(N q^{\frac{N}{3}}) .$$

Similarly, using the Riemann hypothesis, we see

$$\left| \sum_{d|N} \mu(d) \left(\sum_{i=1}^{2g} \pi^{\frac{N}{d}}\right) \right| \leq 2gq^{\frac{N}{2}} + 2gN q^{\frac{N}{4}} .$$

Putting the last three equations together, we find

$$Na_N = q^N + O(q^{\frac{N}{2}}) \ .$$

The theorem follows upon dividing both sides by N.

Note that, in this proof, it was crucial to know the size of the zeros of the zeta function. The proof of Proposition 2.3 was so easy because the zeta function of $A = \mathbb{F}[T]$ has no zeros!

We wish to derive yet another expression for the zeta function. To this end we consider once more the equation

$$Z_K(u) = \prod_{d=1}^{\infty} (1 - u^d)^{-a_d} \ .$$

Take the logarithm of both sides and write the result as a power series in u using the identity $-\log(1 - u) = \sum_{m=1}^{\infty} u^m/m$. The result is

$$\log Z_K(u) = \sum_{m=1}^{\infty} \frac{N_m}{m} u^m \ ,$$

where the numbers N_m are defined by $N_m = \sum_{d|m} d a_d$. These numbers have a very appealing geometric interpretation, which we shall explain in more detail later. Roughly speaking, what is going on is that the function field $K/\mathbb{F}$ is associated to a complete, non-singular curve X defined over $\mathbb{F}$. The number N_m is the number of rational points on X over the unique field extension $\mathbb{F}_m$ of $\mathbb{F}$ of degree m. In any case, using these numbers, the zeta function can be given by

$$Z_K(u) = \exp\left(\sum_{m=1}^{\infty} \frac{N_m}{m} u^m \right) \ .$$

In the course of the proof of Theorem 5.12, we showed that

$$N_m = q^m + 1 - \sum_{i=1}^{2g} \pi_i^m \ .$$

This equality plays an important role in the proof of the Riemann hypothesis for function fields. If we assume the Riemann hypothesis, another consequence is

$$|N_m - q^m - 1| \leq 2gq^{\frac{m}{2}} \ .$$

We will interpret this inequality in Chapter 8 when we discuss constant field extensions of function fields (see Proposition 8.18).

We conclude this chapter by showing how to obtain a weaker result than the Riemann hypothesis, which nevertheless is strong enough to give a proof that

$$a_N = \#\{P \mid \deg(P) = N\} \sim \frac{q^N}{N} \ .$$

Before the statement and proof of the next proposition, we need to deal with an important technical point. Since $\zeta_K(s)$ is a rational function of q^{-s}, it is a periodic function of s with period $2\pi i / \log(q)$. Since it has a pole at $s = 0$ and at $s = 1$ it has infinitely many poles on both the line $\Re(s) = 0$ and the line $\Re(s) = 1$. We will be concerned with the latter line. From Theorem 5.9, we see that all the poles on this line are at the points $1 + 2\pi m i / \log(q)$ for $m \in \mathbb{Z}$.

Proposition 5.13. *Let K be a global function field. The zeta function of K, $\zeta_K(s)$, does not vanish on the line $\Re(s) = 1$.*

Proof. The proof of this proceeds, for the most part, exactly as in the case where K is a number field. It is based on the trigonometric inequality

$$3 + 4\cos\theta + \cos 2\theta \geq 0 \ .$$

The proof of this inequality consists of nothing more than noticing that the left hand side is $2(1 + \cos\theta)^2$.

Write $s = \sigma + it$ where σ and t are real. Assume that $\sigma > 1$. Then a short calculation with the Euler product for $\zeta_K(s)$ yields

$$\Re \ \log \zeta_K(s) = \sum_{P,m} m^{-1} NP^{-m\sigma} \cos(t \log NP^m) \ .$$

Now, replace t with $0, t,$ and $2t$ and use the above identity to derive

$$3\Re\log\zeta_K(\sigma) + 4\Re\log\zeta_K(\sigma + it) + \Re\log\zeta_K(\sigma + 2it) \geq 0 \ .$$

Exponentiating, we find

$$|\zeta_K(\sigma)|^3 |\zeta_K(\sigma + it)|^4 |\zeta_K(\sigma + 2it)| \geq 1 \ .$$

This inequality holds for $\sigma > 1$ and all real t. Suppose t is such that $\zeta_K(1 + it) = 0$. Of course, such a t cannot be zero. It follows that $\zeta_K(\sigma + it)/(\sigma - 1)$ is bounded as $\sigma \to 1$. We know that $(\sigma - 1)\zeta_K(\sigma)$ is bounded as $\sigma \to 1$ since by Theorem 5.9, $\zeta_K(s)$ has a simple pole at $s = 1$. Finally, $\zeta_K(\sigma + 2it)$ is bounded as $\sigma \to 1$ provided that t is not an odd multiple of $\pi / \log(q)$ (see the remarks preceeding the Proposition). Assume this for now. Putting everything together shows that the left-hand side of the above inequality tends to zero as $\sigma \to 1$, which contradicts the fact that it is always greater than or equal to 1.

Now suppose that t is an odd multiple of $2\pi / \log(q)$. In this case, $q^{-(1+it)} = -q^{-1}$. We must show that $\zeta_K(1 + it) = Z_K(-q^{-1}) \neq 0$. By the functional equation, $Z_K(-q^{-1})$ is related to $Z_K(-1)$, which in turn is not zero if and only if $L_K(-1) \neq 0$. To show this we must, unfortunately, use a result from a later chapter, namely, Theorem 8.15.

Let $\mathbb{F}_2$ be a quadratic extension of the constant field $\mathbb{F}$. One can form a new function field from K by extending the field of constants from $\mathbb{F}$ to $\mathbb{F}_2$. Call this new field K_2. Using Theorem 8.15 we can derive the following relation between $L_{K_2}(u)$ and $L_K(u)$.

$$L_{K_2}(u^2) = L_K(u)L_K(-u) .$$

Substitute $u = 1$ into this relationship and use Theorem 5.9 once again. We find that $h_{K_2} = h_K L_K(-1)$ from which it is clear that $L_K(-1) \neq 0$.

Corollary. There is a real number $\theta < 1$ such that $\zeta_K(s)$ does not vanish in the region $\{s \in \mathbb{C} \mid \Re s > \theta\}$.

Proof. The zeta function is represented by a convergent Euler product in the region $\{s \in \mathbb{C} \mid \Re(s) > 1\}$ and so doesn't vanish there. By the functional equation (see Theorem 5.9) it doesn't vanish in $\{s \in \mathbb{C} \mid \Re(s) < 0\}$ either. From the Proposition it doesn't vanish on the boundary of these regions.

The key point that makes the function field case different from the number field case is that $\zeta_K(s)$ is a function of q^{-s} and so it is periodic with period $2\pi i / \log q$. Thus we may confine our search for zeros to the compact region $\{s \in \mathbb{C} \mid 0 \leq \Re(s) \leq 1, \ 0 \leq \Im(s) \leq 2\pi i / \log q\}$. The zero set of an analytic function is discrete, so the number of zeros in this region is finite. The corollary follows immediately.

The Riemann hypothesis for the function field case is that θ can be taken to be $1/2$. It is worth pointing out that nothing as strong as the above corollary is known to be true in the number field case. Zero free regions to the left of the line $\Re(s) = 1$ are known to exist, but the boundary of these regions approach the line as $|\Im(s)| \to \infty$.

Translating the above corollary into a result about $L_K(u) = \prod_{i=1}^{2g}(1 - \pi_i u)$, we see that the assertion is that $|\pi_i| \leq q^{\theta}$ for all $1 \leq i \leq 2g$. If we use this estimate instead of the Riemann hypothesis and follow the steps of the proof of Theorem 5.12, we arrive at the following result.

$$a_N = \#\{P \mid \deg(P) = N\} = \frac{q^N}{N} + O\left(\frac{q^{\theta N}}{N}\right) .$$

As promised, this is good enough to show $a_N \sim q^N / N$ as $N \to \infty$, a result which is much weaker than Theorem 5.12, but is still very interesting.

Exercises

1. Suppose K/F has genus zero. For a divisor D with $\deg D \geq -1$ show that $l(D) = \deg D + 1$.

2. Suppose K/F has genus zero and that C is a divisor in the canonical class. Show $l(-C) = 3$ and conclude that there is a prime P of degree less than or equal to 2.

3. Suppose K/F has genus zero and that there is a prime P of degree 1. Show $K = F(x)$ for some element $x \in K$.

4. Suppose K/F has genus zero and that P is a prime of degree 2. By Exercise 1, $l(P) = 3$. Let $\{1, x, y\}$ be a basis for $L(P)$. Show $K = F(x, y)$. Show further that $\{1, x, y, x^2, y^2, xy\} \subset L(2P)$ and conclude that x and y satisfy a polynomial of degree 2 over F.

5. Suppose that K/F has genus 1. Show that $l(D) = \deg D$ for all divisors D with $\deg D \geq 1$.

6. Suppose K/F has genus 1 and that P is a prime of degree 1. By the last exercise we know $l(2P) = 2$ and $l(3P) = 3$. Let $\{1, x\}$ be a basis of $L(2P)$ and $\{1, x, y\}$ be a basis of $L(3P)$. Show that $K = F(x, y)$. Show also that x and y satisfy a cubic polynomial with coefficients in F of the form

$$Y^2 + a_1 XY + a_3 Y = X^3 + a_2 X^2 + a_4 X + a_6 .$$

Hint: Consider $L(6P)$.

7. Let K/F be of positive genus and suppose there is a prime P of degree 1. Suppose further that $L(2P)$ has dimension 2. Let $\{1, x\}$ be a basis. If char $F \neq 2$, show that there is an element $y \in K$ such that $K = F(x, y)$ and such that x and y satisfy a polynomial equation of the form $Y^2 = f(X)$ where $f(X)$ is a square-free polynomial of degree at least three.

8. Use the Riemann-Roch theorem to show that if B and D are divisors such that $B + D$ is in the canonical class, then $|l(B) - l(D)| \leq \frac{1}{2} |\deg B) - \deg(D)|$.

9. Suppose P is a prime of degree 1 of a function field K/F. For every positive integer n show $l((n+1)P) - l(nP) \leq 1$.

10. Let K/F be a function field of genus $g \geq 2$, and P a prime of degree 1. For all integers k we have $l(kP) \leq l((k+1)P)$. If we restrict k to the range $0 \leq k \leq 2g - 2$ show there are exactly g values of k where $l(kP) = l((k+1)P)$. These are called Weierstrass gaps. Assume F has characteristic zero. If all the gaps are less than or equal to g we say P is a non-Weierstrass point, if not, we say P is a Weierstrass point. It can be shown that there are only finitely many Weierstrass points. In characteristic p there is a theory of Weierstrass points (due to H. Schmid), but the definition is somewhat different.

11. Suppose K/F has genus 1 and that P_∞ is a prime of degree 1, also called a rational point. Let $E(F)$ denote the set of rational points. If $P, Q \in E(F)$, show there is a unique element $R \in E(F)$ such that $P + Q \sim R + P_\infty$. (Recall that for two divisors A and B, $A \sim B$ means that $A - B$ is a principal divisor). Denote R by $P \oplus Q$. Show that $(P, Q) \to P \oplus Q$ makes $E(F)$ into an abelian group with P_∞ as the zero element.

12. With the same assumptions as Exercise 11, map $E(F) \to Cl_K^o$ by sending P to the class of $P - P_\infty$. Show that this map is an isomorphism of abelian groups.

13. Let K/F be a function field and σ an automorphism of K, which leaves F fixed. If $(\mathcal{O}, P)$ is a prime of K, show that $(\sigma\mathcal{O}, \sigma P)$ is also a prime of K. Show, further, that for all $a \in K$, we have $\mathrm{ord}_{\sigma P}(a) = \mathrm{ord}_P(\sigma^{-1}a)$.

14. (Continuation). The map $P \to \sigma P$ on primes extends to an action of σ on divisors. If $a \in K^*$, show that $\sigma(a) = (\sigma a)$.

15. (Continuation). If D is a divisor of K, show $a \to \sigma a$ induces a linear isomorphism from $L(D) \to L(\sigma D)$. In particular, if σ fixes D, i.e., $\sigma D = D$, then σ induces an automorphism of $L(D)$.

16. (Continuation). Suppose P is a prime of degree 1 and that $\sigma P = P$. Then, σ induces an automorphism of $L((2g + 1)P)$. If this induced map is the identity, show that σ is the identity automorphism. (Hint: Find two elements $x, y \in K^*$ fixed by σ such that $K = F(x, y)$).

17. Let A be a divisor and P a prime divisor. Suppose $g \in L(A + P) - L(A)$. If $f \in L(A + P)$ show $f/g \in \mathcal{O}_p$. Use this to prove $l(A + P) \le l(A) + \deg(P)$.

18. Use Exercise 17 to show $l(A) \le \deg(A) + 1$ if A is an effective divisor. Show further that this inequality holds in general. Thus, $l(A)$ is finite for any divisor A.

6
Weil Differentials
and the Canonical Class

In the last chapter we gave some definitions and then the statement of the
Riemann-Roch theorem for a function field K/F. In this chapter we will
provide a proof. In the statement of the theorem an integer, g, enters which
is called the genus of K. Also, a divisor class, $\mathbb{C}$, makes an appearance,
the canonical class of K. We will provide another interpretation of these
concepts in terms of differentials. Thus, differentials give us the tools we
need for the proof and, as well, lead to a deeper understanding of the
theorem. In addition, the use of differentials will enable us to prove two
important results: the strong approximation theorem and the Riemann-
Hurwitz formula. The first of these will be proven in this chapter, the second
in Chapter 7, where we will also prove the ABC conjecture in function fields
and give some of its applications.

We will use a notion of differential which is due to A. Weil. It is somewhat
more abstract than the usual definition but has the advantage of requiring
no special assumptions about the constant field. Also, it leads to very short,
conceptual proofs of the two theorems mentioned in the last paragraph. We
will motivate the definition by first discussing some properties of differentials on compact Riemann surfaces. If the reader is unfamiliar with this
theory, he or she can skip directly to the definition of Weil differential in
the purely algebraic setting.

Let X be a compact Riemann surface of genus g, M the field of meromorphic functions on X, and Ω the space of meromorphic differentials on
X. Fix a non-zero differential $\omega \in \Omega$ and a point $x \in X$. Let t be a local uniformizing parameter at x. In some neighborhood around x we can

express ω in the following form:

$$\omega = \sum_{i=-N}^{\infty} a_i t^i \, dt. \tag{1}$$

If $f \in M_x$, the field of germs of meromorphic functions at x, then we can integrate $f\omega$ around a small circle about x to get $2\pi i \operatorname{Res}_x(f\omega)$. If $f = \sum_{j=-M}^{\infty} bjt^j$, then

$$\operatorname{Res}_x(f\omega) = \sum_{i+j=-1} a_i b_j. \tag{2}$$

Let ω_x be defined to be the $\mathbb{C}$-linear map $f \to \operatorname{Res}_x(f\omega)$ from M_x to $\mathbb{C}$. We now look into the question of what restrictions are placed on the collection of linear functionals $\{\omega_x \mid x \in X\}$ by the fact that they arise from a differential in the manner indicated.

We recall the definition of the order of ω at a point x, $\operatorname{ord}_x(\omega)$. Write ω locally in terms of a uniformizing parameter as in Equation 1. Then the order of ω at x is defined to be the smallest index i such that $a_i \neq 0$. This number is independent of the choice of uniformizing parameter. If $a_{-N} \neq 0$, then $\operatorname{ord}_x(\omega) = -N$. It is well known that $\operatorname{ord}_x(\omega) = 0$ for all but finitely many points $x \in X$ and thus we can associate to $\omega \neq 0$ a divisor:

$$(\omega) = \sum_{x \in X} \operatorname{ord}_x(\omega) \, x.$$

This definition will be useful as we go along. For the moment we will show how to characterize the number $\operatorname{ord}_x(\omega)$ in a different way. Let $O_x \subset M_x$ be the ring of germs of holomorphic functions at x. Each element of O_x has a power series expansion in terms of a uniformizing paramenter t_x at x with all coefficients of negative index zero. O_x is a discrete valuation ring. Its unique maximal ideal P_x is generated by t_x. Every non-zero fractional ideal of O_x is a power P_x^m of P_x where m can be any integer. With this notation we show—

Lemma 6.1. *Let ω be a non-zero meromorphic differential, $x \in X$, and ω_x the linear functional on M_x described above. There is an integer N such that ω_x vanishes on P_x^N but not on P_x^{N-1}. This integer is characterized by*

$$\operatorname{ord}_x(\omega) = -N.$$

Proof. Since we are fixing x in our considerations we set $t_x = t$ and suppose ω is expressed in terms of t as in Equation 1. Assume $a_{-N} \neq 0$ so that $\operatorname{ord}_x(\omega) = -N$. From equation (2) it is then clear that ω_x vanishes on P_x^N. On the other hand, $t^{N-1} \in P_x^{N-1}$ and $\omega_x(t^{N-1}) = a_{-N} \neq 0$.

Corollary. *ω_x is zero on O_x but not on P_x^{-1} for all but finitely many $x \in X$.*

Proof. This follows from the lemma and the fact that $\text{ord}_x(\omega) = 0$ for all but finitely many $x \in X$.

Lemma 6.1 shows that the linear functionals ω_x must satisfy certain vanishing properties. These are all local properties only involving the knowledge of the behavior of ω in the neighborhoods of points. In addition, there is at least one global constraint.

Lemma 6.2. *For every $f \in M$ we have*

$$\sum_{x \in X} \omega_x(f) = 0.$$

Proof. Note that $f \in M$ implies $f \in M_x$ for all $x \in X$, so the terms in the sum make sense. Also, $f \in O_x$ for all but finitely many x (on a compact Riemann surface a meromorphic function has at most finitely many poles). By the corollary to Lemma 6.1, $\omega_x(f) = 0$ for all but finitely many $x \in X$. Thus, the sum is finite.

Now, $f\omega$ is also a meromorphic differential. It is a well-known theorem that on a compact Riemann surface the sum of the residues of a meromorphic differential is zero. Thus,

$$\sum_{x \in X} \omega_x(f) = \sum_{x \in X} \text{Res}_x(f\omega) = 0.$$

We now have all the background we need to set up the notion of a Weil differential. Let $A(X)$ be the subset of $\prod_x M_x$ consisting of elements with all but finitely many coordinates in O_x. It is clear that $A(X)$ is a ring with addition and multiplication defined coordinatewise. We will denote the elements of $A(X)$ by $\phi = (f_x)$; i.e., the x-th component of ϕ is f_x. $A(X)$ is a vector space over $\mathbb{C}$ in the obvious way, $a\phi = a(f_x) = (af_x)$. If $\omega \in \Omega$, define $\tilde{\omega} : A(X) \to \mathbb{C}$ by

$$\tilde{\omega}(\phi) = \sum_{x \in X} \omega_x(f_x).$$

By the corollary to Lemma 6.1 and the definition of $A(X)$, the sum on the right-hand side of the above equation is finite.

Let's map M into $A(X)$ by sending f to (f_x), where $f_x = f$ for all $x \in X$. Clearly, M is isomorphic to its image under this map and from now on we identify M with its image. Lemma 6.2 can now be interpreted as asserting that $\tilde{\omega}$ vanishes in M.

Let $D = \sum_x n_x x$ be any divisor on X. We associate to D a subset of $A(X)$, namely, $A(D) = \{(f_x) \in A(X) \mid \text{ord}_x(f_x) \geq -n_x, \forall x \in X\}$ (we use the convention that $\text{ord}_x(0) = \infty$, which is greater than any integer). Recall that a divisor D is said to be effective, $D \geq 0$, if all its coefficients are non-negative, and one divisor is bigger than another if their difference is effective, i.e., $D \leq C$ iff $C - D \geq 0$. One then checks easily that $D \leq C$ implies $A(D) \subseteq A(C)$ and that $\bigcup A(D) = A(X)$.

With these definitions, the definition of $\tilde{\omega}$, Lemma 6.1, and Lemma 6.2 we can easily prove—

Lemma 6.3. *The functional $\tilde{\omega}$ vanishes on both M and $A((\omega))$. Moreover, if $\tilde{\omega}$ vanishes on $A(D)$, then $A(D) \subseteq A((\omega))$.*

It is possible to show, although we shall not do so here, that if λ is a linear functional on $A(X)$ and λ vanishes on both M and $A(D)$ for some divisor D, then there is a unique differential $\omega \in \Omega$ such that $\tilde{\omega} = \lambda$. For the case when X is the Riemann sphere see Chevalley [1], pp. 29–30. This being so, in the abstract case we shall, following Weil, define differentials to be linear functionals on a certain space, the adele ring of the function field, having properties analogous to those we have seen to be true for the functionals $\tilde{\omega}$ on $A(X)$.

For the remainder of this chapter, let K/F be a function field with constant field F. We make no assumptions about F. Other notations will be the same as those in Chapter 5, except that we now introduce the new notation $\mathcal{S}_K$ for the set of prime divisors of K.

For $P \in \mathcal{S}_K$ let $|a|_P = 2^{-\mathrm{ord}_P(a)}$ for $a \neq 0$ and $|0|_P = 0$ (2 is chosen for convenience, any number greater than one will do). Define a metric on K by $\rho_P(a, b) = |a - b|_P$. We denote by $\hat{O}_P$ and $\hat{K}_P$ the completions of the local ring O_P and the field K with respect to this metric. We assume that the reader is familiar with standard facts about completions. See, for example, Lang [5], Chapter II . The adele ring of K is defined as

$$A_K = \left\{ (\alpha_P) \in \prod_P \hat{K}_P \mid \alpha_P \in \hat{O}_P \text{ for all but finitely many } P \in \mathcal{S}_K \right\}.$$

The analogy between the adele ring A_K of the function field K and the ring $A(X)$ which we attached to a compact Riemann surface is clear.

We imbed K into A_K by taking $x \in K$ to (x_P) where for all P, $x_P = x$. Since for any element $x \in K$, either $x = 0$ or $\mathrm{ord}_P(x) = 0$ for all but finitely many P, the image of x is indeed in A_K. K is isomorphic to its image and we identify K with its image under this map.

If $D = \sum_P n(P)P$ is a divisor of K, define $A_K(D)$ as the set of all $(x_P) \in A_K$ such that $\mathrm{ord}_P(x_P) \geq -n(P)$ for all $P \in \mathcal{S}_P$ (notice the minus sign!). Then, as in the Riemann surface case, it is easy to see that $D \leq C$ implies $A_K(D) \subseteq A_K(C)$ and that $\bigcup A_K(D) = A_K$. It is also useful to notice that $A_K(D) \cap A_K(C) = A_K((D, C))$ and $A_K(C) + A_K(D) = A_K([C, D])$, where (C, D) and $[C, D]$ denote the infimum of C and D and the supremum of C and D, respectively. More concretely,

$$\mathrm{ord}_P((C, D)) = \min(\mathrm{ord}_P(C), \mathrm{ord}_P(D)) \text{ and}$$
$$\mathrm{ord}_P([C, D]) = \max(\mathrm{ord}_P(C), \mathrm{ord}_P(D)).$$

A very important remark for our further considerations is that $A_K(D) \cap K = L(D)$, the vector space whose dimension over F, $l(D)$, is the focus of interest in the Riemann-Roch theorem. This equality follows directly from the definitions of $A_K(D)$, $L(D)$, and the way K is imbedded in $A_K(D)$.

We note that A_K and the subsets $A_K(D)$ are all vector spaces over F under the obvious operation; for $\alpha \in F$, $\alpha(x_P) = (\alpha x_P)$. With all these definitions in place, we can now define a Weil differential.

Weil Differential. *An F-linear map ω from A_K to F is called a Weil differential if it vanishes on K and on $A_K(D)$ for some divisor D. We denote the set of Weil differentials on K by Ω_K and the set of all Weil differentials which vanish on $A_K(D)$ by $\Omega_K(D)$.*

A number of remarks are in order. To begin with, many authors define a somewhat smaller ring than the adele ring, namely, the ring of repartitions, and define Weil differentials using it. The advantage is that one avoids going to the completion at all the primes P of K. While this is more elementary, some of the proofs become more difficult. In particular, the proof of the Riemann-Hurwitz formula is more transparant using the full ring of adeles and this is the principal reason we have used adeles in the above definition.

It is usual to define a topology on A_K by declaring the subsets $A_K(D)$ to be the open neighborhoods of the identity (the adele, all of whose coordinates are zero). We can then say that a Weil differential is a continuous F-linear functional on A_K which vanishes on K. We will not, however, make much use of topological considerations.

A_K is a vector space over K and all the sets $A_K(D)$ are vector spaces over F as we have seen. Ω_K also can be made into a vector space over K by means of the following definition. Let $\xi \in A_K$ and $x \in K$. Define

$$(x\omega)(\xi) = \omega(x\xi).$$

It is clear that $x\omega$ is an F-linear functional on A_K and that it vanishes on K. It requires but a short calculation to see that $\omega \in \Omega_K(D)$ implies $x\omega \in \Omega_K((x) + D)$. Thus, $x\omega$ is a Weil differential.

From now on we will refer to the elements of Ω_K simply as differentials rather than Weil differentials. We will show (Proposition 6.7) that the spaces $\Omega_K(D)$ are finite dimensional over F. Before doing that, we need some preliminary material. In particular, we need Riemann's inequality, the precursor to the Riemann-Roch theorem.

Lemma 6.4. *Let $D \leq C$ be divisors of K. Then,*

$$\dim_F A_K(C)/A_K(D) = \deg C - \deg D.$$

Proof. If $C = D$ the result is clear. Otherwise, C is obtained from D by adding finitely many primes, so it suffices to show that

$$\dim_F A_K(D + P)/A_K(D) = \deg P$$

for any prime P.

Let $\hat{P} = P\hat{O}_P$. Let $n = \mathrm{ord}_P D$. If $\xi = (a_P) \in A_K(D + P)$, then $\mathrm{ord}_P(a_P) \geq -n-1$ which is the same as $a_P \in \hat{P}^{-n-1}$. Now map $A_K(D+P)$ to $\hat{P}^{-n-1}/\hat{P}^{-n}$ by taking $\xi = (a_P)$ to the coset of a_P modulo $\hat{P}^{-n}$. This is clearly an epimorphism and from the definitions the kernel is seen to be $A_K(D)$. Thus,

$$A_K(D + P)/A_K(D) \cong \hat{P}^{-n-1}/\hat{P}^{-n} \cong P^{-n-1}/P^{-n} \cong O_P/P.$$

All these isomorphisms preserve F-vector space structure. Since

$$\dim_F O_P/P = \deg P,$$

the result follows.

Lemma 6.5. *Let $D \leq C$ be divisors of K. Then,*

$$\dim_F \frac{A_K(C) + K}{A_K(D) + K} = (\deg C - l(C)) - (\deg D - l(D)).$$

Proof. Recall that $A_K(C) \cap K = L(C)$. Using the first and second laws of isomorphism, the space on the left-hand side of the above equation is seen to be isomorphic to

$$\frac{A_K(C)}{A_K(D) + L(C)} \cong \frac{A_K(C) \,/\, A_K(D)}{(A_K(D) + L(C)) \,/\, A_K(D)}.$$

Using the first law once again, we see that $(A_K(D) + L(C)) \,/\, A_K(D) \cong L(C)/L(D)$. Thus,

$$\dim_F \frac{A_K(C) + K}{A_K(D) + K} = \dim_F A_K(C)/A_K(D) - \dim_F L(C)/L(D).$$

Using Lemma 6.4, the right-hand side is equal to $(\deg C - \deg D)) - (l(C) - l(D)) = (\deg C - l(C)) - (\deg D - l(D))$ as asserted.

Corollary. *For a divisor D, define $r(D) = \deg D - l(D)$. If $D \leq C$, then $r(D) \leq r(C)$.*

Proof. This is an immediate consequence of the Lemma since the dimension of a vector space is a non-negative integer.

Since both $\deg D$ and $l(D)$ only depend on the linear equivalence class of D, the same is true of $r(D)$. We will use this remark in a moment.

Theorem 6.6. (Riemann's Theorem) *Let K/F be an algebraic function field with field of constants F. There is a unique integer $g \geq 0$ with the following two properties. For all divisors D, we have $l(D) \geq \deg D - g + 1$. Also, there is a constant c such that for all divisors D with $\deg D \geq c$ we have $l(D) = \deg D - g + 1$. (g will turn out to be the constant in the Riemann-Roch theorem, i.e., the genus of K).*

Proof. Choose an element $x \in K^* - F^*$. Then, $K/F(x)$ is a finite extension of degree n, say. Let $B = (x)_\infty$ be the divisor of poles of x. The primes P which occur in the support of B are the only ones for which $\mathrm{ord}_P(x) < 0$. By Proposition 5.1, $\deg B = [K : F(x)] = n$.

Consider the integral closure, R, of $F[x]$ in K. If $\rho \in R$, the only poles of ρ are among the poles of x. Thus, $\rho \in L(m_0 B)$ for some positive integer m_0. It is a standard fact that we can find a basis $\{\rho_1, \rho_2, \ldots, \rho_n\}$ for $K/F(x)$ such that $\rho_i \in R$ for $1 \leq i \leq n$. Choose a positive integer m_0 such that $\rho_i \in L(m_0 B)$ for all $1 \leq i \leq n$. For any integer $m \geq m_0$ the elements $x^j \rho_i$ with $0 \leq j \leq m - m_0$ and $1 \leq i \leq n$ are all in $L(mB)$ and are linearly independent over F. We conclude from this that

$$l(mB) \geq n(m - m_0 + 1) .$$

It follows that for any $m \geq m_0$ we have

$$r(mB) = \deg mB - l(mB) \leq mn - n(m - m_0 + 1) = nm_0 - n .$$

This shows that $r(mB)$, which is an increasing sequence by the Corollary to Lemma 6.5, is bounded above and so must remain constant from some point on. Call this maximum value $g - 1$. Since $O < mB$, $-1 = r(O) \leq r(mB) \leq g - 1$. It follows that $g \geq 0$.

Let D be any divisor. Write $-D = D_1 + D_2$ where the support of D_1 is disjoint from the support of B and the support of D_2 is a subset of the support of B. Let P be in the support of D_1. Then, $F[x] \subset O_P$ and $P \cap F[x] = (g(x))$ where $g(x)$ is a monic, irreducible polynomial. It follows that for some positive integer h, $(g(x)^h) + D_1$ has no pole at P. Multiplying together the polynomials of this type at each P in the support of D_1 and we wind up with a polynomial $f(x)$ with the property that $(f(x)) + D_1$ only has poles among those of x. The same is true of D_2 and so, the same is true of $(f(x)) - D$. It follows that there is a positive integer m such that

$$(f(x)) - D + mB \geq O .$$

By the corollary to Lemma 6.5, we deduce $r(D) \leq r((f(x)) + mB) = r(mB)$. It follows that $r(D) \leq g - 1$ for all divisors D. From the definition of $r(D)$, this is equivalent to

$$l(D) \geq \deg D - g + 1 ,$$

which concludes the proof of Riemann's inequality.

We now have to produce a constant c such that $l(D) = \deg D - g + 1$ whenever $\deg D \geq c$. Let m_1 be a positive integer large enough so that $r(m_1 B) = g - 1$. Define $c = m_1 n + g$. If D is a divisor with $\deg D \geq c$, then by Riemann's inequality we find

$$l(D - m_1 B) \geq \deg(D - m_1 B) - g + 1 \geq 1 .$$

It follows that there is a $y \in K^*$ such that $(y) + D - m_1 B \geq O$ or $m_1 B \leq D + (y)$. Once again invoking the corollary to Lemma 6.5, we find $g - 1 = r(m_1 B) \leq r(D)$. We have already shown that for all divisors D, $r(D) \leq g - 1$. Thus, $r(D) = g - 1$, which is the same as $l(D) = \deg D - g + 1$.

The constant c can, in fact, be taken to be $2g - 1$. This follows from the full Riemann-Roch theorem, as we saw in the last chapter.

The Riemann-Roch theorem replaces the Riemann inequality with an equation. The following proposition is an approximation to what we want.

Proposition 6.7. *For any divisor D of K, the space $\Omega_K(D)$ is finite dimensional over F and*

$$l(D) = \deg D - g + 1 + \dim_F \Omega_K(D).$$

Proof. In Lemma 6.5, we are going to fix D and let C vary over divisors greater than or equal to D. By Riemann's theorem $l(C) \geq \deg C - g + 1$ or what is the same $\deg C - l(C) \leq g - 1$. So, by Lemma 6.5

$$\dim \frac{A_K(C) + K}{A_K(D) + K} \leq g - 1 + l(D) - \deg D.$$

The second part of Riemann's theorem asserts that there is a constant c such that equality holds for all divisors C with $\deg C \geq c$. Let C_o be any divisor greater than or equal to D and with degree bigger than c. Then,

$$\dim_F \frac{A_K(C) + K}{A_K(D) + K} = g - 1 + l(D) - \deg D$$

for all divisors C bigger than C_o. It follows that $A_K(C) + K = A_K(C_o) + K$ for all $C \geq C_o$. However, it is easily seen that for any adele ξ there is a divisor $C \geq C_o$ such that $\xi \in A_K(C)$. Thus, $A_K(C_o) + K = A_K$ and we have shown

$$l(D) = \deg D - g + 1 + \dim_F \frac{A_K}{A_K(D) + K}.$$

To finish the proof one has only to notice that $\Omega_K(D)$ is the F-dual of the vector space $A_K/(A_K(D) + K)$.

Corollary 1. *Let c be the constant in Riemann's theorem. Then if D is a divisor with $\deg D \geq c$, we have $A_K = A_K(D) + K$.*

Proof. We have just shown that $\dim_F(A_K / A_K(D) + K) = l(D) - \deg D + g - 1$, which is zero if $\deg D \geq c$ by Riemann's theorem. Thus $A_K = A_K(D) + K$ in this case.

Corollary 2. *The genus of K, g, can be characterized as the dimension over F of the space $\Omega_K(0)$.*

Proof. The zero divisor, 0, has degree zero and dimension 1. From the proposition we derive $1 = 0 - g + 1 + \dim_F \Omega_K(0)$. This gives the result.

The interested reader can easily show that if on a compact Riemann surface a meromorphic differential ω is such that $\tilde{\omega}$ is zero on $A(0)$, then ω has no poles and conversely. Thus the space $\Omega_K(0)$ is the analogue of the space of holomorphic differentials and this appellation is sometimes used even in the abstract case.

We have now given the promised characterization of the genus in terms of differentials. The next task is to give an interpretation of the canonical class. To do so we have to show how to assign a divisor to a non-zero differential. Since we don't have (yet) local expressions for a differential at a point as in the classical case, we proceed by, in essence, using the result of Lemma 6.3 as a definition. That is, if $\omega \in \Omega_K$, we want to define (ω) as the largest divisor D such that ω vanishes on $A_K(D)$. First we need to show there is such a divisor.

Lemma 6.8. Let $\omega \in \Omega_K$ be a non-zero differential. Then, there is a unique divisor D with the property that ω vanishes on $A_K(D)$ and if D' is any divisor such that ω vanishes on $A_K(D')$, then $D' \leq D$.

Proof. Let $\mathcal{T} = \{D' \mid \omega(A_K(D')) = 0\}$. Since ω is a differential, $\mathcal{T}$ is non-empty. By Corollary 1 to Proposition 6.7, we see that $\deg D' < c$ for all $D' \in \mathcal{T}$, since $\omega \neq 0$. Let D be a divisor of maximal degree in $\mathcal{T}$. We claim that D has the desired properties. Clearly, ω vanishes on $A_K(D)$. Suppose ω vanishes on $A_K(D')$. Then ω vanishes on $A_K(D) + A_K(D') = A_K([D, D'])$; i.e., $[D, D'] \in \mathcal{T}$. Since $\deg [D, D'] \geq \deg D$, it follows that the degrees must be equal and so $[D, D'] = D$, which implies $D' \leq D$ as required. The uniqueness is clear.

We now define the divisor of a differential ω to be the unique divisor D with the properties stated in the Lemma. We use the notation (ω) for the divisor of ω.

Lemma 6.9. Let $\omega \in \Omega_K$ and $x \in K^*$. Then,

$$(x\omega) = (x) + (\omega).$$

Proof. Suppose $\omega \in \Omega_K(D)$. If $\xi \in A_K$, then $x\omega$ vanishes on ξ if $x\xi \in A_K(D)$, which is equivalent to $\xi \in A_K((x) + D)$. Thus, ω vanishes on $A_K(D)$ implies $x\omega$ vanishes on $A_K((x) + D)$. The converse also holds as one can see by observing that $\omega = x^{-1}(x\omega)$. Thus, ω vanishes on $A_K(D)$ if and only if $x\omega$ vanishes on $A_K((x) + D)$ and the result follows easily from this.

In the classical case of compact Riemann surfaces $\Omega(X)$ is one dimensional over the field of meromorphic functions on X. To see this, let $\omega, \omega' \in$

$\Omega(X)$ with $\omega \neq 0$. Suppose $\omega = f(t)dt$ and $\omega' = g(t)dt$ in a neighborhood U of a point $x \in X$. Here, $f(t)$ and $g(t)$ are Laurent series in a uniformizing parameter t about x. Then $h_U(t) = g(t)/f(t)$ is a meromorphic function on U which is well defined in that it is independent of the choice of uniformizing parameter. This follows easily by use of the chain rule. These functions h_U fit together to give a meromorphic function h on X and $\omega' = h\omega$.

A similar proof cannot be given in the abstract case, but nevertheless an analogous result is true. This will follow by the use of Riemann's theorem (once again) together with some elementary linear algebra.

Proposition 6.10. *The space of Weil differentials, Ω_K, is of dimension one when considered as a vector space over K.*

Proof. Let $0 \neq \omega \in \Omega_K$ and $x \in L((\omega) - D)$ where D is some divisor. We claim that $x\omega \in \Omega_K(D)$. By the proof of the previous Lemma, we know that $x\omega$ vanishes on $A_K((x) + (\omega))$. Since $x \in L((\omega) - D)$ we have

$$(x) + (\omega) \geq -((\omega) - D) + (\omega) = D,$$

and so $x\omega$ vanishes on $A_K(D)$. This establishes the claim.

Now let $\omega, \omega' \in \Omega_K$ be non-zero differentials. By the previous paragraph we see that $L((\omega) - D)\omega$ and $L((\omega') - D)\omega'$ are both F-subspaces of $\Omega_K(D)$. If we could show that these subspaces have a non-zero intersection, the proposition would follow immediately. The idea of the proof is to force this to happen by a suitable choice of D.

Let P be any prime, and set $D = -nP$, where n is large and positive (how large will be determined shortly). By Proposition 6.7,

$$\dim_F \Omega(-nP) = l(-nP) + n \deg P + g - 1 = n \deg P + g - 1.$$

Recall $L(-nP) = (0)$ since any element in it would have no pole but would have a zero at P.

Using Riemann's inequality we find

$$\dim_F L((\omega) + nP) \geq \deg(\omega) + n \deg P - g + 1,$$

and similarly for ω'. Thus,

$$\dim_F L((\omega) + nP)\omega \quad + \quad \dim_F L((\omega') + nP)\omega'$$
$$\geq 2n \deg P + \deg(\omega) + \deg(\omega') - 2g + 2.$$

It follows that for large enough n the sum of the dimensions of the two subspaces $L((\omega) + nP)\omega$ and $L((\omega') + nP)\omega'$ exceeds the dimension of the ambient space $\Omega_K(-nP)$. By linear algebra, they must have a non-zero intersection. Thus, there exist $x, y \in K^*$ such that $x\omega = y\omega'$ and so, $\omega' = xy^{-1}\omega$.

Corollary 1. *Let $0 \neq \omega \in \Omega_K$ and let D be a divisor. Then there is an F-linear isomorphism between $L((\omega) - D)$ and $\Omega_K(D)$.*

Proof. In the proof of the proposition we showed that $L((\omega) - D)\omega \subseteq$ $\Omega_K(D)$. So it just remains to show that this inclusion is an equality. Let $\omega' \in \Omega_K(D)$. By the proposition, there is an element $x \in K$ such that $\omega' = x\omega$. Since ω' vanishes on $A_K(D)$ we must have $D \leq (\omega') = (x) + (\omega)$ by Lemma 6.9. Thus, $(x) \geq D - (\omega) = -((\omega) - D)$; i.e., $x \in L((\omega) - D)$.

Corollary 2. *All the divisors of non-zero differentials fill out a single divisor class. This class is called the canonical class of K.*

Proof. If $\omega, \omega' \in \Omega_K$ are non-zero, there exists an $x \in K^*$ such that $\omega' = x\omega$ by the proposition. By Lemma 6.9 we have $(\omega') = (x) + (\omega)$ so that (ω') and (ω) are in the same class. Conversely, if D is in the class of (ω), $D = (x) + (\omega)$ for some $x \in K^*$. Thus, $D = (x\omega)$, the divisor of a differential.

Proof of the Riemann-Roch Theorem. By Corollaries 1 and 2 to Proposition 6.10 and Proposition 6.7, we find

$$l(D) = \deg D - g + 1 + l((\omega) - D).$$

This is the assertion of the Riemann-Roch theorem given in the last chapter, Theorem 5.4. We see that the divisor C in the statement of that theorem can be taken to be any divisor of a non-zero differential. We now have a complete proof of the Riemann-Roch theorem!

Using Theorem 5.4 and its corollaries, we see that the constant c in the statement of Riemann's theorem can be taken to be $2g - 1$ and for any differential ω the degree of (ω) is $2g - 2$.

Finally, we want to decompose a differential into a sum of local pieces analogous to the sum of the residues construction in the classical case. To this end, let's define a map $i_P : \hat{K}_P \rightarrow A_K$. If $x_P \in \hat{K}_P$ let $i_P(x_P)$ be the adele with all components zero except the P-th component which is equal to x_P. Clearly, i_P is an F vector space isomorphism of $\hat{K}_P$ with its image.

Let $\omega \in \Omega_K$. We define $\omega_P \in \operatorname{Hom}_F(\hat{K}_P, F)$ by $\omega_P(x_P) = \omega(i_P(x_P))$. This process associates a family of local functionals $\{\omega_P \mid P \in S_K\}$ to a differential. Knowing this family, we would like to reconstruct the differential and its divisor.

The functionals ω_P are not arbitrary. They must vanish on some power of the maximal ideal $\hat{P} \subset \hat{O}_P$. Indeed, $\omega_P(x_P) = 0$ if $i_P(x_P) \in A_K((\omega))$ and this inclusion holds if $\operatorname{ord}_P(x_P) \geq -\operatorname{ord}_P(\omega)$. Thus, ω_P vanishes on $\hat{P}^{-\operatorname{ord}_P(\omega)}$. This shows the functionals ω_P are continuous in the P-adic topology.

Proposition 6.11. *Let $\omega \in \Omega_K$ and $\xi = (x_P) \in A_K$. Then, for all but finitely many P we have $\omega_P(x_P) = 0$ and*

$$\omega(\xi) = \sum_P \omega_P(x_P).$$

Proof. Let S be the finite set of primes where either $\text{ord}_P(\omega) < 0$ or $\text{ord}_P(x_P) < 0$. If $P \notin S$ then $x_P \in \hat{O}_P$ and so $\omega_P(x_P) = 0$ by the remark preceding the Proposition. Define a new adele ξ' whose P'th component is x_P if $P \notin S$ and 0 if $P \in S$. Then, $\xi' \in A_K((\omega))$ and $\xi = \xi' + \sum_{P \in S} i_P(x_P)$. Thus,

$$\omega(\xi) = \omega(\xi') + \sum_{P \in S} \omega(i_P(x_P)) = \sum_{P \in S} \omega_P(x_P) = \sum_P \omega_P(x_P).$$

The next proposition provides the abstract analogue of Lemma 6.1. It enables one to recover the divisor of ω from properties of the local functionals ω_P. This will be very useful in the proof of the Riemann-Hurwitz formula.

Proposition 6.12. *Let $0 \neq \omega \in \Omega_K$. Then, $N = \text{ord}_P(\omega)$ is determined by the following two properties; ω_P vanishes on $\hat{P}^{-N}$ but does not vanish on $\hat{P}^{-N-1}$.*

Proof. We have already seen in the remarks preceding Proposition 6.10 that ω_P vanishes on $\hat{P}^{-\text{ord}_P(\omega)}$. It remains to show that ω_P doesn't vanish on $\hat{P}^{-\text{ord}_P(\omega)-1}$. We know, from Lemma 6.8 and the definition, that ω does not vanish on $A_K((\omega) + P)$. Let $\xi \in A_K((\omega) + P)$ be such that $\omega(\xi) \neq 0$. As usual, write $\xi = (x_Q)$ with Q varying over all primes. By Proposition 6.11,

$$0 \neq \omega(\xi) = \omega_P(x_P) + \sum_{Q \neq P} \omega_Q(x_Q) = \omega_P(x_P).$$

The last equality follows from the fact that $\text{ord}_Q((\omega) + P) = \text{ord}_Q(\omega)$ for $Q \neq P$.

Since $\xi \in A_K((\omega) + P)$, we must have $x_P \in \hat{P}^{-\text{ord}_P(\omega)-1}$. This concludes the proof.

Corollary. *A differential ω is completely determined by any local component ω_P. That is, if $\omega, \omega' \in \Omega_K$ and $\omega_P = \omega'_P$ then $\omega = \omega'$.*

Proof. If $\omega_P = \omega'_P$ then $(\omega - \omega')_P = 0$. The proposition shows that if $\omega - \omega'$ were a non-zero differential no local component could be the zero map. Thus, $\omega - \omega' = 0$; i.e., $\omega = \omega'$.

We have now accomplished all the goals set out for this chapter except the statement and proof of the strong approximation theorem. This important theorem, strictly speaking, has nothing to do with differentials. However, its proof is an easy consequence of material developed earlier, namely, Corollary 1 to Proposition 6.7.

Let's first recall a version of the weak approximation theorem. Suppose K is a field and $O_1, O_2, \ldots, O_t$ a collection of subrings of K which are discrete valuation rings with quotient field $\dot{K}$. Let $P_i \subset O_i$ be the maximal ideal of O_i. Finally, suppose we are given a set of elements $a_i \in K$ and a set of positive integers n_i with i varying from 1 to t. The weak approximation theorem asserts that there is an element $a \in K$ such that $\mathrm{ord}_{P_i}(a - a_i) \geq n_i$ for $i = 1, 2, \ldots, t$. The proof, which is not hard, can be found in many sources, e.g. see Lang [5]. The strong approximation theorem in function fields is an assertion of similar type, but with much greater constraints on the element $a \in K$.

Theorem 6.13. *Let K/F be a function field and $S \subset \mathcal{S}_K$ a finite set of primes. For each $P \in S$ let an element $a_P \in \hat{K}_P$ and a positive integer n_P be given. Finally, let's specify a prime $Q \notin S$. Then, there is an element $a \in K$ such that $\mathrm{ord}_P(a - a_P) \geq n_P$ for all $P \in S$ and $\mathrm{ord}_P(a) \geq 0$ for all $P \notin S \cup \{Q\}$.*

This theorem is called the strong approximation theorem. Before beginning the proof, two remarks are in order. First, the added generality of choosing the $a_P \in \hat{K}_P$ is very small. If we prove the theorem with the $a_P \in K$, then the full theorem takes just a trivial extra step. The main point is that in addition to the conditions at the primes in S we have added the infinitely many conditions that the element a be integral at all primes not in S with the one exception of Q.

Proof. Define an adele $\xi = (x_P)$ by the conditions that $x_P = a_P$ for $P \in S$ and $x_P = 0$ for $P \notin S$. Next, define a divisor $D = mQ - \sum_{P \in S} n_P \, P$. Choose the integer m so large that the degree of D exceeds the constant c in Riemann's theorem. Then, by Corollary 1 to Proposition 6.7, we have $A_K = K + A_K(D)$. In particular, $\xi = a + \eta$ where $a \in K$ and $\eta \in A_K(D)$. In other words, $\xi - a \in A_K(D)$. Examining this relation, component by component, shows that a has the desired properties.

Exercises

1. Let ω be a meromorphic differential on a compact Riemann surface X. Show that $\tilde{\omega}$ is zero on $A(O)$ if and only if ω has no poles.

2. Let M be the field of meromorphic functions on a compact Riemann surface X and Ω the space of meromorphic differentials on X. Show in detail that Ω is a one-dimensional vector space over M.

3. Show directly (i.e., arguing only with differentials) that $\dim \Omega_K(D) = 0$ if $\deg D > 2g - 2$.

4. Suppose that D is a divisor of degree zero, but that D is not principal. Show $\dim_F \Omega_K(D) = g - 1$.

5. If D is a divisor, and $\deg D < g - 1$, show that $\dim_F \Omega_K(D) > 0$.

6. Suppose that $\omega \in \Omega_K(O)$ and has a zero P of degree 1 and that $\text{ord}_P\omega \geq g$. Show that P is a Weierstrass point (see Exercise 10 of Chapter 5).

7. In this and the following two exercises, we assume that F is algebraically closed. Let P be a prime. Assume the genus g is greater than 0. Show $\Omega_K(P)$ is properly contained in $\Omega_K(O)$.

8. (Continuation) Suppose $g \geq 1$ and $0 < n \leq g$. Show there exist primes $\{P_1, P_2, \ldots, P_n\}$ with the property, $\dim \Omega_K(P_1+P_2+\cdots+P_n) = g-n$.

9. (Continuation) Suppose $g \geq 1$. Show there are primes $\{P_1, P_2, \ldots, P_g\}$ such that $P_1 + P_2 + \cdots + P_g$ is not the polar divisor of any element of K^*.

10. Suppose ω_1 and ω_2 are two Weil differentials with the same divisor. Show $\omega_1 = \alpha\omega_2$ for some $\alpha \in F^*$.

11. Let σ be an automorphism of K which leaves F fixed. Let P be a prime of K and σP the prime obtained by applying σ to P (see Exercise 13 of Chapter 5). Show that σ extends to an isomorphism of $\hat{K}_P$ with $\hat{K}_{\sigma P}$. Show further that σ induces an automorphism of A_K which is F-linear and maps K to itself.

12. (Continuation) If ω is a Weil differential, define $\sigma\omega : A_K \to A_K$ by the equation $\sigma\omega(a) = \omega(\sigma^{-1}a)$ for all $a \in A_K$. Show that $\sigma\omega$ is a differential.

13. (Continuation) Let D be a divisor of K. If $\omega \in \Omega_K(D)$, show that $\sigma\omega \in \Omega_K(\sigma D)$.

14. (Continuation) From the last exercise we see that σ induces an automorphism of $\Omega_K(O)$. If F is algebraically closed and $g \geq 1$, show there is a differential of the first kind ω such that $\sigma(\omega) = (\omega)$.

15. (Continuation) Assume F is algebraically closed and that the genus g of K is ≥ 2. Show there is an integer k with $1 \leq k \leq 2g - 2$ and a prime P such that σ^k leaves P fixed. (This series of exercises was inspired by the paper of Iwasawa and Tamagawa [1], where it was proved that the automorphism group of a function field of genus 2 or greater is finite. In characteristic zero this result is due to A. Hurwitz. In charactersitic p the first proof was given by H. Schmid [1]).

7
Extensions of Function Fields, Riemann-Hurwitz, and the ABC Theorem

Having developed all the basic material we will need about function fields we now proceed to discuss extensions of function fields. This material can be presented in a geometric fashion. Function fields correspond to algebraic curves and finite extensions of function fields correspond to ramified covers of curves. In this chapter, however, we will continue to use a more arithmetic point of view which emphasizes the analogy of function fields with algebraic number fields.

Let K/F be a function field with constant field F and let L be a finite algebraic extension of K. Let E be the algebraic closure of F in L. It is then clear that L is a function field with E as its field of constants. Recall that in this book, a "function field" over F refers to a field which is finitely generated over F and of transcendence degree one. If $L = EK$, we say that L is a constant field extension of K. We will discuss such extensions in detail in the next chapter. If $E = F$, we say that L is a geometric extension of K. In the general case, we have a tower $K \subseteq EK \subseteq L$, where EK/K is a constant field extension, and L/EK is a geometric extension.

Let p denote the characteristic of F. In the characteristic zero case, all extensions are separable and this considerably simplifies the theory. Since we will be especially interested in the case where the constant field F is finite, we must also deal with the theory when $p > 0$ and thus with questions of inseparability. Instead of working in complete generality we will often compromise by assuming that F is perfect, i.e., that all algebraic extensions of F are separable. This holds if F has characteristic zero or is algebraically closed or is finite. These cover all the cases of interest in this book.

This chapter falls naturally into three parts. In the first part we recall some basic facts about extensions of discrete, rank one valuations and, also, the theory of the different and its application to questions of ramification. Here we will assume the reader is somewhat familiar with this material so that the proofs will only be sketched. In the second part we will discuss how differentials behave in extensions. This will lead to the proof of the Riemann-Hurwitz theorem, one of the most important and useful theorems in the subject. Finally, we will discuss the so-called ABC-conjecture of Oesterlé-Masser and give a very simple proof in function fields using an idea the author learned from W. Fulton. Several applications of this result will be given, e.g., a proof of Fermat's last theorem for polynomial rings.

Let L/K be a finite algebraic extension of fields. We will use the abbreviation "dvr" for a discrete valuation ring. Let O_P be a dvr in K having K as its quotient field. Denote its maximal ideal by P. Let $O_{\mathfrak{P}}$ be a dvr in L with maximal ideal $\mathfrak{P}$. We say that $O_{\mathfrak{P}}$ lies above O_P or that $\mathfrak{P}$ lies above P if $O_P = K \cap O_{\mathfrak{P}}$ and $P = \mathfrak{P} \cap O_P$. The notation $\mathfrak{P}|P$ for this relation is often useful. There are two integers associated to this situation, $f = f(\mathfrak{P}/P)$, the relative degree, and $e = e(\mathfrak{P}/P)$, the ramification index. To define f, note that $O_{\mathfrak{P}}/\mathfrak{P}$ is a vector space over O_P/P. The relative degree is defined to be the dimension of this vector space. We shall see shortly that it is finite. Next, $PO_{\mathfrak{P}}$ is a non-zero ideal of $O_{\mathfrak{P}}$ contained in $\mathfrak{P}$. Thus, $PO_{\mathfrak{P}} = \mathfrak{P}^e$ for some integer $e \geq 1$. This integer is called the ramification index. It is easy to see that e is characterized by the following condition; for all $a \in K$, $\operatorname{ord}_{\mathfrak{P}}(a) = e \operatorname{ord}_P(a)$.

The ramification index and the relative degree behave transitively in towers. More precisely, let $K \subseteq L \subseteq M$ be a tower of function fields with L/K and M/L finite, algebraic extensions. If $\mathfrak{P}$ is a prime of M and $\mathfrak{p}$ and P are the primes lying below $\mathfrak{P}$ in L and K respectively, then, $e(\mathfrak{P}/P) = e(\mathfrak{P}/\mathfrak{p})e(\mathfrak{p}/P)$ and $f(\mathfrak{P}/P) = f(\mathfrak{P}/\mathfrak{p})f(\mathfrak{p}/P)$. Both relations follow easily from the definitions.

Proposition 7.1. *With the above notations, $ef \leq n = [L : K]$, the dimension of L over K.*

Proof. Let Π be a generator of $\mathfrak{P}$ and choose $\omega_1, \omega_2, \ldots, \omega_m$ such that their reductions modulo $\mathfrak{P}$ are linearly independent over O_P/P. We will show that the em elements $\omega_i\Pi^j$ with $1 \leq i \leq m$ and $0 \leq j < e$ are linearly independent over K. This is sufficient to establish the result.

Suppose

$$\sum_{j=0}^{e-1} \sum_{i=1}^{m} a_{ij}\omega_i\Pi^j = 0$$

is a linear dependence relation over K. If the $a_{ij} \in K$ are not all zero we can assume they are all in O_P and at least one of them is not in P (since

K is the quotient field of O_P and O_P is a dvr). Consider the elements

$$A_j = \sum_{i=1}^{m} a_{ij}\omega_i.$$

If some $a_{ij} \notin P$, then A_j is a unit in $O_{\mathfrak{P}}$ since its reduction modulo $\mathfrak{P}$ is not zero. Otherwise, A_j is divisible by π, the generator of P, and so $\operatorname{ord}_{\mathfrak{P}}(A_j) \geq e$. Thus, $\operatorname{ord}_{\mathfrak{P}}(\sum_{j=0}^{e-1} A_j\Pi^j) = j_o$ for some $j_o < e$. This is a contradiction since $\sum_{j=0}^{e-1} A_j\Pi^j = 0$.

If we assume L/K is a finite and separable extension, then we can construct all the $\mathfrak{P}$ lying over P as follows. Let O_P be as above, and let R be the integral closure of O_P in L. R is a Dedekind domain. Let $PR = \mathfrak{p}_1^{e_1}\mathfrak{p}_2^{e_2}\ldots\mathfrak{p}_g^{e_g}$ be the prime decomposition of PR in R. The set $\{\mathfrak{p}_1, \mathfrak{p}_2, \ldots, \mathfrak{p}_g\}$ is the complete set of non-zero prime ideals of R. For each i, the localization $R_{\mathfrak{p}_i}$ is a discrete valuation ring with maximal ideal $\mathfrak{P}_i = \mathfrak{p}_i R_{\mathfrak{p}_i}$. Define $O_{\mathfrak{P}_i} = R_{\mathfrak{p}_i}$. Then $\{O_{\mathfrak{P}_1}, O_{\mathfrak{P}_2}, \ldots, O_{\mathfrak{P}_g}\}$ is the complete set of dvrs in L lying above O_P. Let f_i and e_i be the relative degree and ramification index of $\mathfrak{P}_i$ over P_i. By standard properties of localization, the exponents in the decomposition of PR are indeed the same as the ramification indices defined earlier.

Proposition 7.2. *Assume L/K is a finite, separable extension of fields. Then, with the above notations, $\sum_{i=1}^{g} e_i f_i = n = [L : K]$.*

Proof. Since L/K is separable, the trace from L to K, $tr_{L/K}$, is a nontrivial K-linear functional on L. Using this, one can prove that R is a free module over O_P of rank equal to $n = [L : K]$ (see Samuel and Zariski [1]). Thus, R/PR is a vector space over O_P/P of dimension n.

Now, $PR = \mathfrak{p}_1^{e_1}\mathfrak{p}_2^{e_2}\ldots\mathfrak{p}_g^{e_g}$ and so, by the Chinese Remainder Theorem,

$$R/PR \cong R/\mathfrak{p}_1^{e_1} \oplus R/\mathfrak{p}_2^{e_2} \oplus \cdots \oplus R/\mathfrak{p}_g^{e_g}.$$

Again, by standard properties of localization, for each index i we have a ring isomorphism

$$R/\mathfrak{p}_i^{e_i} \cong O_{\mathfrak{P}_i}/\mathfrak{P}_i^{e_i}.$$

The latter ring is a vector space over O_P/P, and we calculate its dimension using the filtration

$$PO_{\mathfrak{P}_i} = \mathfrak{P}_i^{e_i} \subset \mathfrak{P}_i^{e_i-1} \subset \cdots \subset \mathfrak{P}_i \subset O_{\mathfrak{P}_i}.$$

Since $\mathfrak{P}_i$ is principal, the successive quotients are one dimensional over $O_{\mathfrak{P}_i}/\mathfrak{P}_i$. This ring is f_i dimensional over O_P/P (by definition). Thus, the total dimension of $O_{\mathfrak{P}_i}/\mathfrak{P}_i^{e_i}$ over O_P/P is $e_i f_i$.

Summing over i gives $n = \sum_{i=1}^{g} e_i f_i$ as asserted.

Having dealt with the separable case we now prove a simple fact about the purely inseparable case.

Lemma 7.3. *Let L/K be a purely inseparable extension of degree p, the characteristic of K. Assume $K = L^p$ (this strange assumption is often correct in function fields). Suppose $O_P \subset K$ is a dvr with quotient field K. Then there is one and only one dvr $O_{\mathfrak{P}} \subset L$ above O_P. Moreover, $e = p$ and $f = 1$ so $ef = p = [L : K]$.*

Proof. Let $R = \{r \in L \mid r^p \in O_P\}$ and $\mathfrak{P} = \{r \in L \mid r^p \in P\}$. It is easy to see that R is a ring, $\mathfrak{P}$ is a prime ideal in R, and $\mathfrak{P} \cap O_P = P$. We will show that R is a dvr.

Let π be a generator of P. Since $L^p = K$, there is an element $\Pi \in L$ with $\Pi^p = \pi$. Clearly, $\Pi \in \mathfrak{P}$. We claim that every element $t \in L$ is a power of Π times a unit in R. Once this is proved, it is almost immediate that Π generates $\mathfrak{P}$ and that R is a dvr.

Now, $t^p \in K$ so that $t^p = u\pi^s$ where u is a unit in O_P and $s \in \mathbb{Z}$. Thus, $(t/\Pi^s)^p = u$ which shows that $t/\Pi^s \in R$. Since $(\Pi^s/t)^p = u^{-1} \in O_P$ it follows that $\Pi^s/t \in R$ as well. Thus, t is a power of Π times a unit as claimed.

If $O_{\mathfrak{P}'} \subset L$ is any other dvr lying over O_P, let t be one of its elements. Then $t^p \in K \cap O_{\mathfrak{P}'} = O_P$ so that $t \in R$. We have shown $O_{\mathfrak{P}'} \subseteq R$. Since, as we shall show in a moment, dvrs are maximal subrings of their quotient fields, we have $R = O_{\mathfrak{P}'}$, which establishes uniqueness.

To prove the maximality property of dvrs, let $O \subset K$ be a dvr with quotient field K and uniformizing parameter π. Let O' be a subring of K containing O. Suppose there is an element $r \in O'$ with $r \notin O$. Then, there is a unit $u \in O$ such that $r = u\pi^{-n}$ with $n > 0$. Then, $\pi^{-1} = u^{-1}\pi^{n-1}r \in O'$ and it follows that all powers of π, both positive and negative, are in O'. Since every element of K is equal to a unit of O times a power of π, we conclude that if $O \neq O'$, then $O' = K$.

Finally, $\text{ord}_{\mathfrak{P}}(\pi) = p$ so $e = p$. By Proposition 7.1, $ef \leq p$, and it follows that $f = 1$, as asserted.

A field F is called perfect if every algebraic extension is separable. This is automatic in characteristic zero. In characteristic $p > 0$, it is well known that F is perfect if and only if $F = F^p$. We use this criterion in the next proposition.

Proposition 7.4. *Let F be a perfect field of positive characteristic p, and K a function field with constant field F. Then, $[K : K^p] = p$.*

Proof. Let x be an element of K not in F. Then $[K : F(x)] < \infty$. Consider $F(x)^p = F^p(x^p) = F(x^p)$. It is clear that $[F(x) : F(x^p)] = p$. For example, one shows easily that $\{1, x, x^2, \ldots, x^{p-1}\}$ is a field basis for $F(x)$ over $F(x^p)$.

Thus, the proposition follows from the equation

$$[K : F(x^p)] = [K : F(x)][F(x) : F(x^p)] = [K : K^p][K^p : F(x^p)],$$

if we can show $[K : F(x)] = [K^p : F(x^p)]$. To show this, let $\{\omega_1, \omega_2, \ldots, \omega_m\}$ be a field basis for K over $F(x)$. We claim that $\{\omega_1^p . \omega_2^p, \ldots, \omega_m^p\}$ is a field basis for K^p over $F(x^p)$. This is a straightforward calculation.

Corollary. *Let K be a function field of characteristic $p > 0$ with perfect constant field F. Let L be a purely inseparable extension of K of degree p. Then, F is the constant field of L and $L^p = K$.*

Proof. Suppose $\alpha \in L$ is a constant. By definition, it is algebraic over F. Since L/K is purely inseparable of degree p, $\alpha^p \in K$ and is algebraic over F. This implies $\alpha^p \in F$. Since $F = F^p$ there is a $\beta \in F$ with $\alpha^p = \beta^p$ which implies $\alpha = \beta \in F$.

Applying the proposition to L we see $[L : L^p] = p$. However, since L/K is purely inseparable, $L^p \subseteq K$. It follows that $[K : L^p] = 1$ and so $K = L^p$.

Proposition 7.5. *Let K be a function field with a perfect constant field F. Let L be a finite extension of K, and M, the maximal separable extension of K in L. Then, the genus of M is equal to the genus of L. Also, for each prime $\mathfrak{p}$ of M there is a unique prime $\mathfrak{P}$ in L lying above it. Finally, $e(\mathfrak{P}/\mathfrak{p}) = [L : M]$ and $f(\mathfrak{P}/\mathfrak{p}) = 1$.*

Proof. The constant field E of M is perfect since it is a finite extension of F, which is perfect by assumption.

Since L/M is purely inseparable, there is a tower of fields

$$K \subseteq M = K_0 \subset K_1 \subset \cdots \subset K_{n-1} \subset K_n = L,$$

where for each $i \geq 1$, K_i/K_{i-1} is purely inseparable of degree p. By the corollary to Proposition 7.4, and an obvious induction, we have $K_{i-1} = K_i^p$ for each $1 \leq i \leq n$. Raising to the p-th power is thus an isomorphism from K_i to K_{i-1}, which shows that all these fields have the same genus. This proves the first assertion.

The remaining part of the Proposition is proven by induction using the corollary to Proposition 7.4, Lemma 7.3, and the fact that both the relative degree and ramification index are multiplicative in towers.

We are now in a position to prove the theorem at which we have been aiming.

Theorem 7.6. *Let K be a function field with perfect constant field F. Let L be a finite extension of K of degree n. Suppose P is a prime of K and $\{\mathfrak{P}_1, \mathfrak{P}_2, \ldots, \mathfrak{P}_g\}$ the set of primes in L lying above P. Then, $\sum_{i=1}^{g} e_i f_i = n$ where, as usual, e_i is the ramification index and f_i the relative degree of $\mathfrak{P}_i$ over P.*

Proof. Let M be the maximal separable extension of K in L. Let $\mathfrak{p}_i$ be the prime of M lying below $\mathfrak{P}_i$ and let e_i' and f_i' be the ramification index and relative degree of $\mathfrak{p}_i$ over P. By Proposition 7.5, $\mathfrak{P}_i$ is the unique prime of L lying over $\mathfrak{p}_i$. By Proposition 7.2, the theorem is true for M/K.

Thus, $\sum_{i=1}^{g} e_i' f_i' = [M : K]$. Invoking Proposition 7.5 once more we see $e_i'[L : M] = e_i$ and $f_i' = f_i$. Substituting into the sum and noticing that $[L : K] = [L : M][M : K]$ finishes the proof.

When L/K is a finite extension of function fields, the conclusion of Theorem 7.6 holds without any assumption about the constant fields (see Chevalley [1] or Stichtenoth [1]). The method uses results which are specific to function fields, e.g., the degree of the zero divisor of a function and Riemann's inequality. We have chosen a different route, which seems more natural, but at the expense of having to assume the constant field is perfect.

Recall the notation $\mathcal{D}_K$ and $\mathcal{D}_L$ for the divisor groups of K and L, respectively. We introduce homomorphisms $N_{L/K}$ and $i_{L/K}$ as follows:

1. $N_{L/K} : \mathcal{D}_L \to \mathcal{D}_K$ is defined by $N_{L/K}(\mathfrak{P}) = f(\mathfrak{P}/P)P$ for all primes $\mathfrak{P} \in \mathcal{S}_L$ and then extended by linearity. Here, P is the prime of K lying below $\mathfrak{P}$, i.e., $P = \mathfrak{P} \cap K$. $N_{L/K}$ is called the norm map on divisors.

2. $i_{L/K} : \mathcal{D}_K \to \mathcal{D}_L$ is defined by $i_{L/K}(P) = \sum_{\mathfrak{P}|P} e(\mathfrak{P}/P) \mathfrak{P}$ for all $P \in \mathcal{S}_K$ and then extended by linearity. $i_{L/K}$ is called the extension of divisors map, or, sometimes, the conorm map.

A simple consequence of these definitions and Theorem 7.6 is that $N_{L/K} \circ i_{L/K}$ is the map "multiplication by $[L : K]$" on $\mathcal{D}_K$. Thus, $i_{L/K}$ is one to one (which is obvious anyway) and the quotient group $\mathcal{D}_K/N_{L/K}(\mathcal{D}_L)$ is annihilated by $[L : K]$ (which is perhaps not so obvious).

It is important to determine how these maps interact with the degree maps. Suppose that F and E are the constant fields of K and L, respectively. Recall that for a prime $\mathfrak{P}$ of L, $\deg_L(\mathfrak{P})$ is the dimension of $O_{\mathfrak{P}}/\mathfrak{P}$ over E. Similarly, for a prime P of K, $\deg_K(P)$ is the dimension of O_P/P over F. These degree maps are then extended by linearity to $\mathcal{D}_L$ and $\mathcal{D}_K$.

Proposition 7.7. Let $\mathfrak{A} \in \mathcal{D}_L$ and $A \in \mathcal{D}_K$. Then

$$\deg_K N_{L/K}(\mathfrak{A}) = [E : F] \deg_L \mathfrak{A} \quad \text{and} \quad \deg_L(i_{L/K}(A)) = \frac{[L : K]}{[E : F]} \deg_K A.$$

Proof. Both facts follow from the calculation

$$[O_{\mathfrak{P}}/\mathfrak{P} : F] = [O_{\mathfrak{P}}/\mathfrak{P} : E][E : F] = [O_{\mathfrak{P}}/\mathfrak{P} : O_P/P][O_P/P : F],$$

which shows that $[E : F] \deg_L \mathfrak{P} = f(\mathfrak{P}/P) \deg_K P$.

To show the first assertion, we see it is sufficient to do it when $\mathfrak{A} = \mathfrak{P}$ is a prime divisor. In that case, $\deg_K N_{L/K}(\mathfrak{P}) = \deg_K f(\mathfrak{P}/P) P = f(\mathfrak{P}/P) \deg_K P = [E : F] \deg_L \mathfrak{P}$.

To show the second assertion, it is again sufficient to consider the case where $A = P$ is a prime. Then,

$$\deg_L i_{L/K}(P) = \sum_{\mathfrak{P} \mid P} e(\mathfrak{P}/P) \deg_L \mathfrak{P}$$

$$= \frac{1}{[E:F]} \sum_{\mathfrak{P} \mid P} e(\mathfrak{P}/P) f(\mathfrak{P}/P) \deg_K P.$$

The result is now immediate from Theorem 7.6.

We also would like to investigate how these two maps behave on the group of principal divisors. Recall that if $a \in K^*$ its divisor is defined to be $(a)_K = \sum_P \mathrm{ord}_P(a) P$, where the sum is over all $P \in S_K$. Similarly, one defines the principal divisor $(b)_L$ for an element $b \in L^*$.

Proposition 7.8.

(i). If $a \in K^*$, then $i_{L/K}(a)_K = (a)_L$.

(ii). If $b \in L^*$, then $N_{L/K}$ $(b)_L = (N_{L/K}(b))_K$.

Proof. To prove the first assertion, one simply computes

$$i_{L/K}(a)_K = i_{L/K} \sum_P \mathrm{ord}_P(a)\, P = \sum_P \mathrm{ord}_P(a) \sum_{\mathfrak{P}|P} e(\mathfrak{P}/P)\mathfrak{P}$$

$$= \sum_{\mathfrak{P}} e(\mathfrak{P}/P)\mathrm{ord}_P(a)\, \mathfrak{P} = \sum_{\mathfrak{P}} \mathrm{ord}_{\mathfrak{P}}(a)\, \mathfrak{P} = (a)_L.$$

The proof of the second assertion is somewhat more difficult, but standard. It follows from general properties of Dedekind domains. A particularly elegant proof is given in Serre [2]. A more conventional treatment is given in Samuel and Zariski [1].

Corollary. *The maps $i_{L/K}$ and $N_{L/K}$ induce homomorphisms on the class groups Cl_K and Cl_L (which we will designate by the same letters).*

Proof. The proposition shows $i_{L/K}$ maps $\mathcal{P}_K \to \mathcal{P}_L$ and so induces a map from $Cl_K = \mathcal{D}_K/\mathcal{P}_K \to Cl_L = \mathcal{D}_L/\mathcal{P}_L$. Similarly for $N_{L/K}$.

The next topic to consider is that of ramification. Let L/K be a finite extension of function fields, suppose $\mathfrak{P}$ is a prime of L lying over a prime P of K. We say that $\mathfrak{P}$ is unramified over P if two conditions hold: $e(\mathfrak{P}/P) = 1$ and the extension of residue class fields is separable. If either condition is not satisfied, we say P is ramified over $\mathfrak{P}$. In a separable extension of function fields, only finitely many primes are ramified. This important result is a consequence of the theory of the different, which we will now sketch without complete proofs. A detailed treatment can be found in the above cited references, Serre [2] and Samuel and Zariski [1].

We begin with considerations of some generality. Let L/K be a separable extension of fields, $A \subset K$ a discrete valuation ring with quotient field K, and B the integral closure of A in L. One can show that B is a Dedekind domain with only finitely many prime ideals. These are the prime ideals which occur in the prime decomposition of PB, where P is the maximal ideal of A. As an A module, B is a finitely generated free module over A of rank equal to $[L : K]$. Also, the trace of any element of B lies in A. Let $\{x_1, x_2, \ldots, x_n\}$ be an A basis for B and let $\mathfrak{d}_{B/A}$ be the ideal in A generated by $\det(\mathrm{tr}_{L/K}(x_i x_j))$. This ideal is called the discriminant of B/A. It is independent of the choice of a basis. Since L/K is separable, the discriminant is not the zero ideal. Let

$$PB = \prod_{\mathfrak{P}_i \mid P} \mathfrak{P}_i^{e_i}$$

be the prime decomposition of PB and consider the A/P algebra

$$B/PB \cong B/\mathfrak{P}_1^{e_1} \oplus B/\mathfrak{P}_2^{e_2} \oplus \cdots \oplus B/\mathfrak{P}_g^{e_g}.$$

A commutative algebra over a field k is said to be separable if it is a direct sum of separable field extensions of k. From general theory, B/PB is a separable A/P algebra if and only if $\det(\mathrm{tr}_{\bar{B}/\bar{A}}(\bar{x}_i \bar{x}_j)) \neq 0$. Here the bar refers to reduction modulo PB and we have set $\bar{B} = B/\mathfrak{P}B$ and $\bar{A} = A/\mathfrak{P}$. It follows easily that every prime in B is unramified over A if and only if $\mathfrak{d}_{B/A} = A$, in other words, B is unramified over A if and only if the discriminant is all of A.

Define $C_{B/A} = \{x \in L \mid \mathrm{tr}_{L/K}(xb) \in A, \forall b \in B\}$. This set is easily seen to be a B submodule of L. In fact, it is the largest B submodule of L whose trace is contained in A. We shall show it is a fractional B ideal. Notice that $B \subseteq C_{B/A}$. $C_{B/A}$ is called the inverse different of B over A. By definition, $\mathfrak{D}_{B/A} = C_{B/A}^{-1} \subseteq B$ is called the different of B over A.

Since, $C_{B/A}$ is a B-module, to show it is a fractional ideal it suffices to produce a non-zero element d of L such that $dC_{B/A} \subseteq B$. Set $d = \det(\mathrm{tr}_{L/K}(x_i x_j))$, the element we used in defining the discriminant. If $c \in C_{B/A}$, then

$$c = \sum_{i=1}^{n} r_i x_i \qquad r_i \in K.$$

Multiply both sides by x_j and take the trace. We get

$$\mathrm{tr}_{L/K}(cx_j) = \sum_{i=1}^{n} r_i \, \mathrm{tr}_{L/K}(x_i x_j).$$

It follows from Cramer's rule that $dr_i \in A$ for all i, and so $dC_{B/A} \subseteq B$ as asserted. This argument tells us a bit more. Since $C_{B/A} \subseteq d^{-1}B$ we must have $\mathfrak{d}_{B/A}B = dB \subseteq \mathfrak{D}_{B/A}$; i.e., the discriminant is contained in the different. The connection between the different and discriminant is even closer as we see from the following proposition.

Proposition 7.9.   *i) Let A be a dvr with maximal ideal P, K its quotient field, L a finite separable extension, and B the integral closure of A in L. Then, some prime above P in B is ramified if and only if $\mathfrak{d}_{B/A} \subset P$.*
ii) $N_{L/K}\mathfrak{D}_{B/A} = \mathfrak{d}_{B/A}$. In words, the norm of the different is the discriminant.

Proof. We have already given the proof of *i*) in the above discussion. For the proof of part *ii*) see Serre [2]. ∎

We will say that B/A is unramified if no prime of B is ramified over A. From the above proposition, it follows that if $\mathfrak{D}_{B/A} = B$, then B/A is unramified. Much more is true, however. A prime $\mathfrak{P}$ of B is ramified over A if and only if $\mathfrak{P}$ divides $\mathfrak{D}_{B/A}$. The easiest way to see this is to pass to completions.

For each prime $\mathfrak{P} \subset B$ lying over P consider the completion $\hat{L}_{\mathfrak{P}}$ of L at $\mathfrak{P}$. The closure of K in $\hat{L}_{\mathfrak{P}}$ is isomorphic to $\hat{K}_P$, the completion of K at P. We also complete A at P and B at $\mathfrak{P}$ to obtain the rings $\hat{A}_P \subseteq \hat{B}_{\mathfrak{P}}$. It is not hard to show that $\hat{B}_{\mathfrak{P}}$ is the integral closure of $\hat{A}_P$ in $\hat{L}_{\mathfrak{P}}$. This local situation has all the ingredients of the "semi-local" situation considered above, so in exactly the same way we can define the local discriminant and the local different, $\mathfrak{d}_{B/A}(P) \subseteq \hat{A}_P$ and $\mathfrak{D}_{B/A}(\mathfrak{P}) \subseteq \hat{B}_{\mathfrak{P}}$.

Lemma 7.10. *We have $\mathfrak{d}_{B/A}(P) = \mathfrak{d}_{B/A}\hat{A}_P$ and $\mathfrak{D}_{B/A}(\mathfrak{P}) = \mathfrak{D}_{B/A}\hat{B}_{\mathfrak{P}}$. In other words, if $\mathfrak{d}_{B/A} = P^t$, then $\mathfrak{d}_{B/A}(P) = \hat{P}^t$ and if δ is the exact power of $\mathfrak{P}$ dividing $\mathfrak{D}_{B/A}$, then $\mathfrak{D}_{B/A}(\mathfrak{P}) = \hat{\mathfrak{P}}^\delta$.*

For the proof of this result we refer the reader to Serre [2], Chapter 3.

Corollary 1. *As in the Lemma, let δ be the exact power of $\mathfrak{P}$ dividing $\mathfrak{D}_{B/A}$. Then δ can be characterized as the largest integer m such that the trace from $\hat{L}_{\mathfrak{P}}$ to $\hat{K}_P$ of $\hat{\mathfrak{P}}^{-m}$ is contained in $\hat{A}_P$.*

Proof. From the definition, if m has the property described in the corollary, the local inverse different is $\hat{\mathfrak{P}}^{-m}$ and so $\mathfrak{D}_{B/A}(\mathfrak{P}) = \hat{\mathfrak{P}}^m$. The result is then immediate from the lemma. ∎

Corollary 2. *With the same notation as Corollary 1, $\delta \geq e(\mathfrak{P}/P) - 1$ with equality holding if and only if the characteristic of F is either zero or does not divide $e(\mathfrak{P}/P)$.*

Proof. (sketch) Neither $e(\mathfrak{P}/P)$ nor δ changes after passing to the completion (for δ this follows from the lemma). So, we can assume A and B are complete. Again, nothing essential changes if we replace K by the maximal unramified extension of K in L and A by its integral closure in this extension. We can thus assume $\mathfrak{P}$ is totally ramified over P. Set

$e = e(\mathfrak{P}/P) = [L : K]$. Let $\pi \in \mathfrak{P}$ be a uniformizing parameter, and $f(x) = \sum_{i=0}^{e} a_i x^i$ be the monic irreducible polynomial for π over K. $f(x)$ is an Eisenstein polynomial (see Serre [2], Chapter 3) and in particular, $a_i \in P$ for $0 \le i < e$. Under these circumstances, $\{1, \pi, \ldots, \pi^{e-1}\}$ is a basis for B over A and one can show that $\mathfrak{D}_{B/A} = (f'(\pi))$. Now,

$$f'(\pi) = e\pi^{e-1} + (e-1)a_{e-1}\pi^{e-2} + \cdots + a_1.$$

Every term in the sum except possibly the first is divisible by π^e and the first is divisible by π^{e-1}. The first assertion of Corollary 2 follows from this. The first term of the sum is exactly divisible by π^{e-1} if and only if either the characteristic of F is zero or does not divide e. This proves the second assertion.

Let L/K be a finite extension of function fields, $\mathfrak{P}$ a prime of L and P the prime lying below it in K. We say that $\mathfrak{P}$ is tamely ramified over P if it is ramified and either the characteristic of the residue class field $O_{\mathfrak{P}}/\mathfrak{P}$ is zero or does not divide $e(\mathfrak{P}/P)$. The second assertion of Corollary 2 can then be reworded to assert that for a tamely ramified prime $\mathfrak{P}$, the exponent to which it divides the different is $e(\mathfrak{P}/P) - 1$.

Theorem 7.11. *With the above notations and hypotheses, a prime $\mathfrak{P}$ of B is ramified over A if and only if $\mathfrak{P} \mid \mathfrak{D}_{B/A}$.*

Proof. (sketch) The definition of unramifiedness is in two parts; the ramification index must be one, and the residue class field extension must be separable. If one ignores the second condition one can refer to the above Corollary 2 to Lemma 7.10 for a proof of the theorem. We proceed somewhat differently and handle both conditions at once.

By standard properties of localization and completion, a prime $\mathfrak{P}$ of B is ramified over P if and only if $\hat{\mathfrak{P}}$ is ramified over $\hat{P}$. So we can work in the local situation $\hat{B}_{\mathfrak{P}}/\hat{A}_P$. The advantage here is that $\hat{B}_{\mathfrak{P}}$ has only one prime ideal, namely, $\hat{\mathfrak{P}}$. Thus, using what we know about discriminants, $\hat{\mathfrak{P}}$ is ramified over $\hat{P}$ if and only if $\mathfrak{d}_{B/A}(P) \ne \hat{A}_P$. By Proposition 7.9, applied to this situation, we see this is true if and only if $\mathfrak{D}_{B/A}(\mathfrak{P}) \ne \hat{B}_{\mathfrak{P}}$. Finally, by Lemma 7.10 this last condition holds if and only if $\mathfrak{D}_{B/A}$ is divisible by $\mathfrak{P}$.

We have now developed enough theory to enable us to return to function fields. We will define the different divisor and explore its properties. From now on, we suppose L/K is a finite separable extension of function fields with E the constant field of L and F the constant field of K. It is easy to see that E/F is also a finite, separable extension. For any prime P of K we let R_P be the integral closure of O_P in L. As we have seen, the primes of L lying above P are in one to one correspondence with the non-zero prime ideals $\mathfrak{p}$ of R_P. If $\mathfrak{p}$ is such an ideal let $O_{\mathfrak{P}}$ be the localization of R_P at $\mathfrak{p}$

and, of course, the maximal ideal of $O_{\mathfrak{P}}$ is $\mathfrak{P} = \mathfrak{p}O_{\mathfrak{P}}$. Note that the pair R_P, O_P is playing the role of the pair B, A in the above considerations.

For any prime $\mathfrak{P}$ of L, let $\mathfrak{p} = R_P \cap \mathfrak{P}$ and let $\delta(\mathfrak{P})$ be the exact power of $\mathfrak{p}$ dividing the different of R_P over O_P. We define the different divisor of L/K as follows:

$$D_{L/K} = \sum_{\mathfrak{P} \in \mathcal{S}_L} \delta(\mathfrak{P})\, \mathfrak{P}.$$

Actually, we must prove that all but finitely many $\delta(\mathfrak{P})$ are zero before we can be sure this definition makes sense. By our previous work, a prime $\mathfrak{P}$ is unramified over P if and only if $\delta(\mathfrak{P}) = 0$. Once we prove the finiteness assertion, it will follow that only finitely many primes in L can be ramified over K.

To prove the finiteness result, let $\{x_1, x_2, \ldots, x_n\}$ be a basis of L over K. Let $S \subset \mathcal{S}_K$ be the set of primes of K which lie below a pole of some x_i. Since there are only finitely many such poles, the set S is finite. For $P \notin S$ we have $x_i \in R_P$ for $1 \leq i \leq n$. This follows from the fact that R_P is the intersection of the valuation rings containing it (R_P is a Dedekind ring and has quotient field L). Let C_P be the inverse different of R_P over O_P. Let $c \in C_P$ and write $c = \sum_{i=1}^{n} a_i x_i$ with $a_i \in K$. Then, for all $1 \leq j \leq n$,

$$\text{tr}_{L/K}(cx_j) = \sum_{i=1}^{n} a_i \text{tr}_{L/K}(x_i x_j).$$

Using Cramer's rule, as we have previously, we find $da_i \in O_P$ for $1 \leq i \leq n$, where $d = \det(\text{tr}_{L/K}(x_i x_j))$. Thus, $dC_P \subseteq R_P$ and so $R_P \subseteq C_P \subseteq d^{-1}R_P$. Now, $\text{ord}_{\mathfrak{P}}(d) = 0$ for all but finitely many $\mathfrak{P}$ and, consequently, d is a unit in R_P for all but finitely many $P \in \mathcal{S}_K$. This implies $C_P = R_P$ and so $\mathfrak{D}_{R_P/O_P} = R_P$ for all but finitely many P. The fact that $\delta(\mathfrak{P}) = 0$ for all but finitely many $\mathfrak{P}$ is now clear.

We summarize our discussion of the different in the following theorem.

Theorem 7.12. *Suppose L/K is a finite separable extension of function fields. The different $D_{L/K}$ defined above is a divisor with the property that a prime $\mathfrak{P}$ of L is ramified over K if and only if it occurs in $D_{L/K}$ with a non-zero coefficient. In particular, only finitely many primes of L are ramified over K.*

We note that separability is crucial for the last assertion of the theorem. If F is perfect and L/K is purely inseparable of degree p, then Lemma 7.3 shows that every prime of L is ramified over K.

Because it is often useful, we record an important property of differents. The proof is not hard, but will be omitted. Suppose $K \subseteq L \subseteq M$ is a tower of function fields with M/K finite and separable. Then,

$$D_{M/K} = D_{M/L} + i_{M/L}(D_{L/K}) .$$

The next topic will be the behavior of differentials in field extensions. Let ω be a Weil differential of a function field K and suppose L is a finite separable extension of K. We want to associate to ω a differential ω^* of L. After this is done the next task will be to relate the divisor of ω^* to the divisor of ω.

We begin by extending of the trace map from L to K to a trace map from the adeles of L, A_L, to the adeles of K, A_K. The key to this is the following important isomorphism

$$L \otimes_K \hat{K}_P \cong \bigoplus_{\mathfrak{P} | P} \hat{L}_{\mathfrak{P}} .$$

The map involved can be described quite easily. Identifying L as a subfield of $\hat{L}_{\mathfrak{P}}$ and $\hat{K}_P$ as a subfield of $\hat{L}_{\mathfrak{P}}$ there is an obvious K-bilinear map from $L \times \hat{K}_P$ to $\hat{L}_{\mathfrak{P}}$, namely, $(\ell, \alpha) \to \ell\alpha$. Thus, for each $\mathfrak{P} | P$, there is a map from $L \otimes_K \hat{K}_P \to \hat{L}_{\mathfrak{P}}$. Now pass to the direct sum. The fact that the resulting homomorphism is an isomorphism is given in Serre [2], Chapter 3. Note that L is embedded diagonally into the right-hand side.

Both sides of the above isomorphism are $\hat{K}_P$ algebras and the isomorphism respects this structure. If $\{x_1, x_2, \ldots, x_n\}$ is a basis for L/K, then $\{x_1 \otimes 1, x_2 \otimes 1, \ldots, x_n \otimes 1\}$ is a basis for the left-hand side over $\hat{K}_p$. On the other hand, choosing a basis for $\hat{L}_{\mathfrak{P}}$ over $\hat{K}_P$ for each $\mathfrak{P} | P$ and putting these together gives a basis for the right-hand side. Using these bases enable one to prove the following result, which connects the global and local traces and norms.

Proposition 7.13. *Let $T_{\mathfrak{P}}$ and $N_{\mathfrak{P}}$ denote the trace and norm from $\hat{L}_{\mathfrak{P}}$ to $\hat{K}_P$, respectively. Then, for $x \in L$ we have*

$$\operatorname{tr}_{L/K}(x) = \sum_{\mathfrak{P}|P} T_{\mathfrak{P}}(x) \quad \text{and} \quad N_{L/K}(x) = \prod_{\mathfrak{P}|P} N_{\mathfrak{P}}(x) .$$

In words, this says that the global trace is the sum of the local traces and the global norm is the product of the local norms. We will be primarily concerned with the traces.

We now define the trace map from A_L to A_K. Let $\alpha = (\alpha_{\mathfrak{P}})$ be an element of A_L. We map it to the adele of K, whose P-th coordinate, a_P, is $\sum_{\mathfrak{P}|P} T_{\mathfrak{P}}(\alpha_{\mathfrak{P}})$. Since for all $\mathfrak{P}$, $T_{\mathfrak{P}} : \hat{O}_{\mathfrak{P}} \to \hat{O}_P$ we see that for all but finitely many P, a_P is in fact in $\hat{O}_P$. Thus, the image of our map is indeed in A_K. We call this trace map $\operatorname{tr}_{L/K}$ because it extends the trace map on the level of fields. To see this, recall L is embedded diagonally into A_L. If $\lambda \in A_L$ is the adele all of whose coordinates are equal to ℓ, then, by Proposition 7.13, $\operatorname{tr}_{L/K}(\lambda)$ is the adele of K all of whose coordinates are equal to $\operatorname{tr}_{L/K}(\ell)$. One also checks easily that $\operatorname{tr}_{L/K}$ is an F-linear map from A_L to A_K (recall that F is the constant field of K).

Having extended the trace map, it is now relatively easy to define the map from $\Omega_K \to \Omega_L$, which we need. Let $\omega \in \Omega_K$. Define ω^* to be the compositum $\omega \circ \mathrm{tr}_{L/K}$, which is an F-linear homomorphism from A_L to F. From now on we assume that F is also the constant field of L, i.e., that L/K is a geometric extension. We claim that ω^* is, in fact, a Weil differential of L. That it vanishes on L follows from what we have just proved; $\mathrm{tr}_{L/K}$ maps L to K and ω vanishes on K by definition. It remains to prove that ω^* vanishes on $A_L(\mathcal{C})$ for some divisor $\mathcal{C}$ of L.

Since $\omega \in \Omega_K$ there is a divisor C of K such that ω vanishes on

$$A_K(C) = \{(a_P) \in A_K \mid \mathrm{ord}_P(a_P) \geq -\mathrm{ord}_P C, \ \forall P \in S_K\}.$$

Let $\alpha = (\alpha_{\mathfrak{P}}) \in A_L$. Fix a prime P of K and suppose $\{\mathfrak{P}_1, \mathfrak{P}_2, \ldots, \mathfrak{P}_g\}$ are the primes of L lying above P. We need to ascertain the conditions which force

$$\mathrm{ord}_P(\sum_{i=1}^{g} T_{\mathfrak{P}_i}(\alpha_{\mathfrak{P}_i})) \geq -\mathrm{ord}_P C.$$

This will follow if for each i individually $\mathrm{ord}_P(T_{\mathfrak{P}_i}(\alpha_{\mathfrak{P}_i})) \geq -\mathrm{ord}_P C$. Let π be a uniformizing parameter at P and for simplicity set $m = \mathrm{ord}_P C$. The last condition is equivalent to the following: for each i, $\mathrm{ord}_P(T_{\mathfrak{P}_i}(\pi^m \alpha_{\mathfrak{P}_i})) \geq 0$. This will happen if $\pi^m \alpha_{\mathfrak{P}_i}$ is in the local inverse different at $\mathfrak{P}_i$. From the definition of the different and Corollary 1 to Lemma 7.10 this condition is equivalent to

$$\mathrm{ord}_{\mathfrak{P}_i}(\pi^m \alpha_{\mathfrak{P}_i}) \geq -\delta(\mathfrak{P}_i) \quad \text{or} \quad \mathrm{ord}_{\mathfrak{P}_i}(\alpha_{\mathfrak{P}_i}) \geq -\delta(\mathfrak{P}_i) - e(\mathfrak{P}_i/P)\mathrm{ord}_P C.$$

It is easy to check from the definitions that

$$\sum_{\mathfrak{P}}(\delta(\mathfrak{P}) + e(\mathfrak{P}/P)\mathrm{ord}_P C)\,\mathfrak{P} = D_{L/K} + i_{L/K} C.$$

To sum up, we have proven—

Proposition 7.14. *Let L/K be a finite, separable, geometric extension of function fields and ω a non-zero differential of K. The $\omega^* = \omega \circ \mathrm{tr}_{L/K}$ is a differential of L. In more detail, if ω vanishes on $A_K(C)$, ω^* is an F-linear homomorphism from $A_L \to F$ which vanishes on L and on $A_L(i_{L/K} C + D_{L/K})$.*

We would like to determine the divisor of ω^*. Recall, by definition, this is the largest L-divisor B such that ω^* vanishes on $A_L(B)$. In light of the previous proposition, a good guess would be $i_{L/K}(\omega)_K + D_{L/K}$. This is, in fact, correct.

Theorem 7.15. *With the hypotheses and notation of Proposition 7.14, we have*

$$(\omega^*)_L = i_{L/K}(\omega)_K + D_{L/K}.$$

Proof. We are going to use Proposition 6.12 of the last chapter, which shows how to determine the divisor of a differential using properties of its local components.

We begin by recalling the definition. Let $\mu \in \Omega_L$. For a prime $\mathfrak{P}$ of L let $i_{\mathfrak{P}} : \hat{L}_{\mathfrak{P}} \to A_L$ be the map that takes an element $\gamma \in \hat{L}_{\mathfrak{P}}$ to the adele all of whose components are zero except the $\mathfrak{P}$-th component which is γ. Then $\mu_{\mathfrak{P}} = \mu \circ i_{\mathfrak{P}}$.

Now, $\omega_{\mathfrak{P}}^* = (\omega \circ \mathrm{tr}_{L/K}) \circ i_{\mathfrak{P}} = \omega \circ (\mathrm{tr}_{L/K} \circ i_{\mathfrak{P}}) := \omega \circ (i_P \circ T_{\mathfrak{P}}) = \omega_P \circ T_{\mathfrak{P}}$. The third equality follows directly from the definition of the trace map on the adeles. In words, the local component of ω^* at $\mathfrak{P}$ is the local component of ω at P composed with the local trace map.

According to Proposition 6.11, $\mathrm{ord}_{\mathfrak{P}}(\omega^*) = N$, where N is the integer N such that $\omega_{\mathfrak{P}}^*$ vanishes on $\hat{\mathfrak{P}}^{-N}$ but not on $\hat{\mathfrak{P}}^{-N-1}$.

$\omega_{\mathfrak{P}}^*$ vanishes on $\hat{\mathfrak{P}}^{-N}$ if and only if ω_P vanishes on $T_{\mathfrak{P}}(\hat{\mathfrak{P}}^{-N})$, which occurs if and only if $T_{\mathfrak{P}}(\hat{\mathfrak{P}}^{-N}) \subseteq P^{-m}$, where $m = \mathrm{ord}_P(\omega)$. This is equivalent to $T_{\mathfrak{P}}(P^m \hat{\mathfrak{P}}^{-N}) \subseteq \hat{O}_P$, which in turn is true if and only if $\hat{\mathfrak{P}}^{e(\mathfrak{P}/P)m-N} \subseteq \hat{\mathfrak{P}}^{-\delta(\mathfrak{P})}$, by Lemma 7.10 and the definition of the different. We conclude that $\omega_{\mathfrak{P}}^*$ vanishes on $\hat{\mathfrak{P}}^{-N}$ if and only if $N \leq e(\mathfrak{P}/P)\mathrm{ord}_P(\omega) + \delta(\mathfrak{P})$. The largest N with this property is clearly the right-hand side of this inequality. The theorem follows.

It might be asked, what is the necessity of Proposition 7.14 since Theorem 7.15 is a more accurate result? The answer is we had to show ω^* is a differential before we could decompose it into its local components and use Proposition 6.11 to determine its divisor.

We are finally in a position to prove the Riemann-Hurwitz Theorem, one of the main goals of this chapter.

Theorem 7.16. (Riemann-Hurwitz) *Let L/K be a finite, separable, geometric extension of function fields. Then,*

$$2g_L - 2 = [L : K](2g_K - 2) + \deg_L D_{L/K}.$$

In particular,

$$2g_L - 2 \geq [L : K](2g_K - 2) + \sum_{\mathfrak{P}} (e(\mathfrak{P}/P) - 1) \deg_L \mathfrak{P}.$$

where the sum is over all primes $\mathfrak{P}$ of L, which are ramified in L/K. The inequality is an equality if and only if all ramified primes are tamely ramified.

Proof. Let ω be a non-zero differential of K. By the remarks following Corollary 2 to Lemma 6.10, $(\omega)_K$ is in the canonical class of K. By Corollary 3 to Theorem 5.4, every divisor in the canonical class of K has degree

$2g_K - 2$. Thus, $\deg_K(\omega)_K = 2g_K - 2$. Similarly, $\deg_L(\omega^*)_L = 2g_L - 2$. From Theorem 7.15 we find

$$2g_L - 2 = \deg_L(\omega^*)_L = \deg_L i_{L/K}(\omega)_K + \deg_L D_{L/K}.$$

From Proposition 7.7 we see that $\deg_L i_{L/K}(\omega)_K = [L : K]\deg_K(\omega)_K = [L : K](2g_K - 2)$. We have used the assumption that L has the same constant field as K. Substituting into the above equality yields the first assertion of the theorem.

The second assertion is an immediate consequence of Corollary 2 to Lemma 7.10, since $\deg_L D_{L/K} = \sum_{\mathfrak{P}} \delta(\mathfrak{P}) \deg_L \mathfrak{P}$.

The Riemann-Hurwitz theorem has a very large number of consequences. We will give some idea of how it is used by giving three corollaries.

Corollary 1. *Suppose L/K is a finite, separable, geometric extension of function fields. Then, $g_K \leq g_L$. (This need not be true for inseparable extensions!)*

Proof. Since the different is an effective divisor (all its coefficients are non-negative) we see $2g_L - 2 \geq [L : K](2g_K - 2) \geq 2g_K - 2$. Thus $g_K \leq g_L$ as asserted.

One can prove this result in another way. It follows immediately from the theorem that ω^* is a holomorphic differential (no poles) if ω is a holomorphic differential. One checks (using the fact that the trace map from A_L to A_K is onto when L/K is separable) that the map $\omega \to \omega^*$ is a one to one F-linear map from $\Omega_K(0)$ to $\Omega_L(0)$. Since these two vector spaces have dimension g_K and g_L, respectively, it follows that $g_K \leq g_L$.

Corollary 2. (Luroth's Theorem) *Let $L = F(x)$ be a rational function field over F and K a subfield properly containing F. Then, there is a $u \in K$ such that $K = F(u)$.*

Proof. Since K properly contains F, it is easy to see that $[L : K] < \infty$. Let M be the maximal separable extension of K contained in L. If the characteristic of F is zero, then $M = L$. If the characteristic of F is $p > 0$, then L/M is purely inseparable of degree p^n for some $n \geq 0$. It follows that $x^{p^n} \in M$. On the other hand, the polar divisor of x in L is a prime (the prime at infinity) of degree 1 and so the polar divisor of x^{p^n} has degree p^n. Consequently, $[L : F(x^{p^n})] = p^n$ by Proposition 5.1. Thus, $M = F(x^{p^n})$ and M is a rational function field. This shows that we can assume to begin with that L/K is separable. Since L has genus zero it follows by Corollary 1 that $g_K = 0$. Since L has a prime of degree 1, e.g., the zero or pole of x, the prime lying below it in K must also have degree 1. It follows that K is a rational function field (see the discussion after the proof of Corollary 5 to Theorem 5.4).

Corollary 3. *Let L/K be a finite, separable, geometric extension of function fields. Assume $g_L = 1$. Then, $g_K \leq 1$ with equality holding if and only if L/K is unramified.*

Proof. The inequality $g_K \leq 1$ follows from Corollary 1. From $g_L = 1$ and the theorem we deduce, $0 = [L : K](2g_K - 2) + \deg_L D_{L/K}$. If $g_K = 1$ the degree of the different is zero and so the different is the zero divisor (recall that the different is an effective divisor). From Theorem 7.12 it follows that L/K is unramified. By the same theorem, if L/K is unramified then $D_{L/K} = 0$ and so $2g_K - 2 = 0$ or, what is the same, $g_K = 1$.

We will conclude this chapter with a beautiful application of the Riemann-Hurwitz theorem to the proof of the ABC theorem in function fields. Let's begin by recalling the ABC conjecture of Masser and Oesterlé in the case of the rational numbers $\mathbb{Q}$. Suppose $A, B, C \in \mathbb{Z}$ and that $A + B = C$. Suppose further that the three integers $A, B,$ and C are pairwise relatively prime. The conjecture states that for each $\epsilon > 0$ there is a constant M_ϵ such that if $A, B,$ and C satisfy the given conditions, we have

$$\max\left(|A|, |B|, |C|\right) \leq M_\epsilon \left(\prod_{p|ABC} p \right)^{1+\epsilon}.$$

This elegant conjecture has many surprisingly powerful consequences. See Lang [4], Chapter IV, Section 7, for a discussion and a number of references. At present the conjecture is not proven and many people consider it to be beyond the range of the available methods.

The ABC conjecture for $\mathbb{Q}$ can be easily generalized to number fields. We omit this formulation here. Instead we reformulate the conjecture over $\mathbb{Q}$ slightly. In this new formulation it becomes clear what the analogous conjecture should be in the function field case.

Rewrite $A + B = C$ as $A/C + B/C = 1$. Write $u = A/C$ and $v = B/C$. Then $u, v \in \mathbb{Q}$ and $u + v = 1$. Let's recall the definition of the height of a rational number r. Write $r = m/n$ where $m, n \in \mathbb{Z}$ and $(m, n) = 1$. Then the height of r, $\mathrm{ht}(r)$, is defined to be $\max(\log|m|, \log|n|)$. With this notation we can recast the ABC conjecture as follows. Suppose $\epsilon > 0$ is given. Then there is a constant m_ϵ such that whenever $u, v \in \mathbb{Q}^*$ and $u + v = 1$, we have

$$\max(\mathrm{ht}(u), \mathrm{ht}(v)) \leq m_\epsilon + (1 + \epsilon) \sum_{p|ABC} \log p.$$

Here, A and B represent the numerators of u and v and C their common denominator.

Now, let's return to the function field case. Let K be a function field and F its field of constants. Suppose $u, v \in K^*$ and $u + v = 1$. We need a substitute for the notion of height. Let A be the zero divisor of u and

C its polar divisor. A good measure of the size of a divisor is its degree, so it is natural to define the height of u to be $\max(\deg A, \deg C)$. This is fine, but it can be stated more simply. We know (Proposition 5.1) that $\deg A = \deg C = [K : F(u)]$. Instead of calling this number the height of u, it is more conventional to call it the degree of u and use the notation $\deg u$. One should be careful though, the degree of the divisor (u) is zero (Proposition 5.1), whereas the degree of the element u is greater than or equal to zero and is zero only when it is a constant.

For those with some algebraic geometry background, the degree of an element has a nice geometric interpretation. The field K is the function field of a smooth, complete curve Γ defined over F. The element u can be thought of as a rational map from Γ to the projective line $\mathbb{P}^1/F$. The degree of u is the degree of this mapping. If F is algebraically closed, all fibers have $\deg u$ elements with (possibly) finitely many exceptions.

Before stating the next theorem, we need two more definitions.

If $D \in \mathcal{D}_K$ is a divisor, recall that $\mathrm{Supp}(D)$ is defined to be the set of primes which occur in D with non-zero coefficient. This set is called the support of D.

Secondly, suppose $u \in K^*$ is not a constant. Let M be the maximal separable extension of $F(u)$ inside of K. Then, the field degree, $[M : F(u)]$, is called the separable degree of u and is denoted by $\deg_s u$. Note that $\deg_s u \le \deg u$ with equality holding if and only if $K/F(u)$ is separable.

We can now state and prove(!) the ABC conjecture for function fields.

Theorem 7.17. *Let K be a function field with a perfect constant field F. Suppose $u, v \in K^*$ and $u + v = 1$. Then,*

$$\deg_s u = \deg_s v \le 2g_K - 2 + \sum_{P \in \mathrm{Supp}(A+B+C)} \deg_K P .$$

Here, A and B are the zero divisors of u and v in K, respectively, and C is their common polar divisor in K. (Note that no "ϵ" appears in the function field version).

Proof. It is convenient to set $k = F(u)$. We first treat the case that K/k is separable and do the general case later. Let $n = \deg u = [K : k]$. The Riemann-Hurwitz theorem implies that

$$2g_K - 2 \ge -2n + \sum (e(P/\mathfrak{P}) - 1) \deg_K P , \tag{1}$$

where the sum is over all primes of K and for any such prime P, $\mathfrak{P}$ denotes the prime of k lying below P. The point is that since F is perfect a prime is ramified if and only if its ramification index is greater than one. If its ramification index is equal to one, it doesn't contribute to the sum.

In the function field $k = F(u)$ we consider three primes $\mathfrak{P}_0$, $\mathfrak{P}_1$, and $\mathfrak{P}_\infty$, which are the zero divisors in k of u, $1 - u = v$, and $1/u$, respectively. It is easy to see that $A = i_{K/k}(\mathfrak{P}_0)$, $B = i_{K/k}(\mathfrak{P}_1)$, and $C = i_{K/k}(\mathfrak{P}_\infty)$.

(Thinking of u as a mapping from the curve Γ to $\mathbb{P}^1$, this says that A, B, and C are the inverse images (as divisors) of 0, 1, and ∞). In the above sum we are only going to consider primes in the support of either A, B, or C. This will only strengthen the inequality. Consider the sum only over the primes in the support of A. We have

$$\sum_{P \in \text{Supp}(A)} (e(P/\mathfrak{P}_0) - 1) \deg_K P = \deg_K(i_{K/k} \mathfrak{P}_0) - \sum_{P \in \text{Supp}(A)} \deg_K P.$$

By Proposition 7.7, $\deg_K(i_{K/k}\mathfrak{P}_0) = [K : k] \deg_k \mathfrak{P}_0 = n$. So, the above sum is simply n minus the sum of $\deg_K P$ over those P in the support of A. The same considerations prove the analogous result for B and C. So, adding the contributions from these three sums and substituting into Equation 1 above, we find

$$2g_K - 2 \geq n - \sum_{P \in \text{Supp}(A+B+C)} \deg_K P \ ,$$

and that concludes the proof in the case where K/k is separable.

Now suppose the characteristic p of F is positive and that K/k is inseparable. Let M be the maximal separable extension of k in K. Then K/M is purely inseparable of degree p^m for some m. Working with the separable extension M/k, we find

$$2g_M - 2 \geq [M : k] - \sum_{P' \in \text{Supp}(A'+B'+C')} \deg_M P' \ ,$$

where $A' = i_{M/k}(\mathfrak{P}_0)$, $B' = i_{M/k}(\mathfrak{P}_1)$, and $C' = i_{M/k}(\mathfrak{P}_\infty)$. By Proposition 7.5, we see that for each prime P' of M there is one and only one prime P of K lying above P' and that $\deg_K P = \deg_M P'$. Since, by definition, $[M : k] = \deg_s u$, the above inequality can be rewritten as

$$2g_M - 2 \geq \deg_s u - \sum_{P \in \text{Supp}(A+B+C)} \deg_K P \ .$$

Invoking Proposition 7.5 once more, we see $g_M = g_K$. This completes the proof.

To show the power of the ABC Theorem, we will give two applications. The first will concern solutions to the Fermat equation $X^N + Y^N = 1$ in function fields and the second will be the statement and proof of the S-unit Theorem, a powerful result with many applications to diophantine problems.

Proposition 7.18. *Let K be a function field with a perfect constant field F. Consider the equation $X^N + Y^N = 1$. We assume that N is not divisible by the characteristic p of F. If $g_K = 0$ and $N \geq 3$, then there is no nonconstant solution to this equation in K. If $g_K \geq 1$ and $N > 6g_K - 3$, then* *there is no non-constant solution to this equation in K. (By a non-constant solution we mean a pair $(u, v) \in K^2 - F^2$ such that $u^N + v^N = 1$.)*

Proof. Suppose that $(u, v) \in K^2$ is a non-constant solution. Invoking the ABC theorem we find

$$\max(\deg_s u^N, \deg_s v^N) \leq 2g_K - 2 + \sum_{P \in \text{Supp}(A+B+C)} \deg_K P , \qquad (2)$$

where A is the zero divisor of u, B is the zero divisor of v, and C is their common polar divisor. We'll return to this equation in a moment.

Let M be the maximal separable extension of $F(u)$ in K. By considering the tower of fields $F(u^N) \subseteq F(u) \subseteq M \subseteq K$, and noting that $F(u)/F(u^N)$ is separable of degree N (it's here we use the hypothesis $(p, N) = 1$), we see that $\deg_s u^N = N \deg_s u$. Similarly, $\deg_s v^N = N \deg_s v$.

Next, by comparing the zero divisor of u in M to the zero divisor of u in K (as we have done in the proof of Theorem 7.17) we see that $\sum_{P \in \text{Supp}(A)} \deg_K P \leq \deg_s u$. Applying the same reasoning to v yields $\sum_{P \in \text{Supp}(B)} \deg_K P \leq \deg_s v$. Since C is the common polar divisor of u and v we have a similar inequality involving C.

Putting all this together and substituting into Equation 2 yields

$$N \sum_{P \in \text{Supp}(A)} \deg_K P \leq 2g_K - 2 + \sum_{P \in \text{Supp}(A+B+C)} \deg_K P ,$$

with similar equations involving B and C on the left-hand side. Adding all three and rearranging terms gives

$$(N - 3) \sum_{P \in \text{Supp}(A+B+C)} \deg_K P \leq 6g_K - 6 .$$

If $g_K = 0$ and $N \geq 3$, the left-hand side of this inequality is non-negative and the right-hand side is -6. This is impossible and this contradiction establishes the first assertion of the Proposition.

Assume now that $g_K \geq 1$ and $N > 6g_K - 3$. Then certainly $N \geq 4$ so $N - 3$ is positive. Dividing both sides of the inequality by $N - 3$ we see that $6g_K - 6 / N - 3$ must be bigger than or equal to one. If $6g_K - 6 / N - 3 < 1$ we get a contradiction. Since this inequality is equivalent to $N > 6g_K - 3$ the proposition is proved.

Actually, one can get a somewhat better result by a different method. Namely, suppose all the hypotheses of the proposition hold and that $(u, v) \in K^2$ is a non-constant solution. Let $F(u, v)$ be the subfield of K generated by u and v over F. We will show in the next chapter that in characteristic zero or when $(p, N) = 1$ the genus of $F(u, v)$ is equal to $(N-1)(N-2)/2$. When the constant field is perfect the genus of a subfield is less than or equal to the genus of the field. Thus, if a solution exists $(N - 1)(N - 2)/2 \leq g_K$.

Put the other way around, if $g_K < (N-1)(N-2)/2$, there are no non-constant solutions. Although this is quite elegant, the solution using the ABC Theorem is applicable in many situations where this method fails.

By the way, the hypothesis about the constant field being perfect is superfluous. In the next chapter we will show that in this problem we could have replaced F by its algebraic closure $\bar{F}$. Since algebraically closed fields are perfect the method applies and gives the result over $\bar{F}$ and a posteriori over F.

The final result of the chapter involves the notion of S-units. Let K be a function field with constant field F and suppose $S = \{P_1, P_2, \ldots, P_t\}$ is a finite set of primes of K. An element $u \in K^*$ is called an S-unit if $\mathrm{Supp}(u) \subseteq S$, i.e., only primes in S enter into the principal divisor (u). The S-units form a group denoted by U_S. The map $u \to (u)$ is a homomorphism from the S-units into the free abelian group of divisors supported on S. Every element in the kernel of this map has zero for its divisor. Thus, the kernel consists precisely of the constants F^*. The degree of a principal divisor is zero. Thus the image of this map is a subgroup of the divisors of degree zero supported on S. The latter group is free of rank $t-1$. We have shown that U_S/F^* is free of rank $\leq t-1$ where t is the number of elements in S. This tells us something about the multiplicative properties of S-units. The next theorem is about an additive property of S-units.

Theorem 7.19. *Let K be a function field with a perfect constant field F. Let S be a finite set of primes of K. Then, there are only finitely many pairs of separable, non-constant S-units (u, v) such that $u + v = 1$. (u is said to be separable if the field extension $K/F(u)$ is separable). If the characteristic of F is zero, then every solution is separable. If characteristic of F is $p > 0$, then the most general solution to $X + Y = 1$ in non-constant S-units is (u^{p^m}, v^{p^m}) where (u, v) is a separable, non-constant solution in S-units and $m \in \mathbb{Z}, m \geq 0$.*

Proof. Assume to begin with that (u, v) is a non-constant, separable solution to $X + Y = 1$ is S-units. By the ABC Theorem we have

$$\deg u \leq 2g_K - 2 + \sum_{P \in \mathrm{Supp}(A+B+C)} \deg_K P \,,$$

with the usual notations. Let $M = \sum_{P \in S} \deg_K P$. Then, the right-hand side is $\leq 2g_K - 2 + M$ since the supports of A, B, and C are in S. Let $A = \sum_{P \in S} a(P)\, P$ with each $a(P) \geq 0$. Then, $\deg u = \deg_K A = \sum_{P \in S} a(P) \deg_K P$. This shows that for each $P \in S$,

$$a(P) \deg_K P \leq \deg_K A \leq 2g_K - 2 + M,$$

and consequently that $a(P)$ is bounded. Since A is a divisor with support in a finite set of primes with bounded coefficients, A must be one of only finitely many divisors. Similarly for B and C. It follows that the number of possibilities for the principal divisor (u) is finite and similarly for (v). For each of these possible principal divisors choose an S-unit u_i and v_j. We suppose that $1 \leq i \leq l$ and $1 \leq j \leq k$. Since any two non-zero elements of K have the same divisor if and only if they differ by a constant, all the non-constant, separable, S-unit solutions to $X + Y = 1$ have the form $(\alpha u_i, \beta v_j)$ with $\alpha, \beta \in F$. If there are more than lk such solutions, then by the pidgeon hole principal we can find a given pair of indices (i, j) and two distinct pairs of constants $(\alpha, \beta), (\alpha', \beta')$ such that

$$\alpha u_i + \beta v_j = 1 \quad \text{and} \quad \alpha' u_i + \beta' v_j = 1 .$$

Subtracting these two equations, we find that u_i is a constant times v_j. Substituting into the first equation shows that u_i is a constant. This is a contradiction, so we have shown there are only finitely many non-constant, separable, S-unit solutions to $X + Y = 1$.

Now suppose u and v are non-constant S-units and $u + v = 1$. If u is not separable, let M be the maximal separable extension of $F(u)$ in K. Then, $[K : M] = p^m$ for some positive integer m. By the corollary to Proposition 7.4 we see that u and $v = 1 - u$ are p-th powers. Write $u = u_1^p$ and $v = v_1^p$ with $u_1, v_1 \in K$. Note that, in fact, $u_1, v_1 \in U_S$. Since p is the characteristic of K, $1 = u + v = u_1^p + v_1^p = 1 = (u_1 + v_1)^p$ which implies $u_1 + v_1 = 1$. If u_1 is separable, we are done. If not, repeat the process and we find two S-units u_2 and v_2 such that $u_1 = u_2^p$, $v_1 = v_2^p$, and $u_2 + v_2 = 1$. Note that $u = u_2^{p^2}$ and $v = v_2^{p^2}$. Thus, if u_2 is separable we are done. If not, continue the process. This must end in finitely many steps since a non-constant in K cannot be a p^m power for infinitely many m. This is easy to see. For example, if u is not a constant, let P be a prime which is a zero of u. If u is a p^m power then p^m divides $\text{ord}_P(u)$ which bounds m. The proof is now complete.

Corollary. *Suppose K is a function field over a finite field $\mathbb{F}$. Suppose N is greater than 3 and is relatively prime to the characteristic of $\mathbb{F}$. Then, $X^N + Y^N = 1$ has at most finitely many non-constant separable solutions in K.*

Proof. Suppose $(u, v) \in K^2$ is a non-constant solution. In the course of proving Proposition 7.18, we proved that

$$(N - 3) \sum_{P \in \text{Supp}(A+B+C)} \deg_K P \leq 6g_K - 6 ,$$

where A is the zero divisor of u, B is the zero divisor of v, and C is their common polar divisor. Assuming $N \geq 4$, this shows that for any prime P in the support of either u or v we must have $\deg_K P \leq (6g_K - 6) / N - 3$. In a function field over a finite field there are at most finitely many primes whose degree is below a fixed bound (in Chapter 5 we gave estimates for the number of such primes). Let S be the set of all primes in K whose degree is less than or equal to $(6g_K - 6) \; / \; N - 3$. Then every solution to $X^N + Y^N = 1$ in K is an S-unit. The corollary now follows from the theorem.

Notice that the assumption that a solution be separable is essential since if (u, v) is a solution, then (u^{p^m}, v^{p^m}) is also a solution for all $m \geq 1$.

We have just given a taste of the possible applications of the ABC Theorem in function fields. For more, see the paper by Silverman [2] and the book by R.C. Mason [1].

The restriction on the constant field in the corollary to Theorem 7.19 is not necessary. One could apply the classical theorem of de Franchis from algebraic geometry, which states, in part, that if K/F is a function field there are only finitely many subfields M such that $F \subset M$, K is separable over M, and the genus of M is greater than 1. In the notation of the corollary, if (u, v) is a non-constant separable solution to $X^N + Y^N = 1$, then $F(u, v)$ is a subfield, which satisfies these three properties (its genus is $(N-1)(N-2)/2 > 1$ since $N > 3$). Thus, we are reduced to worrying about how many solutions (u, v) and (u', v') can exist with $F(u, v) = F(u', v')$. If this happens, there is an automorphism of $F(u, v)$, which takes u to u' and v to v'. A function field with genus greater than 2 has only finitely many automorphisms (see Iwasawa and Tamagawa [1]). It follows that there are only finitely many non-constant, separable solutions to $X^N + Y^N = 1$ in K.

The theorem of de Franchis is not easy to prove. The paper by E. Kani [1] contains a proof of an effective version of the theorem. The bibliography of that paper gives a number of relevant references to both the classical and more modern treatments.

Exercises

1. Let $K = F(x, y)$ be a function field where x and y satisfy an equation of the form $Y^2 = (X - a_1)(X - a_2) \cdots (X - a_n)$. We assume the a_i are distinct elements of F. Let the divisor of $x - a_i$ in $F(x)$ be denoted by $P_i - P_\infty$. For each i show that $i_{K/F(x)}P_i = 2\mathfrak{P}_i$ where $\mathfrak{P}_i$ is a prime of K of degree 1. Use this information to compute the genus of K (don't forget the role of the prime at infinity).

2. With the same notation as in Exercise 1, suppose that $n \geq 5$. Show that each prime of K which is ramified over $F(x)$ is a Weierstrass point (see Exercise 10 of Chapter 5).

3. Let l be a prime not equal to the characteristic of F and $K = F(x, y)$ a function field where x and y satisfy $Y^l = (X - a_1)^{n_1}(X - a_2)^{n_2} \cdots (X - a_m)^{n_m}$. We assume that the a_i are all distinct and that for each i, $l \nmid n_i$. Compute the genus of K.

4. Assume that F contains a primitive N-th root of unity and that N is not divisible by the characteristic of F. Consider a function field $K = F(x, y)$ where x and y satisfy an equation of the form $X^N + Y^N = 1$. Compute the genus of K.

5. Let K be a function field of genus 0 and L/K a finite geometric extension. If L/K is unramified, show that $L = K$. (Assuming the constant field is algebraically closed, this is the algebraic equivalent of the statement that the projective line is simply connected).

6. Let L/K be a finite, tamely ramified, geometric extension of the rational function field. Let P be a prime of K of degree 1. Suppose that L/K is unramified except possibly at primes lying above P. Show that $L = K$.

7. Let L/K be a finite, separable, geometric extension of function fields. Set $[L : K] = n$. Suppose that $\deg \mathcal{D}_{L/K} > 4(n - 1)$. Show that $g_L + 1 > n(g_K + 1)$.

8. With the same notation and assumptions as Exercise 7, suppose $\mathfrak{P}$ is a prime of L of degree 1 and that $\mathfrak{P}$ is totally ramified over K. Show that $\mathfrak{P}$ is a Weierstrass point.

9. Let L/K be a finite, separable, geometric extension of function fields with five or more totally ramified primes all of degree 1. Show that each of them is a Weierstrass point. (The results contained in Exercises 7, 8, and 9 are due to J. Lewittes.)

10. Let S be a finite set of primes of the function field K. Let $a, b \in K^*$. Show that the equation $aX + bY = 1$ has only finitely many solutions in S-units.

11. Assume $\mathbb{F}$ is finite and let $K/\mathbb{F}$ be a function field, $a, b \in K^*$, and $N \geq 5$ an integer not divisible by the characteristic of F. Show that the equation $aX^N + bY^N = 1$ has only finitely many separable solutions in K. If at least one of the two elements a and b is not a constant, there are only finitely many solutions altogether. (Hint: Pass to the extension field $L = K(\sqrt[N]{a}, \sqrt[N]{b})$.)

8
Constant Field Extensions

In this chapter we investigate a very important class of extensions of function fields, namely, constant field extensions. Let K/F be a function field with constant field F. For every field extension E of F we want to define a function field KE over E and investigate its properties. We shall confine ourselves to the special case where E/F is algebraic, which is substantially easier and which will suffice for most of the applications we have in mind. However, the general case is both interesting and important. Expositions of the general case can be found in Chevalley [1] and Deuring [1].

Let $\bar{K}$ be an algebraic closure of K and $\bar{F} \subset \bar{K}$ the algebraic closure of F in $\bar{K}$. If E is any field intermediate between F and $\bar{F}$, we set KE equal to the compositum of K and E inside $\bar{K}$. By definition, K is finitely generated and of transcendence degree 1 as a field extension of F, and it is clear from this that KE is finitely generated and of transcendence degree 1 as a field extension of E. Thus, KE is a function field over E. It is called the constant field extension of K by E. It is not true, in general, that E is the constant field of KE, but as we shall see shortly, this is often the case. The genus of KE is always less than or equal to the genus of K. It can be shown by example that the genus can decrease. Once again, though, it is often the case that the genus remains unchanged under constant field extension. The magic hypothesis which tends to eliminate all "pathological" behavior is that F is a perfect field. We shall make that assumption throughout this chapter, except when explicitly stated to the contrary. *For emphasis —* unless otherwise stated we shall assume for the rest of this chapter that F is a perfect field. As a consequence E/F will always be separable algebraic and thus KE/K is also separable algebraic.

The last topic we will consider in this chapter is the theory of constant field extensions when the constant fields involved are finite. This will involve interesting questions. Among other things we will consider how primes, the zeta function, and the class number behave under constant field extension. In a later chapter, Chapter 11, we will consider the behavior of the class group and the class number of constant field extensions in greater detail.

Proposition 8.1. *Assume* $[E : F] < \infty$. *Then,* $[KE : K] = [E : F]$. *Any basis for* E/F *is also a basis for* KE/K.

Proof. Suppose first that E/F is a finite, Galois extension. Then, by a standard theorem in Galois theory, KE/K is also Galois and $\mathrm{Gal}(KE/K) \cong \mathrm{Gal}(E/K \cap E)$. Since F is the constant field of K, $E \cap K = F$. It follows that $\mathrm{Gal}(KE/K)$ and $\mathrm{Gal}(E/F)$ have the same number of elements, which implies $[KE : K] = [E : F]$.

Now suppose E/F is finite and separable. Let E_1 be the smallest extension of E in $\bar{F}$ which is Galois over F. Then $[E_1 : F] = [KE_1 : K] = [KE_1 : KE][KE : K] \leq [E_1 : E][E : F] = [E_1 : F]$. The inequality in the middle comes about because, obviously, $[KE_1 : KE] \leq [E_1 : E]$ and $[KE : K] \leq [E : F]$. We conclude that both inequalities are in fact equalities. This proves the first assertion.

Suppose $\{\alpha_1, \alpha_2, \cdots, \alpha_n\}$ is a basis for E/F. It is easy to see that this set also generates KE as a vector space over K. By the first part of the proposition, it follows that the set is also linearly independent since otherwise $[KE : K] < n = [E : F]$.

We will need the following lemma in several of the following proofs.

Lemma 8.2. *(a) Suppose* L/K *is a finite extension of fields and that* K *contains a field* F *which is algebraically closed in* K. *If* $\beta \in L$ *is algebraic over* F, *then* $\mathrm{tr}_{L/K}(\beta) \in F$. *(b) Suppose* L/K *is a finite extension of fields and that* $O \subset K$ *is a subring of* K *which is integrally closed in* K. *If* $b \in L$ *is integral over* O, *then* $\mathrm{tr}_{L/K}(b) \in O$.

Proof. This is fairly standard so we merely sketch the proofs.

For part (a) one considers the minimal polynomial for β over K and shows that all its roots (in some extension field) are algebraic over F. Thus the sum of the roots is algebraic over F and in K, so the sum of the roots is in F. The trace is an integer multiple of the sum of the roots, so it is in F as well.

Part (b) is similar. One shows that all the roots of the minimal polynomial for β over K are integral over O. This implies that the sum of the roots is integral over O. The sum is also in K. Since O is integrally closed, the sum of the roots is in O. The trace is an integral multiple of the sum of the roots so it is also in O.

Proposition 8.3. *E is the exact constant field of* KE.

Proof. We have to show that any element of KE which is algebraic over E is actually in E.

Assume first that $[E : F] < \infty$, and that $\{\alpha_1, \alpha_2, \cdots, \alpha_n\}$ is a basis for E/F. Suppose $\beta \in KE$ is algebraic over E. By Proposition 1, we may write $\beta = \sum_{i=1}^{n} x_i \alpha_i$ where the $x_i \in K$. Multiply this relation by α_j and take the trace of both sides. We find

$$\mathrm{tr}_{KE/K}(\alpha_j \beta) = \sum_{i=1}^{n} \mathrm{tr}_{KE/K}(\alpha_j \alpha_i)\, x_i \quad 1 \leq j \leq n\,.$$

Since β is algebraic over E and E is algebraic over F, it follows that β is algebraic over F. By Lemma 8.2, part (a), $\mathrm{tr}_{KE/K}(\alpha_j \beta) \in F$. Thus, by Cramer's rule, we find each $x_i \in F$. (We have used $\det(\mathrm{tr}_{KE/K}(\alpha_i \alpha_j)) = \det(\mathrm{tr}_{E/F}(\alpha_i \alpha_j)) \neq 0$, which is true because E/F is separable). It follows that $\beta = \sum_{i=1}^{n} x_i \alpha_i \in E$.

Now suppose that E/F is algebraic but not necessarily a finite extension. Since $\beta \in KE$ we must have $\beta \in KE_1$ for some $F \subseteq E_1 \subseteq E$ with $[E_1 : F] < \infty$. By enlarging E_1, if necessary, we can suppose β is algebraic over E_1. By the first part of the proof, $\beta \in E_1$, which is contained in E. The proof is complete.

Our next task is to show that constant field extensions of function fields are unramified extensions. This will be an easy consequence of the next lemma.

Lemma 8.4. *Let E/F be a finite extension with $\{\alpha_1, \alpha_2, \cdots, \alpha_n\}$ a basis for E over F. Let P be a prime of K and O_P the corresponding valuation ring. Let R_P be the integral closure of O_P in KE. Then $\{\alpha_1, \alpha_2, \cdots, \alpha_n\}$ is a free basis for R_P considered as an O_P module.*

Proof. Since $F \subset O_P$ by definition, and each α_i is algebraic over F, it follows that each α_i is integral over O_P.

Suppose $b \in R_P$. By Proposition 8.1, we can write $b = \sum_{i=1}^{n} x_i \alpha_i$, where each $x_i \in K$. Multiply this relation by α_j and take the trace of both sides. One finds

$$\mathrm{tr}_{KE/K}(\alpha_j b) = \sum_{i=1}^{n} \mathrm{tr}_{KE/K}(\alpha_j \alpha_i)\, x_i \quad 1 \leq j \leq n\,.$$

The left-hand side of these equations are in O_P by Lemma 8.2, part (b). Again invoking Cramer's rule and using the fact that the determinant of the coefficient matrix is a non-zero element of F we conclude that each x_i is in O_P. Thus, $\{\alpha_1, \alpha_2, \cdots, \alpha_n\}$ spans R_P over O_P. It is linearly independent over O_P (being a basis for KE over K) so it is a free basis for R_P over O_P as asserted.

Proposition 8.5. *Suppose E/F is a finite extension. Then, KE/K is unramified at all primes.*

Proof. Let P be a prime of K, O_P its valuation ring, and R_P the integral closure of O_P in KE. By Lemma 8.4, any field basis $\{\alpha_1, \alpha_2, \cdots, \alpha_n\}$ for E/F is a free basis for R_P considered as an O_P module. The discriminant ideal $\mathfrak{d}_{R_P/O_P}$ is generated by $\det(\operatorname{tr}_{KE/K}(\alpha_i\alpha_j))$, which is a non-zero element of F. Thus, $\mathfrak{d}_{R_P/O_P} = R_P$. It follows by Proposition 7.9 of the last chapter that KE/K is unramified at every prime above P. Since P was arbitrary, the proof is complete.

It is possible to talk about infinite algebraic extensions being unramified. Once these definitions are given it can be shown that Proposition 8.5 remains valid even without the restriction that E/F be a finite extension.

Now that we know KE/K is unramified, we want to find out how the degree and dimension of a divisor behaves in constant field extensions. For notational convenience, set $L = KE$. Let A be a divisor of K. We want to compare $\deg_L i_{L/K}(A)$ with $\deg_K A$. This will turn out to be fairly easy. More difficult will be the comparison of $l(i_{L/K}(A))$ with $l(A)$. We begin with two lemmas.

Lemma 8.6. *Let* $\{x_1, x_2, \cdots, x_m\} \subset K$ *be linearly independent over* F. *Then, considered as a subset of* KE, *it remains linearly independent over* E.

Proof. Suppose $\sum_{i=1}^m \beta_i x_i = 0$ with each $\beta_i \in E$. Assuming E/F is a finite extension, let $\{\alpha_1, \alpha_2, \cdots, \alpha_n\}$ be a basis for E/F. Then, $\beta_i = \sum_{j=1}^n c_{ji}\alpha_j$ with $c_{ji} \in F$. Substituting and interchanging the order of summation yields

$$\sum_{j=1}^n \left(\sum_{i=1}^m c_{ji}x_i \right) \alpha_j = 0 \ .$$

Using Proposition 8.1, once again, we find $\sum_{i=1}^m c_{ji}x_i = 0$ for each j with $1 \leq j \leq n$. Since the x_i's are linearly independent over F by assumption, it follows that all the c_{ji} are 0 which implies that all the $\beta_i = 0$.

If E/F is not finite, suppose $\sum_{i=1}^m \beta_i x_i = 0$ with each $\beta_i \in E$. Let E_1 be the field obtained from F by adjoining the elements of the set $\{\beta_1, \beta_2, \cdots, \beta_m\}$. E_1 is a finite extension of F. Write $L_1 = KE_1$. Working in this field, and using the first part of the proof, we conclude that all the $\beta_i = 0$.

Lemma 8.7. *Let* L/K *be a finite extension of function fields and* P *a prime of* K. *Suppose that* $\{\mathfrak{P}_1, \mathfrak{P}_2, \cdots, \mathfrak{P}_g\}$, *the primes above* P *in* L, *are all unramified over* P. *Let* $n \in \mathbb{Z}$ *be a given integer. Finally, suppose* $\operatorname{ord}_{\mathfrak{P}_i}(b) \geq -n$ *for all* i *with* $1 \leq i \leq g$. *Then* $\operatorname{ord}_P(\operatorname{tr}_{L/K}(b)) \geq -n$.

Proof. Let $\pi \in K$ be a uniformizing parameter at P. Then, since each $\mathfrak{P}_i$ is unramified over P we have $1 = \operatorname{ord}_P(\pi) = \operatorname{ord}_{\mathfrak{P}_i}(\pi)$ for $1 \leq i \leq n$. The inequalities $\operatorname{ord}_{\mathfrak{P}_i}(b) \geq -n$ are equivalent to $\operatorname{ord}_{\mathfrak{P}_i}(\pi^n b) \geq 0$. It follows that $\pi^n b$ is in the intersection of the valuation rings $O_{\mathfrak{P}_i}$, where

$1 \leq i \leq n$. This intersection is precisely R_P, the integral closure of O_P in L. Thus, $\pi^n b$ is integral over O_P and by Lemma 8.2, part (b), we have $\operatorname{tr}_{L/K}(\pi^n b) \in O_P$. It follows that $\operatorname{ord}_P(\pi^n \operatorname{tr}_{L/K}(b)) \geq 0$ and this is equivalent to $\operatorname{ord}_P(\operatorname{tr}_{L/K}(b)) \geq -n$ as asserted.

We are now in a position to answer the questions raised earlier.

Proposition 8.8. *Let E/F be a finite algebraic extension, K a function field with constant field F, and $L = KE$. Let A be a divisor of K. Then,*

$$(a) \quad \deg_L i_{L/K}(A) = \deg_K A \ .$$

$$(b) \quad l(i_{L/K}(A)) = l(A) \ .$$

Proof. By Proposition 8.3, E is the exact constant field of $L = KE$. By Proposition 8.1, $[L : K] = [E : F]$. Part (a) now follows immediately from Proposition 7.7.

To prove part (b), we recall that $l(i_{L/K}(A))$ is the dimension over E of the vector space $L(i_{L/K}(A)) = \{v \in L \mid (v)_L + i_{L/K}(A) \geq 0\}$. By Proposition 7.8, $i_{L/K}(x)_K = (x)_L$ and it follows immediately that $L(A) \subseteq L(i_{L/K}(A))$. Let $\{x_1, x_2. \cdots, x_d\}$ be a basis for $L(A)$. This set is linearly independent over F, so by Lemma 8.6, it is linearly independent over E. Consequently,

$$l(A) \leq l(i_{L/K}(A)) \ .$$

The reverse inequality will follow if we can show that $\{x_1, x_2, \cdots, x_d\}$ generates $L(i_{L/K}(A))$ over E, and this is what we will prove.

Let $z \in L(i_{L/K}(A))$ and let, as usual by now, $\{\alpha_1, \alpha_2, \cdots, \alpha_n\}$ be a basis for E over F. By Proposition 8.1, we can write $z = \sum_{i=1}^{n} y_i \alpha_i$ where $y_i \in K$ for $1 \leq i \leq n$. Multiply both sides by α_j and take traces to arrive at

$$\operatorname{tr}_{L/K}(\alpha_j z) = \sum_{i=1}^{n} \operatorname{tr}_{L/K}(\alpha_j \alpha_i)\, y_i \quad 1 \leq j \leq n \ .$$

Suppose we can show that the trace of any element in $L(i_{L/K}(A))$ is in $L(A)$. Then the left-hand side of these equations are in $L(A)$ and by Cramer's rule, each $y_i \in L(A)$. It follows that each y_i is in the F-linear span of $\{x_1, x_2, \cdots, x_d\}$ and so z is in the E-linear span of $\{x_1, x_2, \cdots, x_d\}$.

It remains to prove that $v \in L(i_{L/K}(A))$ implies $\operatorname{tr}_{L/K}(v) \in L(A)$. The main tool in doing this will be Lemma 8.7. We begin by recalling that $v \in L(i_{L/K}(A))$ if and only if for every prime $\mathfrak{P}$ of L the following inequality holds:

$$\operatorname{ord}_{\mathfrak{P}}(v) \geq -\operatorname{ord}_{\mathfrak{P}}(i_{L/K}(A)) \ .$$

Let P be the prime of K lying below $\mathfrak{P}$. Since L/K is unramified by Proposition 8.5, we have $\operatorname{ord}_{\mathfrak{P}}(i_{L/K}(A)) = \operatorname{ord}_P(A)$. The condition for v to belong to $L(i_{L/K}(A))$ can be rephrased as follows. For all primes P of

K let $\{\mathfrak{P}_1, \mathfrak{P}_2, \cdots, \mathfrak{P}_g\}$ be the set of primes of L lying above P. Then, for each i with $1 \le i \le g$ we have

$$\mathrm{ord}_{\mathfrak{P}_i}(v) \ge -\mathrm{ord}_P(A) \ .$$

By Lemma 8.7, this implies that $\mathrm{ord}_P(\mathrm{tr}_{L/K}(v)) \ge -\mathrm{ord}_P(A)$ for all primes P of K. These are exactly the conditions for $\mathrm{tr}_{L/K}(v)$ to belong to $L(A)$, so the proof is complete.

Proposition 8.8 provides us with all the background we need to determine how the genus behaves in constant field extensions.

Proposition 8.9. *Let E/F be a finite extension and $L = KE$. Then the genus of L, considered as a function field over E, is equal to the genus of K. (Once more we emphasize that, by hypothesis, F is perfect).*

Proof. Let g be the genus of K and g' the genus of L. Choose a divisor A of K such that $\deg_K(A) \ge \max(2g - 1, 2g' - 1)$, e.g., $A = nP$, where P is a prime divisor and n is a sufficiently large positive integer. By Proposition 8.8, part (a), we have $\deg_L(i_{L/K}(A)) = \deg_K(A)$. By Corollary 4 to Theorem 5.4 we have

$$l(A) = \deg_K(A) - g + 1 \quad \text{and} \quad l(i_{L/K}(A)) = \deg_L(i_{L/K}(A)) - g' + 1 \ .$$

By Proposition 8.8, part (b), we have $l(i_{L/K}(A)) = l(A)$. It follows that $-g + 1 = -g' + 1$ and so $g = g'$.

In the last proposition we could have assumed that E/F is algebraic, but not necessarily finite. The conclusion makes sense and is correct. To prove this in complete generality necessitates a discussion of extension of divisors in infinite, algebraic constant field extensions. We will sketch how this goes after we investigate the way in which primes split in finite constant field extensions.

Proposition 8.10. *Let E/F be a finite extension and $L = KE$. Let $\mathfrak{P}$ be a prime of L and P the prime lying below it in K. Define $E_{\mathfrak{P}} = O_{\mathfrak{P}}/\mathfrak{P}$ and $F_P = O_P/P$. Then, $E_{\mathfrak{P}}$ is the compositum of F_P and E.*

Proof. Let $\bar{\omega} \in E_{\mathfrak{P}}$ and let ω be an element of $O_{\mathfrak{P}}$ representing $\bar{\omega}$. Let's consider $\{\mathfrak{P}_1 = \mathfrak{P}, \mathfrak{P}_2, \cdots, \mathfrak{P}_g\}$, the primes in L lying over P. By the weak approximation theorem, we may find an element $\omega' \in L$ such that $\omega' \equiv \omega \pmod{\mathfrak{P}}$ and $\omega' \equiv 0 \pmod{\mathfrak{P}_i}$ for $2 \le i \le g$. Then $\omega' \in R_P$, the integral closure of O_P in L. By Lemma 8.4, any basis $\{\alpha_1, \alpha_2, \cdots, \alpha_n\}$ of E/F is automatically a free basis of R_P considered as a module over O_P. Thus,

$$\omega' = \sum_{i=1}^{n} x_i\, \alpha_i \quad \text{with } x_i \in O_P \ .$$

Now reduce both sides modulo $\mathfrak{P}$ and we see that $\bar{\omega}$ is in the compositum of F_P and E.

By a small variation of this proof, we can give a very explicit way of understanding how primes split in a constant field extension.

Proposition 8.11. *With the notation of the previous proposition, suppose* $F_P = F[\theta]$ *and that* $h(T) \in F[T]$ *is the irreducible polynomial for* θ *over* F. *Let*

$$h(T) = h_1(T)h_2(T) \cdots h_g(T)$$

be the prime decomposition of $h(T)$ *in* $E[T]$. *There are exactly* g *primes* $\{\mathfrak{P}_1, \mathfrak{P}_2, \cdots, \mathfrak{P}_g\}$ *of* L *lying above* P. *The numbering can be chosen in such a way that for* $1 \leq i \leq g$ *we have* $\deg_L \mathfrak{P}_i = \deg h_i(T)$. *Moreover,*

$$\deg_K P = \sum_{i=1}^{g} \deg_L \mathfrak{P}_i \ .$$

Proof. Lemma 8.4 can be restated to say that $R_P \cong O_P \otimes_F E$. Reducing both sides modulo P yields, $R_P/PR_P \cong F_P \otimes_F E$. By hypothesis, $F_P = F[\theta] \cong F[T]/(h(T))$. Thus,

$$R_P/PR_P \cong F_P \otimes_F E \cong E[T]/(h(T)) \cong \bigoplus_{I=1}^{g} E[T]/(h_i(T)) \ .$$

The right-hand side is a direct sum of fields. Let M_i be the maximal ideal which is the kernel of projection on the i-th factor. Let $\mathfrak{p}_i$ be the maximal ideal of R_P which goes to M_i under $R_P \to R_P/PR_P$ followed by the above sequence of isomorphisms. Set $\mathfrak{P}_i$ equal to the maximal ideal of the ring "R_P localized at $\mathfrak{p}_i$," i.e., the ring $O_{\mathfrak{P}_i}$. A simple check shows that the set of primes of L given by $\{\mathfrak{P}_1, \mathfrak{P}_2, \cdots, \mathfrak{P}_g\}$ has all the properties asserted, except perhaps the last one about the sum of the degrees. To prove this, simply notice that $\sum_{i=1}^{g} \deg_L \mathfrak{P}_i = \sum_{i=1}^{g} \deg h_i(T) = \deg h(T) = \deg_K P$.

Corollary. *Suppose* $F_P \cong F[T]/(h(T))$ *and that* E *is an extension of* F *of degree* n *in which* $h(T)$ *decomposes as a product of linear factors. Then in* KE *the prime* P *splits into* n *primes of degree* 1

Proof. Clear.

The proof of the above proposition was most easily accomplished by choosing a primitive element (which exists since F_P/F is separable) for the field extension F_P/F. However, the situation can be described in a more canonical way without having to make any choices. Consider the algebra $F_P \otimes_F E$ over F. The proof shows this algebra is a direct sum of fields, L_i, say, each of which is a field extension of E.

$$F_P \otimes_F E \cong \bigoplus_{i=1}^{g} L_i \ .$$

Then there is a one-to-one correspondence between primes $\mathfrak{P}_i$, lying above P in KE and the fields L_i with the property that the residue class field of $\mathfrak{P}_i$ is isomorphic to the field L_i.

We have been assuming that the constant field extension E/F is finite, but this is not necessary. Let E/F be an algebraic, but possibly infinite, extension of fields. Using properties of tensor product and Lemma 8.4, one can prove that the integral closure R_P of O_P in KE is isomorphic to $O_P \otimes_F E$, the map being $\omega \otimes \alpha \to \alpha\omega \in R_P$. The statement and proof of Proposition 8.11 can now be repeated without change.

As a special case, let $E = \bar{F}$, an algebraic closure of F. Every polynomial in $F[T]$ splits into a product of linear factors in $\bar{F}[T]$. Consequently, every prime P of K splits into $\deg_K P$ primes of degree 1 in $K\bar{F}$. This will be very useful later when we discuss how to use results about the geometry of algebraic curves to give us information about the arithmetic of algebraic function fields.

As an illustration of the material developed in this chapter we will now discuss the particular case when the constant field $F = \mathbb{F}$, a finite field with q elements. Let $\bar{\mathbb{F}}$ be an algebraic closure of $\mathbb{F}$ and $\mathbb{F}_n$ the unique intermediate extension such that $[\mathbb{F}_n : \mathbb{F}] = n$. Set $K_n = K\mathbb{F}_n$.

We recall some definitions from Chapter 5. We set $a_m(K)$ equal to the number of primes of degree m, $b_m(K)$ equal to the number of effective divisors of degree m, and $h(K)$ equal to the number of divisor classes of degree zero, i.e., the class number of K. The latter number was denoted h_K in Chapter 5. These numbers are all finite. We would like to compare them with the numbers $a_m(K_n)$, $b_m(K_n)$, and $h(K_n)$. We also want to compare the zeta function of K_n with that of K. Of course, all these questions are interrelated. There are connections between this material and Iwasawa's theory of cyclotomic number fields. We will discuss these connections in more detail in Chapter 11.

The first thing to do is to make precise the way in which primes of K split in K_n. Let P be a prime of K and $\mathfrak{P}$ a prime lying above it in K_n. By Proposition 8.10, the residue class field of $\mathfrak{P}$, is the compositum of O_P/P and $\mathbb{F}_n$ inside $O_{\mathfrak{P}}/\mathfrak{P}$. To compute the compositum and its dimension over $\mathbb{F}$ we can invoke the following simple lemma.

Lemma 8.12. *The compositum of $\mathbb{F}_n$ and $\mathbb{F}_m$ is $\mathbb{F}_{[n,m]}$ where $[n,m]$ is the least common multiple of n and m.*

Proof. Let $\mathbb{F}_h$ be the compositum of $\mathbb{F}_n$ and $\mathbb{F}_m$ inside $\bar{\mathbb{F}}$. Since $\mathbb{F}_n, \mathbb{F}_m \subseteq \mathbb{F}_h$, we have $n, m \mid h$, which implies $[n,m] \mid h$. Thus, $\mathbb{F}_{[n,m]} \subseteq \mathbb{F}_h$. On the other hand, since $n, m \mid [n,m]$, we have $\mathbb{F}_n, \mathbb{F}_m \subseteq \mathbb{F}_{[n,m]}$, and so $\mathbb{F}_h \subseteq \mathbb{F}_{[n,m]}$.

Proposition 8.13. *Let P be a prime of K. Then P splits into $(n, \deg_K P)$ primes in K_n. Let $\mathfrak{P}$ be a prime of K_n lying over P. Then*

$$\deg_{K_n} \mathfrak{P} = \frac{\deg_K P}{(n, \deg_K P)} \quad \text{and} \quad f(\mathfrak{P}/P) = \frac{n}{(n, \deg_K P)} \ .$$

Proof. By definition, the dimension of O_P/P over $\mathbb{F}$ is $\deg_K P$. Thus, by the above lemma, the compositum of O_P/P and $\mathbb{F}_n$ inside $O_{\mathfrak{P}}/\mathfrak{P}$ has dimension $[n, \deg_K P]$ over $\mathbb{F}$. By Proposition 8.10, this compositum is equal to $O_{\mathfrak{P}}/\mathfrak{P}$, and so

$$\deg_{K_n} \mathfrak{P} = [O_{\mathfrak{P}}/\mathfrak{P} : \mathbb{F}_n] = \frac{[n, \deg_K P]}{n} = \frac{\deg_K P}{(n, \deg_k P)} \; .$$

The last equality follows from elementary number theory — for any two non-zero integers n and m, $nm = (n, m)[n, m]$.

The relative degree $f(\mathfrak{P}/P)$ is the dimension of $O_{\mathfrak{P}}/\mathfrak{P}$ over O_P/P, which in this case is $[n, \deg_K P]/\deg_K P = n/(n, \deg_K P)$.

Finally, we recall that K_n/K is unramified. Since each prime $\mathfrak{P}$ over P has relative degree $n/(n, \deg_K P)$ and $[K_n : K] = n$ we see the number of primes above P is $(n, \deg_K P)$ by Theorem 7.6.

Corollary. *A prime P of K splits into $\deg_K P$ primes of degree 1 in K_n if and only if $\deg_K P$ divides n.*

Proof. This is immediate from the proposition.

Proposition 8.13 is the key to comparing the zeta function of K with that of K_n. The only other piece of information needed is provided by the following elementary lemma.

Lemma 8.14. *Let $\zeta_n \in \mathbb{C}$ be a primitive n-th root of unity and m a positive integer. Then*

$$\prod_{i=0}^{n-1}(1 - \zeta_n^{im} u^m) = (1 - u^{[n,m]})^{(n,m)} \; .$$

Proof. First consider the case where $m = 1$. The result in this case follows from the identity $T^n - 1 = \prod_{i=1}^n (T - \zeta_n^i)$ by making the substitution $T = u^{-1}$ and simplifying.

In the general case, let $m' = m/(n, m)$ and $n' = n/(n, m)$. It is easy to see that ζ_n^m is a primitive n' root of unity. Call it $\zeta_{n'}$. Every i in the range $0 \le i < n$ can be uniquely represented in the form $i = kn' + r$, where $0 \le k < (n, m)$ and $0 \le r < n'$. Thus,

$$\prod_{i=1}^{n}(1 - \zeta_n^{im} u^m) = \prod_{h=0}^{(n,m)-1} \prod_{r=0}^{n'-1}(1 - \zeta_{n'}^r u^m) = (1 - u^{mn'})^{(n,m)} \; .$$

Finally, $mn' = mn/(n, m) = [n, m]$.

Theorem 8.15. *Let $u = q^{-s}$ and $\zeta_K(s) = Z_K(u)$. Then,*

$$\zeta_{K_n}(s) = Z_{K_n}(u^n) = \prod_{i=0}^{n-1} Z_K(\zeta_n^i u) \; .$$

Proof. Since the constant field of K_n is $\mathbb{F}_n$ which has q^n elements, we have $\zeta_{K_n}(s) = Z_{K_n}(u')$, where $u' = (q^n)^{-s} = q^{-ns} = u^n$. Thus, $\zeta_{K_n}(s) = Z_{K_n}(u^n)$, which proves the first equality in the statement of the theorem.

Setting $d_P = (n, \deg_K P)$ for each $P \in \mathcal{S}_K$, we have

$$\zeta_{K_n}(s) = \prod_{\mathfrak{P}}(1 - N\mathfrak{P}^{-s})^{-1} = \prod_{P}(1 - NP^{-s\frac{n}{d_P}})^{-d_P} ,$$

where $\mathfrak{P}$ ranges over $\mathcal{S}_{K_n}$ and P ranges over $\mathcal{S}_K$. We have used Proposition 8.13 and the fact that $N\mathfrak{P} = NP^{f(\mathfrak{P}/P)}$. Since $NP = q^{\deg_K P}$, by definition, the last product can be rewritten as

$$\prod_{P}(1 - u^{n\deg_K P/d_P})^{-d_P} = \prod_{P}\prod_{i=0}^{n-1}(1 - \zeta_n^{i\deg_K P}u^{\deg_K P})^{-1} .$$

Here we have used Lemma 8.14 with $m = \deg_K P$. Recall that $Z_K(u) = \prod_P(1 - u^{\deg_K P})^{-1}$ Now, interchanging the order of the products on the right-hand side of the above identity completes the proof.

By the proof of Theorem 5.9, $Z_K(u) = L_K(u)/(1-u)(1-qu)$ where $L_K(u)$ is a polynomial of degree equal to twice the genus g of K. By Proposition 8.9, K_n has the same genus as K. Let $u' = u^n$, as above, and we have

$$Z_{K_n}(u') = \frac{L_{K_n}(u')}{(1-u')(1-q^n u')} .$$

Proposition 8.16. *Let $L_K(u) = \prod_{j=1}^{2g}(1 - \pi_j u)$ be the factorization of $L_K(u)$ in $\mathbb{C}[u]$. Then,*

$$L_{K_n}(u') = \prod_{j=1}^{2g}(1 - \pi_j^n u') .$$

Proof. Using the definitions and Theorem 8.15, we find

$$\frac{L_{K_n}(u^n)}{(1-u^n)(1-q^n u^n)} = \prod_{i=0}^{n-1}\frac{L_K(\zeta_n^i u)}{(1-\zeta_n^i u)(1-q\zeta_n^i u)} .$$

For any complex number π we have the identity $\prod_{i=0}^{n-1}(1-\zeta_n^i \pi u) = 1-\pi^n u^n$. Thus,

$$\frac{L_{K_n}(u^n)}{(1-u^n)(1-q^n u^n)} = \frac{\prod_{j=1}^{2g}(1-\pi_j^n u^n)}{(1-u^n)(1-q^n u^n)} .$$

The proposition follows upon canceling the denominators.

Corollary. $h(K_n) = \prod_{i=1}^{2g}(1 - \pi_i^n)$.

Proof. Applying Theorem 5.9 with K replaced by K_n we find $L_{K_n}(1) = h(K_n)$. The result follows upon substituting $u' = 1$ in the proposition.

By the Riemann Hypothesis, $|\pi_i| = \sqrt{q}$ for $1 \leq i \leq 2g$. Using this and the corollary we get the following lower bound for $h(K_n)$:

$$h(K_n) \geq (q^{n/2} - 1)^{2g} .$$

This shows that $h(K_n)$ grows rapidly with n. A more precise investigation of how $h(K_n)$ varies with n is possible. The results one obtains lead the way, by analogy, to Iwasawa theory in algebraic number fields (see Iwasawa [3]). As stated earlier, we will discuss these matters in greater detail in Chapter 11.

The next proposition gives some insight into how the numbers $b_m(K_n)$ grow with n.

Proposition 8.17. $b_m(K_n) = h(K_n)\frac{q^{nm-g+1}-1}{q^n-1}$, provided $m > 2g - 2$.

Proof. We proved earlier (see the remarks following Lemma 5.8) that if $m > 2g - 2$, where g is the genus of K, that

$$b_m(K) = h(K) \frac{q^{m-g+1} - 1}{q - 1} .$$

In this equality, replace K by K_n and q by q^n (since the constant field of K_n has q^n elements). The resulting equation is valid because the genus of K_n is the same as the genus of K (Proposition 8.9).

In Chapter 5 we provided three different description of $Z_K(u)$, namely,

$$Z_K(u) = \sum_{m=0}^{\infty} b_m(K)u^m = \prod_{d=1}^{\infty}(1 - u^d)^{-a_d(K)} = \exp\left(\sum_{m=1}^{\infty} \frac{N_m(K)}{m}u^m\right).$$

By definition, $N_m(K) = \sum_{d|m} da_d(K)$ and we showed in Chapter 5 that $N_m(K) = q^m + 1 - \sum_{i=1}^{2g} \pi_i^m$. Although these numbers are clearly very important we did not give an interpretation of them. We can do so now.

Proposition 8.18. $N_m(K)$ *is equal to the number of prime divisors of* K_m *of degree* 1.

Proof. It is interesting to note that $N_1(K) = a_1(K)$, so the result is certainly true when $m = 1$.

To prove the general case, we invoke Proposition 8.16.

$$\frac{\prod_{j=1}^{2g}(1 - \pi_j^m u')}{(1 - u')(1 - q^m u')} = Z_{K_m}(u') = \sum_{k=0}^{\infty} b_k(K_m)u'^k .$$

Compute $Z'_{K_m}(0)/Z_{Km}(0)$ in two different ways using these expressions. We find

$$q^m + 1 - \sum_{j=1}^{2g} \pi_j^m = b_1(K_m) \ .$$

The left-hand side is just $N_m(K)$ and the right-hand side is the number of effective divisors of degree 1 in K_m, which is the same as the number of prime divisors of K_m of degree 1.

We have proved $N_m(K) = b_1(K_m) = a_1(K_m) = N_1(K_m)$. Assuming some algebraic geometry we can reword the result as follows. Let X be an absolutely irreducible, non-singular curve defined over a finite field $\mathbb{F}$ with q elements. For each $m \geq 1$ let $N'_m(X)$ be the number of rational points on X over $\mathbb{F}_m$, i.e., $N'_m(X) = \#X(\mathbb{F}_m)$. It can be shown that $N'_m(X)$ is equal to the number of prime divisors of degree 1 belonging to the function field $\mathbb{F}_m(X)$ of X over $\mathbb{F}_m$. This means that $N'_m(X) = N_m(K)$ by the above proposition, where $K = \mathbb{F}(X)$. Thus, the zeta function of the function field of X, $K = \mathbb{F}(X)$, is equal to $\exp(\sum_{m=1}^{\infty} N'_m(X)/m \ u^m)$. This approach enables one, in a fairly obvious manner, to define the zeta function of a variety X of any dimension over a finite field by using the numbers $\#X(\mathbb{F}_m)$. A beautiful exposition of the general theory is given in Serre [1].

Consider the identity $N_m(K) = N_1(K_m)$. Let's apply this to the field K_n rather than K. It is easy to see that $(K_n)_m = K_{nm}$ and it follows that

$$N_m(K_n) = N_1(K_{nm}) = N_n(K_m) \ .$$

This identity allows us to derive an interesting expression for the number of primes of degree m in the field K_n, i.e., $a_m(K_n)$.

Proposition 8.19. $a_m(K_n) = m^{-1} \sum_{d|m} \mu(d) a_1(K_{nm/d}) \ .$

Proof. From the definition, $N_m(K_n) = \sum_{d|m} d a_d(K_n)$. Using Möbius inversion, we see that $m a_m(K_n) = \sum_{d|m} \mu(d) N_{m/d}(K_n)$. From the relation $N_{m/d}(K_n) = N_1(K_{nm/d}) = a_1(K_{nm/d})$ the result follows.

There is much more to be said about the fascinating sequences of numbers we have introduced, but it is time to break off this development for now and to pass on to other matters.

Exercises

1. Let $K = F(x, y)$ be the function field associated to the curve $Y^2 = f(X)$, where $f(X)$ is a square-free polynomial of degree n. Assume that $\operatorname{char}(F) \neq 2$. Compute the genus. (Hint: Reduce to the case where $f(X)$ is a product of linear factors and apply Exercise 1 of Chapter 7).

2. Generalize Exercise 1 to the case where l is a prime, $f(X)$ is l-th power-free, $l \nmid \operatorname{char}(F)$, and the curve is $Y^l = f(X)$.

3. Consider the curve $X^N + Y^N = 1$ and the associated function field $F(x, y)$. Assume $\operatorname{char}(F) \nmid N$. Compute the genus of K.

4. Let F_0 be a field of characteristic $p > 0$ and set $F = F_0(T)$, the rational function field over F_0. Consider the function field $K = F(x, y)$ over F, where x and y satisfy the equation $Y^2 = X^p - T$. Prove that the genus of K is $(p - 1)/2$ (use the Riemann-Hurwitz theorem arguing via the extension $K/F(x)$). Now, extend the constant field to $F' = F(\sqrt[p]{T})$. Show that $K' = F'(x, y)$ has genus 0. This does not contradict Proposition 8.9 since the extension F'/F is purely inseparable.

5. Let E/F be a finite Galois extension with group G. Identify G with the Galois group of KE/K. Let $\mathcal{B}$ be a divisor of KE which is invariant under G, i.e., $\sigma\mathcal{B} = \mathcal{B}$ for all $\sigma \in G$. Show that $L(\mathcal{B})$ has a basis consisting of elements of K. (Hint: Use Propostion 9.2 of the next Chapter to show that $\mathcal{B} = i_{KE/K}B$ for some divisor B of K. Then invoke the proof of Proposition 8.8).

6. Let $K/\mathbb{F}$ be a function field over a finite field and let $L_K(u) = \prod_{i=1}^{2g}(1 - \pi_i u)$ be the numerator of the zeta function of K. Assume that there is a positive constant C such that for all $r \geq 1$ we have $|N_r(K) - q^r - 1| \leq Cq^{\frac{r}{2}}$. Prove that $|\pi_i| = \sqrt{q}$ for all i. (Hint: Expand $L'_K(u)/L_K(u)$ in a power series about $u = 0$ and consider the radius of convergence).

7. Let $K/\mathbb{F}$ be a function field of genus 1 over a finite field. Show that $N_1(K)$ determines all the other numbers $N_r(K)$.

8. Generalize the last exercise as follows. Let $K/\mathbb{F}$ be a function field of genus $g \geq 1$ over a finite field. Show that the numbers $N_1(K)$, $N_2(K), \ldots, N_g(K)$ determine all the other numbers $N_r(K)$.

9
Galois Extensions - Hecke and Artin L-Series

In Chapters 7 and 8 we discussed finite extensions L/K of algebraic function fields. We propose to continue that discussion here under the special assumption that the extension L/K is Galois. To simplify the discussion we continue to assume that the constant field F of K is perfect.

After proving a number of basic results in the general case, i.e., F being perfect but otherwise arbitrary, we specialize to the case where $F = \mathbb{F}$, a finite field with q elements. Then, for every prime $\mathfrak{P}$ of L unramified over K we associate an automorphism $(\mathfrak{P}, L/K)$ in $G = \mathrm{Gal}(L/K)$ called the Frobenius automorphism of $\mathfrak{P}$. This is one of the most fundamental notions in the number theory of local and global fields. It will be seen that if P is a prime of K, unramified in L, the set of automorphisms $(P, L/K) =: \{(\mathfrak{P}, L/K) \mid \mathfrak{P}$ above $P\}$ fill out a conjugacy class in G. Suppose $C \subset G$ is a conjugacy class. One can ask how big is the set of primes $P \in \mathcal{S}_K$ such that $(P, L/K) = C$? The answer to this question is given by the Tchebotarev density theorem. We will discuss two forms of this important result, one involving Dirichlet density and the other involving natural density. The key tool will be Artin L-series and their properties.

Let χ be a complex character of the group $G = \mathrm{Gal}(L/K)$. E. Artin showed how to associate an L-function, $L(s, \chi)$, with such a character (see Artin [2]). It is defined and analytic in the half plane $\{s \in \mathbb{C} \mid \Re(s) > 1\}$. Artin was able to show that $L(s, \chi)$ can be analytically continued to a neighborhood of $s = 1$ and that if χ is irreducible and $\chi \neq \chi_o$, the trivial character, then $L(1, \chi) \neq 0$. It is this property which will enable us to prove the version of the Tchebotarev density theorem formulated in terms of Dirichlet density.

Artin conjectured that his L-series can be meromorphically continued to the entire plane and that if χ is non-trivial and irreducible then $L(s,\chi)$ can, in fact, be continued to an entire function on the whole plane. R. Brauer was able to prove the first part of this by means of a deep theorem about group characters. His proof works in both number fields and function fields. The second part of Artin's conjecture is still an open question in number fields. It is one of the most important open questions in that area. In the function field case, the matter was resolved by A. Weil [1] in the same small book in which he first proved the Riemann hypothesis for curves over a finite field. He showed, subject to a small technical restriction (which we will discuss), that if χ is non-trivial and irreducible, then $L(s,\chi)$ is a polynomial in q^{-s}. On the basis of this result we will give a proof of the version of the Tchebotarev density theorem formulated in terms of natural density.

The reader will not fail to notice that the above discussion has the same flavor as the material in Chapter 4 where we discussed Dirichlet L-series and the Dirichlet theorem about primes in an arithmetic progression. However, in that chapter there was no discussion of Galois groups and characters on them. We considered the groups $(A/mA)^*$ and to a character on such a group we associated an L-series. Is there any connection between the two types of L-series? The answer is yes, but the explanation is extremely subtle and difficult. It is by trying to answer this question in the most general context that Artin was led to the famous Artin reciprocity law, perhaps the deepest and most far-reaching theorem in all of algebraic number theory. We will attempt a general discussion of these matters, but mainly without proofs. We will investigate what happens when $\mathrm{Gal}(L/K)$ is an abelian group. This will lead to a rough statement of Artin's reciprocity law. When $G(L/K)$ is abelian and L/K is unramified we will give a proof of Weil's result using Artin's reciprocity law. In general, we will define Hecke L-functions for characters of finite order (Dirichlet L-series are a special case of these) and state some of their properties without proof. Artin reciprocity allows one to show that for one-dimensional characters Artin L-series $L(s,\chi)$ "are" Hecke L-series and, in the abelian case, Artin's conjecture about his L-series being entire will follow from this.

It is time to begin!

We assume L/K is a finite, Galois extension of function fields and denote the Galois group by $G = \mathrm{Gal}(L/K)$. As usual, let F be the constant field of K and E the constant field of L.

Proposition 9.1. *The field extension E/F is Galois and the map $G \to \mathrm{Gal}(E/F)$ obtained by restriction of automorphisms to E is onto. Let $N \subseteq G$ be the kernel of this map. Then the fixed field of N is KE, the maximal constant field extension of K contained in L.*

Proof. If $\sigma \in G$ and $\alpha \in E$, then the fact that α is a root of a polynomial with coefficients in F shows that $\sigma\alpha$ must be a root of the same polynomial since σ fixes F. Thus, $\sigma\alpha$ is algebraic over F, which implies $\sigma\alpha \in E$. This shows that the restriction map takes G to the group of automorphisms of E which leave F fixed, namely, $\text{Aut}(E/F)$. Let G' be the image of this map. Then, the fixed field of G' is $E \cap K = F$. This proves E/F is Galois and that G' is its Galois group.

It is clear that N leaves KE fixed, so to prove KE is the fixed field of N it suffices to show $|N| = [L : KE]$. Since $G/N \cong G'$ we see $|G| = |N||G'|$ which by Galois theory is the same as $[L : K] = |N|[E : F]$. By Proposition 8.1, $[E : F] = [KE : K]$ and it follows that $|N| = [L : KE]$ as required.

Let $\mathfrak{P}$ be a prime of L lying over a prime P of K. Recall that $\mathfrak{P}$ is the maximal ideal of a discrete valuation ring $O_{\mathfrak{P}}$ which contains the constant field E and whose quotient field is L. Let $\sigma \in G$. Then $\sigma O_{\mathfrak{P}}$ is a dvr with the same properties and its maximal ideal is $\sigma\mathfrak{P}$. Thus, $\sigma\mathfrak{P}$ is another prime of L and it is easy to verify that it also lies above P. The group G acts as a group of permutations on the set of primes above P.

Proposition 9.2. *Let $\{\mathfrak{P}_1, \mathfrak{P}_2, \cdots, \mathfrak{P}_g\}$ be the set of primes of L lying above P. The Galois group G acts transitively on this set.*

Proof. For each i with $1 \leq i \leq g$ we need to show there is a $\sigma \in G$ such that $\sigma\mathfrak{P}_1 = \mathfrak{P}_i$.

Consider the set $\{\sigma\mathfrak{P}_1 \mid \sigma \in G\}$. Suppose some $\mathfrak{P}_i$ is not in this set, $\mathfrak{P}_g$ say. We will derive a contradiction.

By the weak approximation theorem we can find an element $x \in L$ such that $x \equiv 0 \pmod{\mathfrak{P}_g}$ and $x \equiv 1 \pmod{\mathfrak{P}_i}$ for $i \neq g$. Since these conditions imply $x \in O_{\mathfrak{P}_i}$ for all $1 \leq i \leq g$, we have $x \subset R_P$ the integral closure of O_P in L. It follows that $\sigma x \in R_P$ for all $\sigma \in G$ and $\prod_{\sigma \in G} \sigma x \in R_P \cap K = O_P$. Since $x \in \mathfrak{P}_g \cap R_P$, we have, $\prod_{\sigma \in G} \sigma x \in \mathfrak{P}_g \cap R_P \cap O_P = P \subset \mathfrak{P}_1$. Since $\mathfrak{P}_1$ is a prime ideal, there is a $\tau \in G$ such that $\tau x \in \mathfrak{P}_1$ and so $x \in \tau^{-1}\mathfrak{P}_1$, which contradicts $x \equiv 1 \pmod{\tau^{-1}\mathfrak{P}_1}$.

Proposition 9.3. *We continue to use the notation introduced above, except that we now denote the number of primes in L lying above P by $g(P)$. We have $f(\mathfrak{P}_i/P) = f(\mathfrak{P}_j/P)$ and $e(\mathfrak{P}_i/P) = e(\mathfrak{P}_j/P)$ for all $1 \leq i, j \leq g$. If we denote by $f(P)$ the common relative degree and by $e(P)$ the common ramification index, then $e(P)f(P)g(P) = n = [L : K]$. In particular, $e(P)$, $f(P)$, and $g(P)$ divide n.*

Proof. For a given pair i and j there is an automorphism $\sigma \in G$ such that $\sigma\mathfrak{P}_i = \mathfrak{P}_j$. Map $O_{\mathfrak{P}_i}/\mathfrak{P}_i \to O_{\mathfrak{P}_j}/\mathfrak{P}_j$ by $\bar{\omega} \to \overline{\sigma\omega}$. It is straightforward to check that this map is well defined and gives a field isomorphism which leaves O_P/P fixed. It follows immediately that $f(\mathfrak{P}_i/P) = f(\mathfrak{P}_j/P)$ as asserted.

Similarly, if $PO_{\mathfrak{P}_i} = \mathfrak{P}_i^e$, applying σ to both sides yields $PO_{\mathfrak{P}_j} = \mathfrak{P}_j^e$. Thus, $e(\mathfrak{P}_i/P) = e(\mathfrak{P}_j/P)$.

By Theorem 7.6, $\sum_{i=1}^{g(P)} e(\mathfrak{P}_i/P) f(\mathfrak{P}_i/P) = n$, so the last two assertions follows from this and the first part of the proof.

Let $\mathfrak{P}$ be a prime of L lying over a prime P of K. We now define two important subgroups of $G = \mathrm{Gal}(L/K)$:

$$Z(\mathfrak{P}/P) = \{\sigma \in G \mid \sigma\mathfrak{P} = \mathfrak{P}\} \text{ and}$$
$$I(\mathfrak{P}/P) = \{\tau \in G \mid \tau\omega \equiv \omega \pmod{\mathfrak{P}}, \forall\omega \in O_{\mathfrak{P}}\}$$

The first is called the decomposition group of $\mathfrak{P}$ over P and the second is called the inertia group of $\mathfrak{P}$ over P.

Lemma 9.4. *The order of $Z(\mathfrak{P}/P)$ is $e(\mathfrak{P}/P)f(\mathfrak{P}/P)$.*

Proof. By Proposition 9.2, the group G acts transitively on the set of primes of L lying above P. The group $Z(\mathfrak{P}/P)$ is the isotropy group for this action. From this it follows that $[G : Z(\mathfrak{P}/P)] = g(P)$, the number of primes in L above P. By Proposition 9.3, we have $e(\mathfrak{P}/P)f(\mathfrak{P}/P)g(P) = [L : K] = \#G$. Thus, $\#Z(\mathfrak{P}/P) = e(P)f(P)$.

Let $M \subseteq L$ be the fixed field of $Z(\mathfrak{P}/P)$ and $\mathfrak{p}$ the prime in M lying below $\mathfrak{P}$. M is sometimes called the decomposition field of $\mathfrak{P}$.

Lemma 9.5. *With the above notation, $\mathfrak{P}$ is the only prime in L lying above $\mathfrak{p}$. Moreover, $e(\mathfrak{p}/P) = f(\mathfrak{p}/P) = 1$ and $[M : K] = g(P)$.*

Proof. The first assertion follows by applying Proposition 9.2 to the Galois extension L/M and using the definition of the decomposition group. By Lemma 9.4, $\#Z(\mathfrak{P}/\mathfrak{p}) = e(\mathfrak{P}/\mathfrak{p})f(\mathfrak{P}/\mathfrak{p})$. On the other hand, $Z(\mathfrak{P}/\mathfrak{p}) = Z(\mathfrak{P}/P)$ and the order of this group is $e(\mathfrak{P}/P)f(\mathfrak{P}/P)$. The fact that $e(\mathfrak{p}/P) = f(\mathfrak{p}/P) = 1$ follows from this since $e(\mathfrak{P}/P) = e(\mathfrak{P}/\mathfrak{p})e(\mathfrak{p}/P)$ and $f(\mathfrak{P}/P) = f(\mathfrak{P}/\mathfrak{p})f(\mathfrak{p}/P)$. Finally, the index relation $[L : K] = [L : M][M : K]$, together with $[L : K] = \#G$ and $[L : M] = \#Z(\mathfrak{P}/P)$, implies the last assertion that $[M : K] = g(P)$.

Let's reintroduce the notation $E_{\mathfrak{P}}$ for the residue class field of $O_{\mathfrak{P}}$ and F_P for the residue class field of O_P.

Theorem 9.6. *Suppose L/K is a Galois extension with $G = \mathrm{Gal}(L/K)$ and that $\mathfrak{P}$ is a prime of L lying over a prime P of K. Then the extension $E_{\mathfrak{P}}/F_P$ is also a Galois extension. There is a natural homomorphism from $Z(\mathfrak{P}/P)$ onto $\mathrm{Gal}(E_{\mathfrak{P}}/F_P)$ and the kernel of this homomorphism is $I(\mathfrak{P}/P)$. The inertia group is a normal subgroup of the decomposition group and $\#I(\mathfrak{P}/P) = e(\mathfrak{P}/P)$.*

Proof. Since F_P is perfect, there is an element $\theta \in O_{\mathfrak{P}}$ such that $E_{\mathfrak{P}} = F_P(\bar{\theta})$ where $\bar{\theta}$ is the residue class of θ modulo $\mathfrak{P}$. By using the weak approximation theorem, if necessary, we can assume that θ is integral over O_P. As above, let M be the fixed field of $Z(\mathfrak{P}/P)$ and $f(X) \in M[X]$ the minimal polynomial for θ over M. Since θ is an integral element, the coefficients of $f(X)$ are in $O_{\mathfrak{p}} = O_{\mathfrak{P}} \cap M$. Since L/M is a Galois extension, $f(X)$ splits into linear factors in L, i.e., $f(X) = \prod_{i=1}^{m}(X - \theta_i)$ where $\theta = \theta_1$. Reducing modulo $\mathfrak{P}$, we have $\bar{f}(X) = \prod_{i=1}^{m}(X - \bar{\theta}_i)$. The coefficients of $\bar{f}(X)$ are in the residue class field of $O_{\mathfrak{p}}$, which is the same as F_P since $f(\mathfrak{p}/P) = 1$ by Lemma 9.5. This shows that $E_{\mathfrak{P}} = F_P(\bar{\theta})$ is the splitting field of $\bar{f}(X)$ and so $E_{\mathfrak{P}}$ is Galois over F_P as asserted.

If $\sigma \in Z(\mathfrak{P}/P)$ and $\bar{\omega} \in E_{\mathfrak{P}}$, define $\bar{\sigma}$ by the equation $\bar{\sigma}(\bar{\omega}) = \overline{\sigma\omega}$. It is easy to check that $\bar{\sigma}$ is a well-defined mapping from $E_{\mathfrak{P}}$ to itself which is, in fact, an automorphism leaving F_P fixed, i.e., $\bar{\sigma} \in \mathrm{Gal}(E_{\mathfrak{P}}/F_P)$. The map σ to $\bar{\sigma}$ is a homomorphism and the kernel of this homomorphism is $I(\mathfrak{P}/P)$. Again, all this is straightforward from the definition. It remains to show that the homomorphism which takes σ to $\bar{\sigma}$ is onto $\mathrm{Gal}(E_{\mathfrak{P}}/F_P)$.

Let $\lambda \in \mathrm{Gal}(E_{\mathfrak{P}}/F_P)$. Let $h(X) \in F_P[X]$ be the irreducible polynomial of $\bar{\theta}$ over F_P. Then $\lambda\bar{\theta}$ is also a root of $h(X)$. Since $\bar{\theta}$ is also a root of $\bar{f}(X)$ (see the first paragraph), $h(X)|\bar{f}(X)$. It follows that $\lambda\bar{\theta} = \bar{\theta}_i$ for some root θ_i of $f(X)$. Since $f(X)$ is irreducible over M, there is a $\sigma \in Z(\mathfrak{P}/P)$ such that $\sigma\theta = \theta_i$. Thus, $\bar{\sigma}\bar{\theta} = \lambda\bar{\theta}$. From this, and the fact that $\bar{\theta}$ generates $E_{\mathfrak{P}}$ over F_P we can conclude that $\lambda = \bar{\sigma}$. This proves the onto-ness. We have shown that the following sequence is exact.

$$(e) \to I(\mathfrak{P}/P) \to Z(\mathfrak{P}/P) \to \mathrm{Gal}(E_{\mathfrak{P}}/F_P) \to (e) \ .$$

The middle term has order $e(\mathfrak{P}/P)f(\mathfrak{P}/P)$ and the end term has order $f(\mathfrak{P}/P)$. One concludes that $\#I(\mathfrak{P}/P) = e(\mathfrak{P}/P)$.

Corollary. *If $\mathfrak{P}/P$ is unramified, then $Z(\mathfrak{P}/P) \cong \mathrm{Gal}(E_{\mathfrak{P}}/F_P)$.*

We continue with two propositions about how the decomposition groups and inertia groups behave "functorially."

Proposition 9.7. *Suppose L/K is a Galois extension of function fields and suppose $\mathfrak{P}$ is a prime of L lying above a prime P of K. Let $\sigma \in \mathrm{Gal}(L/K)$. Then, $Z(\sigma\mathfrak{P}/P) = \sigma Z(\mathfrak{P}/P)\sigma^{-1}$ and $I(\sigma\mathfrak{P}/P) = \sigma I(\mathfrak{P}/P)\sigma^{-1}$. In particular, all the decomposition groups of primes above P in L are conjugate and similarly for the inertia groups.*

Proof. By definition, $\tau \in Z(\sigma\mathfrak{P}/P)$ if and only if $\tau\sigma\mathfrak{P} = \sigma\mathfrak{P}$. This is so if and only if $\sigma^{-1}\tau\sigma\mathfrak{P} = \mathfrak{P}$, i.e., if and only if $\sigma^{-1}\tau\sigma \in Z(\mathfrak{P}/P)$, which holds if and only if $\tau \in \sigma Z(\mathfrak{P}/P)\sigma^{-1}$. This proves the first assertion. The proof of the second is entirely similar.

To prove the last assertion, it is enough to recall that by Proposition 9.2, all the primes above P are of the form $\sigma\mathfrak{P}$ for $\sigma \in \mathrm{Gal}(L/K)$.

Proposition 9.8. *Let L/K be a Galois extension of function fields and M an arbitrary intermediate field. Let $\mathfrak{P}$ be a prime of L and $\mathfrak{p}$ and P the primes of M and K respectively which lie below $\mathfrak{P}$. Set $H = \mathrm{Gal}(L/M)$. Then,*

$$(i) \quad Z(\mathfrak{P}/\mathfrak{p}) = H \cap Z(\mathfrak{P}/P) \quad \text{and} \quad I(\mathfrak{P}/\mathfrak{p}) = H \cap I(\mathfrak{P}/P) \ .$$

Now, assume H is a normal subgroup and that ρ is the restriction map from $\mathrm{Gal}(L/K) \to \mathrm{Gal}(M/K)$. Then,

$$(ii) \quad \rho(Z(\mathfrak{P}/P)) = Z(\mathfrak{p}/P) \quad \text{and} \quad \rho(I(\mathfrak{P}/P)) = I(\mathfrak{p}/P) \ .$$

Proof. Part (i) of the proposition follows directly from the definitions.

To prove part (ii) we first remark that from the definitions it is easy to prove that ρ maps $Z(\mathfrak{P}/P)$ to $Z(\mathfrak{p}/\mathfrak{p})$. The kernel of the this map is $Z(\mathfrak{P}/P) \cap H = Z(\mathfrak{P}/\mathfrak{p})$. Thus, the order of the image is

$$e(\mathfrak{P}/P)f(\mathfrak{P}/P)/e(\mathfrak{P}/\mathfrak{p})f(\mathfrak{P}/\mathfrak{p}) = e(\mathfrak{p}/P)f(\mathfrak{p}/P) = \#Z(\mathfrak{p}/P).$$

This proves the map is onto.

The proof for the inertia groups is entirely similar.

Let L/K be a finite extension of function fields, and P a prime of K. We say that P splits completely in L if there are $[L : K]$ primes above it in L. From the relation $\sum_{i=1}^{g} e_i f_i = n$ it follows that if a prime splits completely in L, every prime above it is unramified and of relative degree 1. Suppose L/K is a Galois extension and that $\mathfrak{P}$ is some prime of L above P. Then, by Proposition 9.3 and Lemma 9.4, we see that P splits completely in L if and only if $Z(\mathfrak{P}/P) = (e)$. More directly, the Galois group acts transitively on the primes above P and the decomposition group of one of them is an isotropy group for this action. Thus, one gets $[L : K]$ primes above P if and only if this decomposition group is trivial.

We recall that a prime P of K is said to be unramified in L if and only if every prime above it in L is unramified.

Proposition 9.9. *Let M_1 and M_2 be two Galois extensions of a function field K and let $L = M_1 M_2$ be the compositum. A prime P of K splits completely in L if and only if it splits completely in M_1 and M_2. A prime P of K is unramified in L if and only if it is unramified in M_1 and M_2.*

Proof. Let $\mathfrak{P}$ be some prime of L lying above P. If P splits completely in L, then by the previous remarks $Z(\mathfrak{P}/P) = (e)$. Let $\mathfrak{p}_1$ and $\mathfrak{p}_2$ be the primes of M_1 and M_2, respectively, which lie below $\mathfrak{P}$. By Proposition 9.8, part (ii), we deduce that $Z(\mathfrak{p}_1/P) = (e)$ and $Z(\mathfrak{p}_2/P) = (e)$. Thus, P splits completely in M_1 and M_2.

Now suppose that P splits completely in M_1 and M_2. Then, $Z(\mathfrak{p}_1/P) = (e)$ and $Z(\mathfrak{p}_2/P) = (e)$. Let $\sigma \in Z(\mathfrak{P}/P)$. By Proposition 9.8, part (ii), we see the restrictions of σ to both M_1 and M_2 are the identity maps. Since M_1 and M_2 generate L, it follows that σ is the identity. Thus, $Z(\mathfrak{P}/P) = (e)$ and so P splits completely in L.

Once again, the proof of the last assertion about unramifiedness is entirely similar. We omit the details.

We conclude this part of the chapter by sketching the behavior of a prime P in the fixed fields of $Z(\mathfrak{P}/P)$ and $I(\mathfrak{P}/P)$, where $\mathfrak{P}$ is a prime of L lying above P. To ease the notation, call the two subgroups Z and I and the corresponding subfields L_Z and L_I of L (we previously denoted L_Z by M). We have $K \subseteq L_Z \subseteq L_I \subseteq L$. The fields L_Z and L_I are called the decomposition field and the inertia field of $\mathfrak{P}$. Let $\mathfrak{p}_Z$ and $\mathfrak{p}_I$ be the primes of L_Z and L_I, respectively, which lie below $\mathfrak{P}$. Then $f(\mathfrak{p}_Z/P) = 1$ and $e(\mathfrak{p}_Z/P) = 1$. If $\mathrm{Gal}(L/K)$ is abelian, it follows that P splits completely in L_Z. It is the case that $\mathfrak{p}_I$ is the only prime of L_I above $\mathfrak{p}_Z$ and we have $e(\mathfrak{p}_I/\mathfrak{p}_Z) = 1$ and $f(\mathfrak{p}_I/\mathfrak{p}_Z) = f(\mathfrak{P}/P) = [L_I : L_Z]$. Finally, $\mathfrak{P}$ is the only prime of L above $\mathfrak{p}_I$ and we have $f(\mathfrak{P}/\mathfrak{p}_I) = 1$ and $e(\mathfrak{P}/\mathfrak{p}_I) = e(\mathfrak{P}/P) = [L : L_I]$. We say that $\mathfrak{p}_Z/P$ is unramified of degree 1, that $\mathfrak{p}_I/\mathfrak{p}_Z$ is inert, and $\mathfrak{P}/\mathfrak{p}_I$ is totally ramified. All this is relatively easy to prove on the basis of our earlier results. We leave the details as an exercise.

This is about as far as we wish to go with the general theory. Although we have been working in function fields, it is clear that most of what we have proven will work in a more general context of Dedekind domains, their quotient fields, and finite extensions thereof.

For the rest of this chapter we will be working with global function fields, i.e., function fields whose field of constants is finite. A key notion in this context is that of the Frobenius automorphism attached to an unramified prime ideal. Our first goal will be to define this object and discuss its properties.

Let K be a function field whose constant field $\mathbb{F}$ is a finite field with q elements. Let L/K be a finite, Galois extension with constant field $\mathbb{E}$. Let G, as usual, denote $\mathrm{Gal}(L/K)$. Suppose P is a prime of K and $\mathfrak{P}$, a prime of L lying above P. The residue class fields are finite and, as is well known, the Galois group, $\mathrm{Gal}(\mathbb{E}_\mathfrak{P}/\mathbb{F}_P)$, is cyclic, generated by ϕ_P which is defined by $\phi_P(x) = x^{NP}$ for all $x \in \mathbb{E}_\mathfrak{P}$ (the point is that $NP = |\mathbb{F}_P|$).

If we suppose that $\mathfrak{P}/P$ is unramified, then by the corollary to Proposition 9.6 we have a canonical isomorphism $Z(\mathfrak{P}/P) \cong \mathrm{Gal}(\mathbb{E}_\mathfrak{P}/\mathbb{F}_P)$. Under these circumstances, there is a unique element $(\mathfrak{P}, L/K) \in Z(\mathfrak{P}/P)$ which corresponds to ϕ_P under this isomorphism. $(\mathfrak{P}, L/K)$ is called the Frobenius automorphism of $\mathfrak{P}$ for the extension L/K. Going back through the definitions we see that the Frobenius automorphism can be characterized by the following condition

$$(\mathfrak{P}, L/K)\, \omega \equiv \omega^{NP} \pmod{\mathfrak{P}} \quad \forall\, \omega \in O_\mathfrak{P} .$$

Proposition 9.10. *Let L/K be a Galois extension of global function fields, $\mathfrak{P}$ a prime of L and P the prime of K lying below it. Suppose $\mathfrak{P}/P$ is unramified. Then, $(\mathfrak{P}, L/K)$ is a cyclic generator of $Z(\mathfrak{P}/P)$ and consequently has order $f(\mathfrak{P}/P)$. Furthermore, if $\sigma \in \text{Gal}(L/K)$, then $(\sigma\mathfrak{P}, L/K) = \sigma(\mathfrak{P}, L/K)\sigma^{-1}$.*

Proof. The first assertion is true by the definition of the Frobenius automorphism via the isomorphism $Z(\mathfrak{P}/P) \cong \text{Gal}(\mathbb{E}_{\mathfrak{P}}/\mathbb{F}_P)$.

To prove the second assertion, recall that for $\sigma \in G$ we have $\sigma O_{\mathfrak{P}} = O_{\sigma\mathfrak{P}}$. Thus, $(\sigma\mathfrak{P}, L/K)$ is characterized by

$$(\sigma\mathfrak{P}, L/K)\, \sigma\omega \equiv (\sigma\omega)^{NP} \pmod{\sigma\mathfrak{P}} \quad \forall\, \omega \in O_{\mathfrak{P}}\,.$$

Applying σ^{-1} to both sides of this congruence we deduce that $\sigma^{-1}(\sigma\mathfrak{P}, L/K)\sigma = (\mathfrak{P}, L/K)$ from which the result follows immediately.

From the second part of this proposition and Proposition 9.2, we see that as $\mathfrak{P}$ varies over the primes above P in L, the Frobenius automorphisms $(\mathfrak{P}, L/K)$ fill out a conjugacy class in G. This leads to the following formal definition.

Definition. Let L/K be a Galois extension of global function fields, and P a prime of K which is unramified in L. The Artin conjugacy class of P, $(P, L/K)$, is defined as the set of all Frobenius automorphisms $(\mathfrak{P}, L/K)$ as $\mathfrak{P}$ varies over the primes in L above P.

The map from $\mathcal{S}_K$ to the conjugacy classes of $\text{Gal}(L/K)$ given by $P \rightarrow (P, L/K)$ is called the Artin map. It is extremely important and we shall discuss it in some detail. First, however, we will record some more "functorial" properties of the Frobenius automorphism.

Proposition 9.11. *Let L/K be a Galois extension of global function fields and M an arbitrary intermediate field. Let $\mathfrak{P}$ be a prime of L and $\mathfrak{p}$ and P the primes lying below it in M and K, respectively. Assume $\mathfrak{P}/P$ is unramified. Then,*

$$(\mathfrak{P}, L/K)^{f(\mathfrak{p}/P)} = (\mathfrak{P}, L/M)\,.$$

If M/K is also a Galois extension, then

$$(\mathfrak{P}, L/K)|_M = (\mathfrak{p}, M/K)\,.$$

Proof. Using the characterization

$$(\mathfrak{P}, L/K)\, \omega \equiv \omega^{NP} \pmod{\mathfrak{P}} \quad \forall\omega \in O_{\mathfrak{P}}\,,$$

we deduce

$$(\mathfrak{P}, L/K)^{f(\mathfrak{p}/P)}\omega \equiv \omega^{NP^{f(\mathfrak{p}/P)}} \pmod{\mathfrak{P}} \quad \forall\omega \in O_{\mathfrak{P}}\,.$$

Since $NP^{f(\mathfrak{p}/P)} = N\mathfrak{p}$, this also characterizes $(\mathfrak{p}, L/M)$. This proves the first assertion.

To prove the second, just recall $O_\mathfrak{p} = O_\mathfrak{P} \cap M$ and $\mathfrak{p} = \mathfrak{P} \cap M$. Thus,

$$(\mathfrak{P}, L/K)\, \omega \equiv \omega^{NP} \pmod{\mathfrak{p}} \quad \forall \omega \in O_\mathfrak{p} \ .$$

This characterizes the automorphism $(\mathfrak{p}, M/K)$ as well, so $(\mathfrak{P}, L/K)|_M = (\mathfrak{p}, M/K)$, which finishes the proof.

One of the main goals of this chapter is to investigate the set of primes P of K which go to a fixed conjugacy class C in $\mathrm{Gal}(L/K)$ via the Artin map $P \to (P, L/K)$. One way to describe the abundance of such primes is via the notion of Dirichlet density. We introduced this notion in a special case in Chapter 4. The next task is to give the general definition and to investigate its properties.

Let $\mathcal{M} \subseteq \mathcal{S}_K$ be a set of primes in K. The Dirichlet density of $\mathcal{M}$, $\delta(\mathcal{M})$, is given by the following limit, provided that the limit exists. If it doesn't exist we say that $\mathcal{M}$ does not have Dirichlet density.

$$\delta(\mathcal{M}) = \lim_{s \to 1^+} \frac{\sum_{P \in \mathcal{M}} NP^{-s}}{\sum_{P \in \mathcal{S}_K} NP^{-s}} \ .$$

The expression "$s \to 1^+$" means that s approaches 1 through real values from above.

It is clear from this definition that when $\delta(\mathcal{M})$ exists we have $0 \leq \delta(\mathcal{M}) \leq 1$.

The denominator in this definition can be replaced with either $\log \zeta_K(s)$ or $-\log(1-s)$. To see this, consider the following calculation where throughout we assume $\Re(s) > 1$, and sums over P mean sums over all P in $\mathcal{S}_K$.

$$\log \zeta_K(s) = \sum_P \sum_{k=1}^\infty k^{-1} NP^{-ks} = \sum_P NP^{-s} + \sum_P \sum_{k=2}^\infty k^{-1} NP^{-ks} \ .$$

Let's call the second sum $R_K(s)$. We claim $R_K(s)$ remains bounded as $s \to 1^+$. To see this, set $x = \Re(s)$ and note that

$$|R_K(s)| < \sum_P \sum_{k=2}^\infty NP^{-kx} = \sum_P NP^{-2x}(1 - NP^{-x})^{-1}$$

$$< 2 \sum_P NP^{-2x} < 2\zeta_K(2x) \ .$$

Since $\zeta_K(s)$ is holomorphic for $\Re(s) > 1$, we see that $\zeta_K(2s)$ is bounded in a neighborhood of 1, which establishes our claim.

Next, by Theorem 5.9, $\zeta_K(s)$ has a simple pole at $s = 1$. Thus, $\lim_{s \to 1^+}(s-1)\zeta_K(s) = \alpha_K \neq 0$. Confining s to a small neighborhood of 1 we see that $\log((s-1)\zeta_K(s)) = \log(s-1) + \log\zeta_K(s)$ is bounded. Thus, we have shown

$$\sum_P NP^{-s} \approx \log\zeta_K(s) \approx -\log(s-1) \ .$$

where $f(s) \approx g(s)$ means that $f(s) - g(s)$ is bounded in a neighborhood of 1 (in particular, on an interval of the form $(1, r)$). It follows that all three functions tend to infinity as $s \to 1^+$ and, also, that the ratio of any two tend to 1 as $s \to 1^+$. So, we have justified the claim that we could have used any of the three functions as the denominator in the definition of Dirichlet density. The function $-\log(s-1)$ is particularly useful for some purposes. It does not depend on K!

Certain properties of Dirichlet density follow easily from the definition. We have already mentioned one of them. We summarize those which will be needed later.

Proposition 9.12. *Let K be a global function field and $\mathcal{M} \subseteq \mathcal{S}_K$ a set of primes. If the Dirichlet density of $\mathcal{M}$ exists, $0 \leq \delta(\mathcal{M}) \leq 1$. If $\mathcal{M}$ is finite, $\delta(\mathcal{M}) = 0$. Also, $\delta(\mathcal{S}_K) = 1$. Suppose $\mathcal{M}_1$ and $\mathcal{M}_2$ both have Dirichlet density. If these two sets differ by only finitely many primes, then $\delta(\mathcal{M}_1) = \delta(\mathcal{M}_2)$. If $\mathcal{M}_1 \subseteq \mathcal{M}_2$ then $\delta(\mathcal{M}_1) \leq \delta(\mathcal{M}_2)$. If $\mathcal{M}_1 \cap \mathcal{M}_2 = \phi$, the empty set, then $\delta(\mathcal{M}_1 \cup \mathcal{M}_2) = \delta(\mathcal{M}_1) + \delta(\mathcal{M}_2)$.*

The property involving disjoint unions extends to finitely many sets, but not to denumerably many sets! For each $P \in \mathcal{S}_K$ let $\{P\}$ be the set consisting of one element, P. Then, $\delta(\{P\}) = 0$ for every P, $\mathcal{S}_K = \bigcup_P\{P\}$, but $\delta(\mathcal{S}_K) = 1 \neq 0$. One must not think of Dirichlet density as a measure (in the technical sense) on the set of primes of K.

We have enough information to prove an important special case of the Tchebotarev density theorem, and we proceed to do so. For much of the rest of this chapter we will fix a global function field K as base field and consider a finite Galois extension L of K. Given such an extension, we define $\{L\} \subseteq \mathcal{S}_K$ to be the set of primes in K which split completely in L. By our previous work, this can be characterized as the set of primes P of K, which are unramified in L and for which $(P, L/K) = (e)$, the conjugacy class of $\mathrm{Gal}(L/K)$ consisting of the identity element.

Proposition 9.13. *Let L/K be a Galois extension of global function fields. The Dirichlet density of the set of primes in K which split in L is given by $\delta(\{L\}) = 1/[L : K]$. If L_1 and L_2 are two Galois extensions of K and $\{L_1\} = \{L_2\}$, then $L_1 = L_2$.*

Proof. We consider the zeta function of L. We have,

$$\log\zeta_L(s) = \sum_{\mathfrak{P} \in \mathcal{S}_L} \sum_{k=1}^{\infty} k^{-1} N\mathfrak{P}^{-ks} = \sum_{\mathfrak{P}} N\mathfrak{P}^{-s} + \sum_{\mathfrak{P}} \sum_{k=2}^{\infty} k^{-1} N\mathfrak{P}^{-ks} \ .$$

The double sum is what we previously labeled $R_L(s)$. This was shown to be bounded in a neighborhood of $s = 1$. In the sum that remains, group the terms lying over a fixed prime P in K and we get

$$\log \zeta_L(s) \approx \sum_{P \in \mathcal{S}_K} \sum_{\mathfrak{P} | P} N\mathfrak{P}^{-s} .$$

We can ignore the finitely many ramified primes. Set $[L:K] = n$. For the remaining primes we have $f(\mathfrak{P}/P)g(\mathfrak{P}/P) = n$. Thus,

$$\sum_P \sum_{\mathfrak{P}|P} N\mathfrak{P}^{-s} = \sum_{f|n} \frac{n}{f} \sum_{\substack{P \in \mathcal{S}_K \\ f(\mathfrak{P}/P) = f}} NP^{-fs} ,$$

where we have used $N\mathfrak{P} = NP^{f(\mathfrak{P}/P)}$. The sum of the terms with $f > 1$ is bounded in a neighborhood of 1 and the sum of the terms with $f = 1$ is exactly $n \sum_{P \in \{L\}} NP^{-s}$. Putting all this together we find

$$\log \zeta_L(s) \approx [L:K] \sum_{P \in \{L\}} NP^{-s} .$$

Finally, divide both sides by $-\log(s-1)$ and take the limit as $s \to 1^+$. We conclude that $1 = [L:K]\delta(\{L\})$ and so $\delta(\{L\}) = [L:K]^{-1}$.

To prove the second part of the Proposition, consider the compositum $L = L_1 L_2$. By Proposition 9.9, a prime splits completely in L if and only if it splits completely in L_1 and L_2. Thus, $\{L\} = \{L_1\} \cap \{L_2\} = \{L_1\} = \{L_2\}$. From the first part of the proposition we conclude that $[L:K] = [L_1:K] = [L_2:K]$. Since $L_1 \subseteq L$ and $L_2 \subseteq L$ we have $L_1 = L = L_2$. The proof is complete.

We note that to get the second part of the Proposition it would have been enough to assume $\{L_1\}$ and $\{L_2\}$ differ by at most a set of Dirichlet density zero. This generalization is sometimes quite useful.

We are now in a position to state the two different forms of the Tchebotarev density theorem that we have promised.

Theorem 9.13A. (Tchebotarev Density Theorem, first version). *Let L/K be a Galois extension of global function fields and set $G = \mathrm{Gal}(L/K)$. Let $C \subset G$ be a conjugacy class in G and $\mathcal{S}'_K$ be the set of primes of K which are unramified in L. Then*

$$\delta(\{P \in \mathcal{S}'_K \mid (P, L/K) = C\}) = \frac{\#C}{\#G} .$$

In particular, every conjugacy class C is of the form $(P, L/K)$ for infinitely many primes P in K.

Theorem 9.13B. (Tchebotarev Density Theorem, second version). *Let L/K be a geometric, Galois extension of global function fields and set*

$G = \mathrm{Gal}(L/K)$. Let $C \subset G$ be a conjugacy class. Suppose the common constant field $\mathbb{F}$ of K and L has q elements. As above let S'_K be the set of primes of K unramified in L. Then, for each positive integer N, we have

$$\#\{P \in S'_K \mid \deg_K P = N, \ (P, L/K) = C\} = \frac{\#C}{\#G}\frac{q^N}{N} + O\left(\frac{q^{N/2}}{N}\right).$$

In particular, for every sufficiently large integer N, there is a prime P of degree N with $(P, L/K) = C$.

In the second theorem, the hypothesis that the extension be geometric is not absolutely necessary, but it simplifies both the statement and the proof. The interested reader can investigate how matters should be modified to handle the general case.

It will be seen that both of these theorems have considerable depth. However, the second is much stronger and much more difficult to prove.

In fact, we will not prove either theorem completely, but will reduce both theorems to facts about Artin L-series. This may not be the easiest way to proceed, but is, perhaps, the most instructive. M. Deuring was able to prove the number field version of the first theorem by reducing to the case where $\mathrm{Gal}(L/K)$ is abelian by means of a very clever trick. The reader may wish to adapt this proof to function fields. See Lang [5], Ch. VIII, Theorem 10, for an exposition of Deuring's proof.

Of course, before we can go forward along the lines indicated toward a proof of either theorem, we have to define Artin L-functions and discuss their properties. So, we do this first, and afterwards sketch the proofs.

Let $G = \mathrm{Gal}(L/K)$ be the Galois group of a Galois extension of global function fields and $\rho : G \to \mathrm{Aut}_{\mathbb{C}}(V)$ a representation of G. Here V is a finite-dimensional vector space over the complex numbers $\mathbb{C}$ of dimension m. By choosing a basis of V over $\mathbb{C}$ we are led to an isomorphism $\mathrm{Aut}_{\mathbb{C}}(V) \cong \mathrm{GL}_m(\mathbb{C})$. Thus, for $\sigma \in G$ we can think of $\rho(\sigma)$ either as an automorphism of V or an $m \times m$ matrix with complex coefficients. The latter way of looking at things is more concrete, but depends on the choice of a basis. However, our definitions will only depend on the determinant and trace of such a matrix and these only depend on the automorphism.

Let P be a prime of K which is unramified in L and let $\mathfrak{P}$ be a prime of L lying above it. We define the local factor $L_P(s, \rho)$ as follows:

$$L_P(s, \rho) = \det\left(I - \rho((\mathfrak{P}, L/K))NP^{-s}\right)^{-1}.$$

Here, I is the identity automorphism on V and $(\mathfrak{P}, L/K)$ is the Frobenius automorphism at $\mathfrak{P}$. By Proposition 9.10, we easily verify that the definition is independent of the choice of $\mathfrak{P}$ above P.

Let $\{\alpha_1(P), \alpha_2(P), \cdots, \alpha_m(P)\}$ be the eigenvalues of $\rho((\mathfrak{P}, L/K))$. Again, Proposition 9.10 can be used to show that these eigenvalues depend only on P and not the choice of $\mathfrak{P}$ above P. In terms of these eigenvalues, we get the following useful expression for the local factor at P:

$$L_P(s,\rho)^{-1} = (1 - \alpha_1(P)NP^{-s})(1 - \alpha_2(P)NP^{-s}) \cdots (1 - \alpha_m(P)NP^{-s}) \ .$$

We remark that since $(\mathfrak{P}, L/K)$ has finite order, these eigenvalues are roots of unity ($f(\mathfrak{P}/P)$-th roots of unity, to be precise).

Next, we must answer the question of what should be the local factors at P in the case when P is ramified in L. Let $\mathfrak{P}$ lie above P and set $Z = Z(\mathfrak{P}/P)$ and $I = I(\mathfrak{P}/P)$ (the context should keep this use of "I" separate from its use as the identity automorphism). We recall the exact sequence:

$$(e) \to I \to Z \to \mathrm{Gal}(\mathbb{E}_{\mathfrak{P}}/\mathbb{F}_P) \to (e) \ .$$

Let $V^I = \{v \in V \mid \rho(\tau)v = v, \ \forall \tau \in I\}$. This is a vector subspace of V. Let $\gamma_{\mathfrak{P}}$ be any element in Z which maps to $\phi_P \in \mathrm{Gal}(\mathbb{E}_{\mathfrak{P}}/\mathbb{F}_P)$, the automorphism defined by raising to the NP power. We define the local factor at P by

$$L_P(s,\rho) = \det\left(I' - \rho(\gamma_{\mathfrak{P}})|_{V^I} NP^{-s}\right)^{-1} \ .$$

Here, I' is the identity automorphism on V^I. Since any two choices of $\gamma_{\mathfrak{P}}$ differ by an element in I, the definition of $L_P(s,\rho)$ is unaffected. As before, the definition is also unaffected by the choice of $\mathfrak{P}$ lying above P.

Let m' be the dimension of V^I and $\{\alpha_1(P), \alpha_2(P), \cdots, \alpha_{m'}(P)\}$ the eigenvalues of $\rho(\gamma_{\mathfrak{P}})$. These, indeed, depend only on P, and we have

$$L_P(s,\rho)^{-1} = (1 - \alpha_1(P)NP^{-s})(1 - \alpha_2(P)NP^{-s}) \cdots (1 - \alpha_{m'}(P)NP^{-s}) \ .$$

We remark that $m' \leq m$ and, once again, the eigenvalues are all roots of unity.

Having defined the local factors for all $P \in S_K$ we now define the Artin L-series associated to the representation ρ by the equation

$$L(s,\rho) = \prod_{P \in S_K} L_P(s,\rho) \ .$$

Suppose $\rho = \rho_o$, the trivial representation. This means that V is one dimensional and $\rho_o(\sigma)$ is the identity for all $\sigma \in G$. It follows easily from the definitions that $L(s,\rho_o) = \zeta_K(s)$.

Another interesting representation of G is the regular representation ρ_{reg}. In this case $V = \mathbb{C}[G]$, the group ring of G over $\mathbb{C}$ and for all $\sigma \in G$, $\rho_{\mathrm{reg}}(\sigma)$ is given by left multiplication by σ. It can be shown that $L(s,\rho_{\mathrm{reg}}) = \zeta_L(s)$. We will return to these matters later.

Suppose (V,ρ) and (V',ρ') are isomorphic representations. This means there is an isomorphism $\mu : V \to V'$, such that for all $v \in V$ and $\sigma \in G$ we have $\mu(\rho(\sigma)(v)) = \rho'(\sigma)(\mu(v))$. It is easy to see that if ρ and ρ' belong to isomorphic representations, then $L(s,\rho) = L(s,\rho')$. It follows that the

L-series depends only on the character χ of the representation. Recall that if (V, ρ) is a representation, the corresponding character is given by $\chi(\sigma) = \text{trace}(\rho(\sigma))$ for all $\sigma \in G$. χ is a complex-valued function on G. It is a class function in the sense that $\chi(\tau^{-1}\sigma\tau) = \chi(\sigma)$ for all $\sigma, \tau \in G$. It is easily seen that two isomorphic representations have the same character. We will write $L(s, \rho) = L(s, \chi)$.

Lemma 9.14. *Let (V, ρ) be a representation of $G = \text{Gal}(L/K)$ where L/K is a Galois extension of global function fields. Let P be a prime of K unramified in L. Then*

$$\log L_P(s, \chi) = \sum_{k=1}^{\infty} \frac{\chi(P^k)}{kNP^{ks}} \ ,$$

where $\chi(P^k)$ means $\chi((\mathfrak{P}, L/K)^k)$ for a prime $\mathfrak{P}$ lying over P.

Proof. If $\{\alpha_1(P), \alpha_2(P), \cdots, \alpha_m(P)\}$ are the eigenvalues of $(\mathfrak{P}, L/K)$, then we showed earlier that $L_P(s, \chi)^{-1} = \prod_{i=1}^{m}(1 - \alpha_i(P)NP^{-s})$. Taking the logarithm of both sides and using the identity $-\log(1-X) = \sum_{k=1}^{\infty} k^{-1}X^k$, we find

$$\log L_P(s, \chi) = \sum_{k=1}^{\infty} \frac{\sum_{i=1}^{m} \alpha_i(P)^k}{k} \frac{1}{NP^{ks}} \ .$$

The sum $\sum_{i=1}^{m} \alpha_i(P)^k$ is equal to the trace of $\rho((\mathfrak{P}, L/K)^k)$, which is $\chi(P^k)$ by definition.

The reader may wish to give a similar expression for $\log L_P(s, \chi)$ when P is ramified (see Artin [2] or Lang [5], Chapter XII).

Up to now we have been treating everything in a formal manner and not worrying about where these new L-series are defined. It is relatively easy to provide some information by using the comparison test.

Proposition 9.15. *With the above notation and conventions, $L(s, \chi)$ converges absolutely in the region $\Re(s) > 1$ and for every $\delta > 0$ it converges absolutely and uniformly in the region $\Re(s) \geq 1 + \delta$. Consequently, $L(s, \chi)$ is holomorphic and non-vanishing for all s with $\Re(s) > 1$.*

Proof. An infinite product $\prod_{n=1}^{\infty}(1 + a_n)$ converges absolutely if and only if $\sum_{n=1}^{\infty} |a_n|$ converges. Using the local decomposition $L_P(s) = \prod_{i=1}^{m}(1 - \alpha_i(P)NP^{-s})^{-1}$, we can use this criterion together with the fact that the zeta function $\zeta_K(s) = \prod_P (1 - NP^{-s})^{-1}$ converges absolutely in the region $\Re(s) > 1$ to show the same holds for $L(s, \chi)$. Since each of the local factors, $L_P(s, \chi)$, is non-vanishing in that region, the same holds for the product, $L(s, \chi)$.

The statement about uniform convergence can be proved in a similar fashion.

To prove the Tchebotarev Density Theorem we need more information on the analytic properties of Artin L-series. We present the next two theorems in parallel with the two versions of the Tchebotarev Theorem, i.e., Theorems 9.13A and 9.13B.

Theorem 9.16A. (E. Artin) *Let L/K be a Galois extension of global fields (either function fields or number fields) and $L(s, \chi)$ a corresponding Artin L-series. Then, for some positive integer n, $L(s, \chi)^n$ has a meromorphic continuation to the whole complex plane. Moreover, if χ is a non-trivial irreducible character, then $L(s, \chi)^n$ is holomorphic and non-vanishing in a neighborhood of $s = 1$.*

Theorem 9.16B. (A. Weil) *Let L/K be a geometric, Galois extension of global function fields. Denote by q the number of elements in the constant field. Let $L(s, \chi)$ be a corresponding Artin L-series and assume that χ is irreducible and non-trivial. Then $L(s, \chi)$ is a polynomial in q^{-s}. In particular, this implies that $L(s, \chi)$ has a holomorphic continuation to the whole complex plane. Moreover, denoting by m the degree of $L(s, \chi)$ in q^{-s}, we have*

$$L(s, \chi) = \prod_{i=1}^{m} (1 - \pi_i(\chi) q^{-s}),$$

where for each i with $1 \le i \le m$, $|\pi_i(\chi)| = \sqrt{q}$.

We will not prove either of these results. In the case of Artin's theorem, we will show later how to reduce the proof to the case of one-dimensional characters and how, via Artin's reciprocity law, the result can be made to follow from Hecke's work on another type of L-series. Our main goal is to show how to use Theorem 9.16A to prove Theorem 9.13A and how to use Theorem 9.16B to prove Theorem 9.13B. First, however, a series of remarks. These remarks are not needed in the proofs, so the impatient reader can simply skip over them.

1.   Artin deduced Theorem 9.16A by means of a theorem on group characters. Namely, he showed that any complex character of a finite group G can be written as a rational linear combination of induced characters from cyclic subgroups of G. See Serre [3] for the definition of induced character and the proof of this theorem (Chapter 9). From this it follows that there is an integer $n > 0$ such that $L(s, \chi)^n$ can be written as a product of Artin L-series corresponding to one-dimensional characters divided by another such product. Since, via Hecke's work and the reciprocity law, he knew the result to be true for one-dimensional characters, the meromorphic continuation follows. Using the same ideas he deduced $L(1, \chi) \ne 0$ by reducing to the case of one-dimensional characters.

2.   Strictly speaking, Theorem 9.16A only gives information about $L(s, \chi)^n$ about $s = 1$, not $L(s, \chi)$ itself. However, the result implies that on any real

interval $(1, t)$, $L(s, \chi)$ is bounded from above and is bounded away from 0. This is all we need to prove Theorem 13A.

3.   In 1947 , R. Brauer proved a much stronger theorem on group characters. Namely, if χ is any complex character on a finite group G, then χ can be written as a $\mathbb{Z}$-linear combination of characters induced from one-dimensional characters on "elementary subgroups" (again, see Serre [3], Chapter 10, for the definitions and proof). This sufficed to show that any Artin L-series has a meromorphic continuation to the whole complex plane, i.e., the troublesome "n" in Artin's theorem can be taken to be 1. Brauer's result did not give Artin's conjecture that $L(s, \chi)$ has a holomorphic continuation to the whole complex plane when χ is irreducible and non-trivial. In the number field case, this remains an open conjecture. In the function field case, Weil proved it in the precise form given by Theorem 9.16B, using algebraic-geometric methods.

4.   In Theorem 9.16B, once the first part of the theorem has been established, the second part, about the size of the inverse roots, follows from an important, but not deep, property of Artin L-series and the Riemann hypothesis for function fields.

Let G be a finite group and χ_{reg} the character of the regular representation described earlier. Let $\{\chi_1, \chi_2, \cdots, \chi_g\}$ be the set of irreducible characters of G. We set $\chi_1 = \chi_o$, the trivial character. Denote by d_i the degree of χ_i, i.e., $d_i = \chi_i(e) =$ the dimension of the representation space corresponding to χ_i. In this language, the one-dimensional characters are those of degree 1. It is well known that $\chi_{\text{reg}} = \sum_{i=1}^{g} d_i \chi_i$ (See Serre [3], Chapter 2).

To avoid awkward notation, we consider a geometric, Galois extension of function fields M/K (not L/K for now). Let $G = \text{Gal}(M/K)$. Then, using the result about group characters given in the last paragraph, formal properties of Artin L-series, and $L(s, \chi_{\text{reg}}) = \zeta_M(s)$, one deduces

$$\zeta_M(s) = \zeta_K(s) \prod_{i=2}^{g} L(s, \chi_i)^{d_i} .$$

Assuming the first part of Weil's result, set, for $2 \leq i \leq g$, $L(s, \chi_i) = P(u, \chi_i)$, a polynomial in $u = q^{-s}$. Now use Theorem 5.9, which describes the form of the zeta function of a global function field. Substituting into the last equation, we get

$$L_M(u) = L_K(u) \prod_{i=2}^{g} P(u, \chi_i)^{d_i} .$$

By the Riemann hypothesis for global function fields (see Theorem 5.10), and the fact that M/K is a geometric extension, the inverse roots of $L_M(u)$ all have size $\sqrt{q}$. The right-hand side of the above equation is a product of polynomials, so each of these polynomials must have inverse roots whose absolute value is $\sqrt{q}$.

5.   Weil was able to determine the exact degree m_i of the polynomials $P(u, \chi_i)$. The answer is that for $2 \leq i \leq g$, $m_i = d_i(2g_K - 2) + \deg_K F(\chi_i)$. Here, $F(\chi_i)$ is an effective divisor of K called the Artin conductor of the character χ_i. We will not define it here, but the interested reader can consult Serre [2], Chapter VI. We will give the definition later in the special case where χ is a linear character. By considering the last relation in Remark 4 above, and taking degrees, we find

$$2g_M - 2 = [M : K](2g_K - 2) + \sum_{i=2}^{g} d_i \deg_K F(\chi_i),$$

a relation which is the function field analogue of the conductor-discriminant theorem of algebraic number theory. (We have used $[M : K] = \sum_{i=1}^{g} d_i^2$ which follows from $\chi_{\text{reg}} = \sum_{i=1}^{g} d_i \chi_i$ by evaluating both sides at the identity element e).

We now return to our main business.

Lemma 9.17. *Let G be a finite group and $C \subset G$ a conjugacy class of G. Let $\sigma \in C$ and $\tau \in G$. Then*

$$\sum_{\chi} \overline{\chi(\sigma)} \chi(\tau) = 0 \; if \; \tau \notin C \; and \; \frac{\#G}{\#C} \; if \; \tau \in C \; ,$$

where the sum is over all irreducible characters of G.

Proof. This is one of the two standard othogonality relations among characters of finite groups. See Lang [4] or Serre [3].

We have all the tools we need to give a proof of the first form of the Tchebotarev Density Theorem.

Proof of Theorem 9.13A. Let χ be any irreducible character and define $L^*(s, \chi) = \prod_{P \in S'_K} L_P(s, \chi)$. We have omitted the finitely many factors from the product defining $L(s, \chi)$ which correspond to primes of K ramified in L. It is still true that $L^*(s, \chi_0)$ has a simple pole at $s = 1$ (since it differs from $\zeta_K(s)$ by a factor which is holomorphic and non-vanishing at $s = 1$) and that $L^*(s, \chi)$ is bounded and bounded away from 0 on any real interval of the form $(1, t)$ if $\chi \neq \chi_0$ is irreducible. This follows from Theorem 9.16A.
   By Lemma 9.14, we have, for $\Re(s) > 1$

$$\log L^*(s, \chi) = \sum_{P \in S'_K} \log L_P(s, \chi) = \sum_{P \in S'_K} \sum_{k=1}^{\infty} \frac{\chi(P^k)}{kNP^{ks}} \; .$$

For any element $\tau \in G$, we have $|\chi(\tau)| \leq d$, the degree of χ. This follows since $\chi(\tau)$ is the sum of d roots of unity. From this, we see

$$\left| \sum_{P \in S_K'} \sum_{k=2}^{\infty} \frac{\chi(P^k)}{kNP^{ks}} \right| \leq d \sum_{P \in S_K'} \sum_{k=2}^{\infty} \frac{1}{NP^{kx}} = dR_K(x) .$$

Just as in a previous discussion, after the definition of Dirichlet density, one shows $R_K(x) \leq 2\zeta_K(2x)$. Since this is bounded in a neighborhood of 1 we deduce

$$\log L^*(s, \chi) \approx \sum_{P \in S_K'} \frac{\chi(P)}{NP^s} .$$

Choose an element $\sigma \in C$ and multiply both sides by $\overline{\chi(\sigma)}$ and add the result over all irreducible characters χ of G. Making use of Lemma 9.17, we obtain

$$\sum_{\chi} \overline{\chi(\sigma)} \log L^*(s, \chi) \approx \frac{\#G}{\#C} \sum_{P, \ (P, L/K)=C} \frac{1}{NP^s} . \qquad (*)$$

Since $L^*(s, \chi_o)$ has a simple pole at $s = 1$ and for the other irreducible characters, $L^*(s, \chi)$ is bounded and bounded away from zero on any interval of the form $(1, t)$ we have

$$\lim_{s \to 1^+} \frac{\log L^*(s, \chi_o)}{-\log(s-1)} = 1 \quad \text{and} \quad \lim_{s \to 1^+} \frac{\log L^*(s, \chi)}{-\log(s-1)} = 0 \ \text{for } \chi \neq \chi_o .$$

In equation $(*)$ above, divide both sides by $-\log(s-1)$ and take the limit as $s \to 1^+$. The result is

$$1 = \frac{\#G}{\#C} \delta(\{P \in S_k' \mid (P, L/K) = C\}) ,$$

which concludes the proof of the theorem.

The exact same proof works equally well in algebraic number fields. The reader will not fail to notice how similar this proof is to the proof of Dirichlet's theorem, Theorem 4.7. We shall indicate below how the two results are connected. In fact, Theorem 9.16A should be thought of as a vast generalization of Dirichlet's theorem. First, however, it is time to prove the second form of Tchebotarev's density theorem.

Proof of Theorem 9.13B. We begin by making a remark about Lemma 9.14. We proved it assuming P is a prime of K unramified in L. We would like to take the ramified primes into account as well. Using the definition of the local factor of an Artin L-series at a ramified prime we found in this case also one can write $L_P(s, \chi)^{-1} = \prod_{i=1}^{m'} (1 - \alpha_i(P)NP^{-s})$ where the α_i

are roots of unity and $m' \leq m =$ the degree of χ. We define $\chi(P^k)$ to be $\sum_{1=1}^{m'} \alpha_i^k$. This coincides with the definition in case P is unramified and with this definition the formula of Lemma 9.14 is valid for all primes.

We next calculate the logarithmic derivative of $L(s, \chi)$ assuming $\Re(s) > 1$. From the Euler factor definition, $L(s, \chi) = \prod_{P \in S_K} L_P(s, \chi)$, we find, using Lemma 9.14, that

$$\log L(s, \chi) = \sum_{P \in S_K} \sum_{k=1}^{\infty} \frac{\chi(P^k)}{k N P^{ks}} .$$

We now switch to the variable $u = q^{-s}$ and, by abuse of notation, write $L(s, \chi) = L(u, \chi)$. The above relation becomes

$$\log L(u, \chi) = \sum_{P \in S_K} \sum_{k=1}^{\infty} \frac{\chi(P^k)}{k} u^{k \deg P} .$$

Take the derivative of both sides and multiply the resulting equation by u. We find

$$u \frac{L'(u, \chi)}{L(u, \chi)} = \sum_{P \in S_k} \sum_{k=1}^{\infty} \deg P \, \chi(P^k) u^{k \deg P} = \sum_{n=1}^{\infty} c_n(\chi) u^n .$$

The coefficient $c_n(\chi)$ of u^n is given by

$$c_n(\chi) = \sum_{P, \, \deg P | n} \deg P \, \chi(P^{n/ \deg P}) . \tag{1}$$

We write this as

$$c_n(\chi) = n \Big(\sum_{P, \, \deg P = n} \chi(P) \Big) + R_n(\chi) , \tag{2}$$

and we will show later that $R_n(\chi) = O(q^{n/2})$.

The main idea of the proof is to express $c_n(\chi)$ in another way using the zeros and poles of the various L-series whose size we know something about because of the Riemann hypothesis for function fields and Theorem 9.16B above. From this it will turn out that

$$c_n(\chi) = q^n \delta(\chi, \chi_o) + O(q^{n/2}) , \tag{3}$$

where $\delta(\chi, \chi_o) = 1$ if $\chi = \chi_o$ and is 0 otherwise.

Assuming these facts about $c_n(\chi)$ we will now show how to complete the proof. Afterwards we will give the details behind these two separate evaluations of $c_n(\chi)$.

Combining equations 2 and 3, we find

$$q^n \delta(\chi, \chi_o) + O(q^{n/2}) = n \Big(\sum_{P, \, \deg P = n} \chi(P) \Big) + O(q^{n/2}) . \tag{4}$$

There are only finitely many ramified primes in L/K so for all n sufficiently large there are no ramified primes of degree n. Thus, from some point on $\chi(P) = \chi((P, L/K))$ for all primes of degree n. We assume n is at least this big. Now, choose an element $\sigma \in C$ and multiply both sides of Equation 4 by $\overline{\chi(\sigma)}$ and sum over all irreducible characters χ. Using Lemma 9.17, we deduce

$$q^n + O(q^{n/2}) = n \frac{\#G}{\#C} \#\{P \in \mathcal{S}_K \mid \deg P = n, \ (P, L/K) = C\} + O(q^{n/2}) .$$

Divide both sides of the equation by $n\#G/\#C$, combine the error terms, and, subject to the proofs of 2 and 3, the theorem follows.

We now proceed to show the validity of the two expressions we have given for $c_n(\chi)$. Consider first Equation 2. From Equation 1 we get the following explicit expression for $R_n(\chi)$:

$$R_n(\chi) = \sum_{\substack{\deg P \mid n \\ \deg P < n}} \deg P \ \chi(P^{n/\deg P}) .$$

If h is the degree of χ, then, as we have seen, $|\chi(\tau)| \leq h$ for all $\tau \in G$. So, taking absolute values and using the triangle inequality, we get

$$|R_n(\chi)| \leq h \sum_{d \mid n, d < n} d a_d(K) .$$

Recall that $a_d(K)$ is the number of primes of K of degree d. We know that $\sum_{d \mid n} d a_d(K) = N_n(K)$. It follows that

$$|R_n(\chi)| \leq h |N_n(K) - n a_n(K)| . \tag{5}$$

By the analogue of the prime number theorem, Theorem 5.12, we know $n a_n = q^n + O(q^{n/2})$. Since $N_n(K) = q^n + 1 - \sum_{i=1}^{2g} \pi_i^n$, where for each i we have $|\pi_i| = \sqrt{q}$, we also have $N_n(K) = q^n + O(q^{n/2})$. It follows that $|N_n(K) - n a_n(K)| = O(q^{n/2})$. The required estimate $R_n(\chi) = O(q^{n/2})$ now follows from Equation 5.

The final step is to prove the estimate for $c_n(\chi)$ in Equation 3. We begin with the trivial character χ_o. As we have seen, the Artin L-series for χ_o is just the zeta function of K. Thus,

$$L(s, \chi_o) = \frac{L_K(u)}{(1 - u)(1 - qu)},$$

where $L_K(u) = \prod_{i=1}^{2g} (1 - \pi_i u)$ and for each i, $|\pi_i| = \sqrt{q}$. Taking the logarithmic derivative of both sides, multiplying by u and equating coefficients, we find

$$c_n(\chi_o) = q^n + 1 - \sum_{i=1}^{2g} \pi_i^n = q^n + O(q^{n/2}) .$$

This verifies the estimate for $c_n(\chi_o)$. We have done this calculation much earlier, in the proof of Theorem 5.12, and we have used the result in a different context in the last paragraph.

If $\chi \neq \chi_o$ is irreducible, then by Weil's result, Theorem 9.16B, we can write $L(u, \chi) = \prod_{i=1}^{m}(1 - \pi_i(\chi)u)$ where m is the degree of $L(u, \chi)$ and each $\pi_i(\chi)$ has absolute value $\sqrt{q}$. Taking logarithmic derivatives, multiplying by u, and comparing coefficients we derive

$$c_n(\chi) = -\sum_{i=1}^{m} \pi_i(\chi)^n .$$

From this it is clear that $c_n(\chi) = O(q^{n/2})$.

Theorem 9.13B is proved.

We have been content to be somewhat careless about the error term. It can be estimated effectively by keeping careful track of constants at each step of the proof. The interested reader can try working this out or he/she can consult Murty and Scherk [1].

The method of proof is often used in analytic number theory. We have an arithmetic L-series which is defined by an Euler product over primes. One then tries to continue the function to be analytic on the whole complex plane. One then writes the same function as a product over its zeros and, when they exist, poles. Taking the logarithmic derivative of both product expansions and comparing the results usually leads to important results. This idea goes back to Riemann. It has been a very fruitful method.

For the rest of the chapter we will treat the case of abelian extensions of global function fields. For the most part we will be content to sketch this beautiful theory, but from time to time complete proofs will be supplied. Our main objective is to set out the connection between Artin L-series associated to abelian extensions and Hecke L-series (to be defined below). This is fundamental to any deeper understanding of the material we have covered up to now.

From now on we assume that L/K is a finite, Galois extension of global function fields and that the Galois group, $G = \mathrm{Gal}(L/K)$, is an abelian group. As before, $\mathbb{E}$ will be the constant field of L and $\mathbb{F}$ the constant field of K. We set $q = \#\mathbb{F}$ and $m = [\mathbb{E} : \mathbb{F}]$. If P is a prime of K and $\mathfrak{P}_1$ and $\mathfrak{P}_2$ are two unramified primes of L lying above P, then by Proposition 9.10 the two Frobenius automorphisms $(\mathfrak{P}_1, L/K)$ and $(\mathfrak{P}_2, L/K)$ are conjugate in G. Since we are assuming G is abelian, these two automorphisms are equal. Thus, the conjugacy class $(P, L/K)$ contains only one element. We identify this conjugacy class consisting of one element with an element of G. The automorphism $(P, L/K)$ is called the Artin automorphism at P. Let $S_K' \subset S_K$ denote the set of primes in K which are unramified in L. Then, $P \to (P, L/K)$ is a well-defined map from S_K' to G. Let $\mathcal{D}_K' \subset \mathcal{D}_K$ be the divisors of K whose support lies in S_K'. Then, by linearity, $P \to (P, L/K)$

extends to a homomorphism from $\mathcal{D}'_K \to G$ which is called the Artin map, $(*, L/K)$. To be explicit, if $D \in \mathcal{D}'_K$, then

$$(D, L/K) = (\sum_{P \in \mathcal{S}'_K} a(P)P, L/K) = \prod_{P \in \mathcal{S}'_K} (P, L/K)^{a(P)} .$$

Proposition 9.18. *The Artin map* $(*, L/K) : \mathcal{D}'_K \to G$ *is onto and the kernel contains the group* $N_{L/K}\mathcal{D}'_L$ *where* $\mathcal{D}'_L$ *is the subgroup of* $\mathcal{D}_L$ *generated by primes of* L *unramified over* K.

Proof. Let G' denote the image of $(*, L/K)$ and $M \subset L$ the fixed field of G'. If $P \in \mathcal{S}'_K$ then, by Proposition 9.11, $(P, L/K)|_M = (P, M/K)$. By definition, $(P, L/K)|_M = e$. Thus, $(P, M/K) = e$ which implies that P splits completely in M. Since $\mathcal{S}'_K$ has Dirichlet density 1, it follows from Theorem 9.13 that $M = K$. Galois theory now yields that $G' = G$, i.e., $(*, L/K)$ is onto.

If $\mathfrak{P}$ is a prime of L lying above P, then by the definition of the norm map (see the discussion following Proposition 7.6) we have $N_{L/K}\mathfrak{P} = f(\mathfrak{P}/P)P$. Thus,

$$(N_{L/K}\mathfrak{P}, L/K) = (P, L/K)^{f(\mathfrak{P}/P)} = e .$$

The last equality is a consequence of Proposition 9.10 which asserts that the Frobenius automorphism $(\mathfrak{P}, L/K)$ (and so $(P, L/K)$) has order $f(\mathfrak{P}/P)$. The second assertion of the proposition follows from this.

The exact nature of the kernel of the Artin map is a very difficult question. We first turn our attention to a much simpler question. Among abelian extensions of K the simplest are the constant field extensions. How does the general theory play out in this special case? The key to answering this question is to determine explicitly the Artin automorphism $(P, K\mathbb{E}/K)$.

Recall that $\text{Gal}(\mathbb{E}/\mathbb{F})$ is cyclic of order m generated by the automorphism ϕ_q which maps $\alpha \to \alpha^q$ for all $\alpha \in \mathbb{E}$. We have shown previously that $\text{Gal}(K\mathbb{E}/K) \cong \text{Gal}(\mathbb{E}/\mathbb{F})$. From now on we identify these two groups.

Proposition 9.19. *Let* $L = K\mathbb{E}$ *where* $\mathbb{E}$ *is an extension of* $\mathbb{F}$ *of degree* m. *Let* P *be any prime of* K. *Then* $(P, L/K) = \phi_q^{\deg_K P}$.

Proof. Every prime of K is unramified in L since L is a constant field extension. See Proposition 8.5.

Suppose $\alpha \in \mathbb{E}$. From the definition, $(P, L/K)\alpha \equiv \alpha^{NP} \pmod{\mathfrak{P}}$, where $\mathfrak{P}$ is a prime of L above P. Both sides of this congruence are in $\mathbb{E}$ and thus the difference is in $\mathbb{E} \cap \mathfrak{P} = (0)$. It follows that

$$(P, L/K)\alpha = \alpha^{NP} \quad \forall \alpha \in \mathbb{E} .$$

Now, $NP = q^{\deg_K P}$. It follows that the right-hand side of the above equality coincides with $\phi_q^{\deg_K P}(\alpha)$. Since $\alpha \in \mathbb{E}$ is arbitrary, we deduce $(P, L/K) = \phi_q^{\deg_K P}$.

Proposition 9.20. *Maintaining the notation of the previous proposition, the Artin map* $(*, L/K) : \mathcal{D}_K \to \mathrm{Gal}(L/K) \cong \mathrm{Gal}(\mathbb{E}/\mathbb{F})$ *is onto and the kernel is the group* $\mathcal{D}_K^o \mathcal{D}_K^m$*. Here,* $\mathcal{D}_K^o$ *denotes the group of divisors of degree zero.*

Proof. We already know that the map is onto. To determine the kernel we note that the Artin map is given by $(D, L/K) = \phi_q^{\deg_K D}$ for $D \in \mathcal{D}_K$. This is true for prime divisors by the previous proposition and it follows in general by linearity. From this we see $\mathcal{D}_K^o$ is in the kernel. Since $\mathrm{Gal}(L/K)$ has order m, it follows that $\mathcal{D}_K^m$ is also in the kernel. Thus, $\mathcal{D}_K^o \mathcal{D}_K^m$ is in the kernel. This group is equal to the kernel since it has index m in $\mathcal{D}_K$. This follows since $\mathcal{D}_K^o \mathcal{D}_K^m$ is the kernel of the map

$$D \to \deg_K D \pmod{m} \quad \text{from} \quad \mathcal{D}_K \to \mathbb{Z}/m\mathbb{Z} \, .$$

We note, for future reference, that $\mathcal{P}_K$, the principal divisors of K, have degree zero and are thus in the kernel of the Artin map for constant field extensions.

We can now determine the Artin L-functions associated to constant field extensions.

Proposition 9.21. *Again maintaining the notations and hypotheses of Proposition 9.19, let* χ *be an irreducible character of* $G = \mathrm{Gal}(L/K)$*. Then,* $L(s, \chi) = Z_K(\chi(\phi_q)u)$ *where, as usual,* $u = q^{-s}$*, and* $Z_K(u) = \zeta_K(s)$*.*

Proof. Since G is abelian, χ is a linear character, i.e., a homomorphism from G to $\mathbb{C}^*$. From the definitions, and Proposition 9.19,

$$
\begin{aligned}
L(s, \chi) &= \prod_{P \in \mathcal{S}_K} (1 - \chi((P, L/K)) N P^{-s})^{-1} \\
&= \prod_{P \in \mathcal{S}_K} (1 - \chi(\phi_q)^{\deg_K P} q^{-s \deg_K P})^{-1} \\
&= \prod_{P \in \mathcal{S}_K} (1 - (\chi(\phi_q)u)^{\deg_K P})^{-1} = Z_K(\chi(\phi_q)u) \, .
\end{aligned}
$$

This proposition gives a meromorphic continuation of $L(s, \chi)$ to the whole plane, which is good. However, all of these functions have poles, which seems to be bad.

This result seems troubling at first sight. If $\chi = \chi_o$, the trivial character, the Artin L-function, is the zeta function, which is as it should be. If χ is not trivial, then $L(s, \chi)$ is not a polynomial in u and in fact has poles (at $s \in \mathbb{C}$ such that $\chi(\phi_q)q^{-s} = q^{-1}$ or 1). This seems to contradict Artin's conjecture and Weil's theorem. It does not contradict Weil's theorem, Theorem 9.16B, because part of the hypothesis was that L/K be a geometric extension, i.e., L and K have the same constant fields. Artin's conjecture has to be modified in the function field case to accommodate characters belonging to constant field extensions. We will explain this more fully later.

It is worth noting that a consequence of Proposition 9.21 is that for constant field extensions $L(1,\chi) \neq 0$ if χ is linear and non-trivial. This follows because

$$L(1,\chi) = Z_K(\chi(\phi_q)q^{-1}) \neq 0 ,$$

since $\zeta_K(s)$ has no zeros on the line $\Re(s) = 1$ by Proposition 5.13, and so $Z_K(u)$ has no zeros on the circle $|u| = q^{-1}$.

Another interesting consequence, which is a special case of a far more general result (which we discussed in Remark 4 following the statement of Theorem 9.16B) is that

$$\zeta_L(s) = \prod_\chi L(s,\chi) ,$$

where the product is over all linear characters of G. This is immediate from Theorem 8.15 and Proposition 9.21.

Having investigated constant field extensions the next question is to see how they fit into the more general situation. Let L/K be a general abelian extension with $\mathbb{E}$ the constant field of L and $\mathbb{F}$ the constant field of K. Then $K\mathbb{E}$ is the maximal constant field extension of K in L. $L/K\mathbb{E}$ is a geometric function field extension.

Proposition 9.22. *Let L/K be an abelian extension of global function fields and $K\mathbb{E}$ be the maximal constant field extension of K in L. Let $G =$ $\mathrm{Gal}(L/K)$ and G' the image of $\mathcal{D}'^o_K$ under the Artin map. Then, $G' =$ $\mathrm{Gal}(L/K\mathbb{E})$. In particular, if L/K is a geometric extension, $\mathcal{D}'^o_K$ maps onto G under the Artin map.*

Proof. Let P be a prime of K which is unramified in L. By Proposition 9.11 and Proposition 9.19 we see $(P, L/K)|_{K\mathbb{E}} = (P, K\mathbb{E}/K) = \phi_q^{\deg_K P}$. By linearity, if $D \in \mathcal{D}'_K$, then $(D, L/K)|_{K\mathbb{E}} = \phi_q^{\deg_K D}$. It follows that $\mathcal{D}'^o_K$ maps to $\mathrm{Gal}(L/K\mathbb{E})$, i.e., $G' \subset \mathrm{Gal}(L/K\mathbb{E})$.

To show the Artin map from $\mathcal{D}'^o_K$ to $\mathrm{Gal}(L/K\mathbb{E})$ is onto is a little tricky. We first need a subsidiary result, namely, that $m = [K\mathbb{E} : K]$ is the greatest common divisor of the degrees of the primes in $\{L\}$, the primes of K which split completely in L. Note first that if $P \in \{L\}$ then $(P, L/K) = e$. Thus, $(P, K\mathbb{E}/K) = e$ and this occurs if and only if $m | \deg_K P$. Let m' be the greatest common divisor of the degrees of primes in $\{L\}$. We have just shown $m|m'$, and we want to show $m = m'$. To do this, consider the finite field $\mathbb{E}' \supseteq \mathbb{E}$ whose degree over $\mathbb{F}$ is m'. Every prime in $\{L\}$ splits completely in $K\mathbb{E}'$ since the degree of any such prime is divisible by m'. The field $L\mathbb{E}'$ is a Galois extension of K since it is the composite of two Galois extensions of K, L and $K\mathbb{E}'$. A moment's reflection shows that $\{L\} = \{L\mathbb{E}'\}$. By Proposition 9.13, this implies that $L = L\mathbb{E}'$, i.e., $\mathbb{E}' \subset L$. This shows that $\mathbb{E}' = \mathbb{E}$ and it follows that $m = m'$.

From what we have proven, it follows that there are primes $P_1, P_2, \cdots,$ $P_t \in \{L\}$ and integers $a_1, a_2, \cdots, a_t \in \mathbb{Z}$ such that

$$\sum_{i=1}^{t} a_i \deg_K P_i = m .$$

Set $C = \sum_{i=1}^{t} a_i P_i$. Then, $C \in \mathcal{D}'_K$ and $\deg_K C = m$. Also, $(C, L/K) = e$ since every prime in the support of C splits completely in L.

To finish the proof, choose $\sigma \in \text{Gal}(L/K\mathbb{E})$. By Proposition 9.18, $\sigma = (D, L/K)$ for some $D \in \mathcal{D}'_K$. Since σ is the identity on $K\mathbb{E}$ it follows from $e = \sigma|_{K\mathbb{E}} = (D, K\mathbb{E}/K) = \phi_q^{\deg_K D}$, that $m | \deg_K D$. Suppose $\deg_K D = km$ with $k \in \mathbb{Z}$. Then $D - kC$ has degree zero, and

$$(D - kC, L/K) = (D, L/K)(C, L/K)^{-k} = \sigma e^{-k} = \sigma .$$

The proof is complete!

Let S be a finite set of primes in a global function field K and $\mathcal{F} = \sum_{P \in S} h(P) P$ an effective divisor of K with support in S. We define the ray modulo $\mathcal{F}$, $\mathcal{P}^{\mathcal{F}}$, to be the set of principal divisors of K generated by elements $x \in K^*$ which satisfy

$$\text{ord}_P(x - 1) \geq h(P) \quad \forall P \in S .$$

Clearly, the ray modulo $\mathcal{F}$ is a subgroup of the group of principal divisors $\mathcal{P}_K$. In fact, it is a subgroup of $\mathcal{P}(S)$, the principal divisors of K whose support is disjoint from S. Let $\mathcal{D}(S) \subset \mathcal{D}_K$ be the group of divisors whose support is disjoint from S.

The ray class group modulo $\mathcal{F}$, $Cl_{\mathcal{F}}$, is defined to be the quotient $\mathcal{D}(S)/\mathcal{P}^{\mathcal{F}}$. This group is not finite. However, there is an exact sequence

$$(0) \to Cl_{\mathcal{F}}^o \to Cl_{\mathcal{F}} \to \mathbb{Z} \to (0)$$

induced by the degree map. It can be shown, using the finiteness of the divisor class group of degree zero, that $Cl_{\mathcal{F}}^o$ is a finite group. The first step in the proof (which we shall not pursue) is to show the following sequence is exact:

$$(0) \to \mathcal{P}(S)/\mathcal{P}^{\mathcal{F}} \to Cl_{\mathcal{F}}^o \to Cl_K^o \to (0) ,$$

where the map from $Cl_{\mathcal{F}}^o$ to Cl_K^o is as follows: given a ray class in $Cl_{\mathcal{F}}^o$, find a divisor representing it and map that divisor to its class in Cl_K^o. We know that this latter group is finite, so it all comes down to showing that $\mathcal{P}(S)/\mathcal{P}^{\mathcal{F}}$ is finite. This is not difficult.

The relevance of these notions comes from the following theorem, which is one form of the Artin reciprocity law.

Theorem 9.23. (E. Artin) *Let L/K be a finite abelian extension of global function fields. Let S be the set of primes' of K which are ramified in L. Then the Artin map, $(*, L/K)$, takes $D(S)$ onto $\mathrm{Gal}(L/K)$ and there is an effective divisor $\mathcal{F}$ supported on S such that the kernel of the map is $\mathcal{P}^{\mathcal{F}} N_{L/K} \mathcal{D}'_L$.*

As we have already mentioned, this is a very deep result whose proof is long and involved. We have proved a portion of the Theorem in Proposition 9.18. The Artin map is onto and the norms of divisors are contained in the kernel. The exisitence of a divisor $\mathcal{F}$ such that $\mathcal{P}^{\mathcal{F}} N_{L/K} \mathcal{D}'_L$ is the kernel is the hard part. We will not prove it here, but, accepting its truth, we will derive some consequences.

Notice that another way to state the same thing is that the Artin map takes the ray class group $Cl_{\mathcal{F}}$ onto $\mathrm{Gal}(L/F)$ and the kernel is generated by the classes of the norms of unramified primes in L. With minor modifications, the same result holds in algebraic number fields. What is required in this case is some attention to the archimedean primes. These do not exist in function fields.

How unique is the divisor $\mathcal{F}$ which plays such a major role in the Theorem? It turns out it is not unique. However, one can show that in the set of all effective divisors with the same property as $\mathcal{F}$ there is a minimum one (recall that one divisor is greater than or equal to another if their difference is effective or zero). This minimum divisor, which we continue to denote by $\mathcal{F}$, is called the conductor of L/K. Sometimes one writes this as $\mathcal{F}_{L/K}$.

Our next goal is to define Hecke L-series and then, using Artin's theorem, connect these to Artin L-series.

Let $\mathcal{F}$ be an effective divisor with support $S \subset \mathcal{S}_K$. A character of finite order on $Cl_{\mathcal{F}}$ is called a Hecke character modulo $\mathcal{F}$. There is a more general notion of Hecke character which is very important, but we will confine our attention to those which satisfy the definition just given.

Another way to phrase the definition is to say a Hecke character modulo $\mathcal{F}$ is a homomorphism from $\mathcal{D}(S) \to \mathbb{C}^*$ whose kernel is a subgroup of finite index containing the ray $\mathcal{P}^{\mathcal{F}}$.

Let λ be a Hecke character modulo $\mathcal{F}$ in the sense just given. We want to define a Hecke L-series, $L(s, \lambda)$. The definition suggests itself. Let $P \notin S$ be a prime of K. Define $\lambda(P)$ to be λ evaluated on the ray class in $Cl_{\mathcal{F}}$ containing P. Then, define

$$L(s, \lambda) = \prod_{P \notin S} (1 - \lambda(P) N P^{-s})^{-1} .$$

Since $|\lambda(P)| = 1$, one sees easily by the comparison test that $L(s, \lambda)$ converges absolutely for $\Re(s) > 1$ and for every $\delta > 0$ it converges absolutely and uniformly in the region $\Re(s) \geq 1 + \delta$. It follows that Hecke L-series are entire function of s in the region $\Re(s) > 1$. Since the terms of the product are non-vanishing in that region, the same is true for $L(s, \lambda)$. The following result is essentially due to Hecke. He proved the analogous result in the case of algebraic number fields. The details of the function field version were first worked out by F.K. Schmidt. Nowadays, one can give a uniform proof of both versions simultaneously (and for the most general Hecke characters). This was done by J. Tate [2] in his thesis. A more classical approach to the function field case can be found in Deuring [1] and Moreno [1]. We will give the proof of a special case below (Proposition 9.26).

Theorem 9.24. *Let λ be a Hecke character modulo $\mathcal{F}$ and assume that λ is not trivial on $\mathcal{D}^o(S)$. Then $L(s, \lambda)$ is an entire function of s. In fact, it is a polynomial in q^{-s}. Moreover, $L(1, \lambda) \neq 0$.*

This is actually a rough version of the full result which includes a beautiful functional equation that is satisfied when the character λ is primitive. We will briefly explain what this means and write down the functional equation.

Suppose $\mathcal{F}' \leq \mathcal{F}$ are two effective divisors. It is easy to see that there is a natural map $\pi : Cl_{\mathcal{F}} \to Cl_{\mathcal{F}'}$. If λ' is a character of $Cl_{\mathcal{F}'}$ then $\lambda = \lambda' \circ \pi$ is a character of $Cl_{\mathcal{F}}$. λ is said to be induced from λ'.

A character λ modulo $\mathcal{F}$ is said to be primitive if it is not induced from a character of any properly smaller modulus. In this case $\mathcal{F}$ is said to be the conductor of λ and we write $\mathcal{F} = \mathcal{F}_\lambda$.

Theorem 9.24A. *Let λ be a primitive Hecke character with conductor $\mathcal{F}_\lambda$ and suppose λ is not trivial on $\mathcal{D}^o(S)$. Then $L(s, \lambda)$ is a polynomial in q^{-s} of degree $2g - 2 + \deg_K \mathcal{F}_\lambda$. Define $\Lambda(s, \lambda) = q^{(g-1)s} N\mathcal{F}_\lambda^{s/2} L(s, \lambda)$. Then*

$$\Lambda(s, \lambda) = \epsilon(\lambda)\Lambda(1 - s, \bar{\lambda}) ,$$

where $\epsilon(\lambda)$ is a complex number of absolute value 1.

We are finally in a position to explain why, for linear (one dimensional) characters, Artin L-series are "the same as" Hecke L-series.

Let L/K be a finite, abelian extension of global function fields, $G = \mathrm{Gal}(L/K)$, and χ a linear character on G. We want to show that the Artin L-series $L(s, \chi)$ is equal to a Hecke L-series. As a first step, let $N_\chi \subset G$ be the kernel of χ and let $K_\chi \subset L$ be the fixed field of N_χ. It is almost immediate that $\mathrm{Gal}(K_\chi/K)$ is cyclic of order equal to the order of χ in the character group of G. We set $\mathrm{Gal}(K_\chi/K) = G_\chi$, and note that χ gives rise to a character on $G_\chi \cong G/N_\chi$. We call this character χ as well. Let $\mathcal{F}_\chi$ be the conductor of the extension K_χ/K. By Artin's theorem, Theorem 9.23, the Artin map, $(*, K_\chi/K)$, gives a homomorphism from $Cl_{\mathcal{F}_\chi}$ onto $\mathrm{Gal}(K_\chi/K)$. Call this homomorphism ρ and set $\lambda = \chi \circ \rho$. Then λ is a homomorphism from $Cl_{\mathcal{F}_\chi}$ to $\mathbb{C}^*$, i.e., a Hecke character modulo $\mathcal{F}_\chi$. It can be shown that λ is a primitive character modulo $\mathcal{F}_\chi$, i.e., $\mathcal{F}_\chi = \mathcal{F}_\lambda$. With all this in place, it now follows directly from the definitions that

$$L(s, \chi) = L(s, \lambda) .$$

This is the long awaited identification of an Artin L-series associated to a linear character with a Hecke L-series. We can now invoke Theorem 9.24 to establish the analytic continuation of $L(s, \chi)$ to the whole complex plane. It remains to explain when this continuation is entire.

Theorem 9.25. *Let L/K be a finite, abelian extension of global function fields, $G = \mathrm{Gal}(L/K)$, and χ a linear character of G. Then, $L(s, \chi)$ has an analytic continuation to an entire function in the whole complex plane if and only if K_χ/K is not a constant field extension.*

Proof. From the discussion preceding the statement of the theorem, $L(s, \chi) = L(s, \lambda)$ where $\lambda = \chi \circ \rho$. Here, $\rho : Cl_{\mathcal{F}_\chi} \to G_\chi$ and $\chi : G_\chi \to \mathbb{C}^*$. Let G'_χ be the image of $Cl^0_{\mathcal{F}_\chi}$ in G_χ. By Proposition 9.22, the fixed field of G'_χ is the maximal constant field extension of K inside of K_χ.

Since ρ is onto λ is not trivial on $Cl^0_{\mathcal{F}_\chi}$ if and only if χ is not trivial on G'_χ. Since χ is one to one on G_χ we see χ is trivial on G'_χ if and only if G'_χ is trivial and this happens if and only if K_χ is a constant field extension. Equivalently, χ is not trivial on G'_χ if and only if K_χ/K is not a constant field extension.

Thus, if K_χ/K is not a constant field extension, λ is not trivial on $Cl^0_{\mathcal{F}_\chi}$ and by Theorem 9.24 this shows $L(s, \lambda) = L(s, \chi)$ is entire. If K_χ/K is a constant field extension, Proposition 9.21 shows $L(s, \chi)$ is meromorphic, but not entire.

Theorem 9.25 gives a precise understanding of when $L(s, \chi)$ fails to satisfy Artin's conjecture in the function field case.

In the exercises we will outline the relationship of Hecke L-series to the Dirichlet L-series, which were introduced and investigated in Chapter 4. As it turns out, the latter are simply a special case of the former.

We will conclude this chapter by proving a portion of Theorem 9.24, namely for those Hecke characters belonging to the trivial modulus. In this case, the ray is just the group of principal ideals, $\mathcal{P}_K$, and the ray class group is $\mathcal{D}_K/\mathcal{P}_K = Cl_K$, the class group of K. So, we will consider L-series attached to characters of finite order of the class group of K. In this case, the analytic continuation and the functional equation follow from the Riemann-Roch theorem by using the same ideas that went into the proof of Theorem 5.9.

Proposition 9.26. (special case of Theorem 9.24A) *Let λ be a character of finite order of Cl_K and suppose that λ is not trivial on Cl^0_K. Then, $L(s, \lambda)$ is a polynomial in q^{-s} of degree $2g - 2$. Set $\Lambda(s, \lambda) = q^{(g-1)s}L(s, \lambda)$. Then, $\Lambda(s, \lambda) = \lambda(C)\Lambda(1 - s, \bar\lambda)$, where C is the canonical class of K.*

Proof. $L(s, \lambda) = \prod_P (1 - \lambda(P)NP^{-s})^{-1} = \sum_A \lambda(A)NA^{-s}$ where the product is over all primes of K and the sum is over all effective divisors of K.

Summing by degrees we find

$$L(s, \lambda) = \sum_{k=0}^{\infty} \left(\sum_{\deg_K A=k} \lambda(A) \right) q^{-ks}$$

We claim $\sum_{\deg_K A=k} \lambda(A) = 0$ for all $k > 2g - 2$. This will show that $L(s, \chi)$ is a polynomial in q^{-s} of degree at most $2g - 2$.

Assume $k > 2g - 2$ and let $\{A_1, A_2, \cdots, A_h\}$ be a set of divisors of degree k representing the divisor classes of degree k. Here, $h = h_K$ is the class number of K. For any two divisors B_1 and B_2 we will use the notation $B_1 \sim B_2$ to mean that B_1 and B_2 are linearly equivalent, i.e., $B_1 - B_2$ is principal. All sums will be over effective divisors A. We have

$$\sum_{\deg_K A=k} \lambda(A) = \sum_{i=1}^{h} \left(\sum_{A \sim A_i} \lambda(A) \right) = \frac{q^{k-g+1} - 1}{q - 1} \sum_{i=1}^{h} \lambda(A_i) .$$

We have used two facts. Since λ takes principal divisors to 1, $A \sim A_i$ implies $\lambda(A) = \lambda(A_i)$. Secondly, if $k > 2g - 2$, the number of effective divisors linearly equivalent to A_i is $(q^{k-g+1} - 1)/(q-1)$. This follows from Lemma 5.7 and the fact that $l(A_i) = k - g + 1$ since $\deg_K A_i = k > 2g - 2$ (see Theorem 5.4, Corollary 4).

Let D be a divisor of K of degree 1. Such a divisor exists by the theorem of F.K. Schmidt. Write $A_i - kD = B_i$ for each i with $1 \leq i \leq h$. The divisors $\{B_1, B_2, \cdots, B_h\}$ have degree zero and, in fact, are a set of representatives for the divisor classes of degree zero. Substituting $A_i = kD + B_i$ in the above sum, we see that

$$\sum_{\deg_K A=k} \lambda(A) = \frac{q^{k-g+1} - 1}{q - 1} \lambda(D)^k \sum_{i=1}^{h} \lambda(B_i) .$$

The latter sum is zero, since it is the sum of the character λ evaluated on all the elements of the group Cl_K^0, and by hypothesis, λ is not trivial on that group. This completes the first part of the proof.

To prove the functional equation for $L(s, \lambda)$ we first ease the notation by setting $u = q^{-s}$ and writing $L(s, \lambda) = f(u, \lambda)$. We have shown that $f(u, \lambda)$ is a polynomial in u of degree at most $2g - 2$. Now,

$$L(s, \lambda) = f(u, \lambda) = \sum_{\deg_K A \leq 2g-2} \lambda(A) u^{\deg_K A}$$

$$= \sum_{\deg_K \mathcal{A} \leq 2g-2} \lambda(\mathcal{A}) \frac{q^{l(\mathcal{A})} - 1}{q - 1} u^{\deg_K \mathcal{A}} .$$

The first sum is over effective divisors A and the second sum is over divisor classes $\mathcal{A}$. The passage from the first sum to the second uses Lemma 5.7 once again.

A simplification occurs because

$$\sum_{\deg_K \mathcal{A} \leq 2g-2} \lambda(\mathcal{A})u^{\deg_K \mathcal{A}} = 0 \ .$$

To see this, simply sum by degrees and check that each coefficient of the resulting polynomial is 0 because the sum of λ evaluated on all divisor classes of a fixed degree is 0 (one reduces to the case of divisor classes of degree zero, as above). We are thus led to the following simple expression

$$(q-1)f(u,\lambda) = \sum_{\deg_K \mathcal{A} \leq 2g-2} \lambda(\mathcal{A})q^{l(\mathcal{A})}u^{\deg_K \mathcal{A}} \ .$$

Multiply both sides by $u^{1-g} = q^{(g-1)s}$ and we get

$$(q-1)u^{1-g}f(u,\lambda) = \sum_{\deg_K \mathcal{A} \leq 2g-2} \lambda(\mathcal{A})q^{l(\mathcal{A})}u^{\deg_K \mathcal{A}-g+1} \ .$$

The key observation is that if $\mathcal{C}$ denotes the canonical class, the map $\mathcal{A} \to \mathcal{C} - \mathcal{A}$ is a permutation of the divisor classes of degree less than or equal to $2g-2$. Thus, in the last summation we can substitute $\mathcal{C} - \mathcal{A}$ for $\mathcal{A}$ and the sum remains the same. Let's investigate how the individual terms change.

The expression $\lambda(\mathcal{A})$ becomes $\lambda(\mathcal{C} - \mathcal{A}) = \lambda(\mathcal{C})\lambda(\mathcal{A})^{-1} = \lambda(\mathcal{C})\overline{\lambda(\mathcal{A})}$.

The expression $q^{l(\mathcal{A})}$ becomes $q^{l(\mathcal{C}-\mathcal{A})} = q^{g-1-\deg_K \mathcal{A}}q^{l(\mathcal{A})}$, since, by the Riemann-Roch theorem, $l(\mathcal{A}) = \deg_K \mathcal{A} - g + 1 + l(\mathcal{C} - \mathcal{A})$.

Finally, $u^{\deg_K \mathcal{A}-g+1}$ becomes $u^{g-1-\deg_K \mathcal{A}}$ since $\deg_K \mathcal{C} = 2g-2$.

Making all these substitutions in the above equation yields

$$(q-1)u^{1-g}f(u,\lambda) = \lambda(\mathcal{C}) \sum_{\deg_K \mathcal{A} \leq 2g-2} \overline{\lambda(\mathcal{A})}q^{l(\mathcal{A})}(q^{-1}u^{-1})^{\deg_K \mathcal{A}-g+1} =$$

$$(q-1)\lambda(\mathcal{C})(q^{-1}u^{-1})^{1-g}f(q^{-1}u^{-1},\bar{\lambda}) \ .$$

If we let $F(u,\lambda) = u^{1-g}f(u,\lambda)$, we have shown that $F(u,\lambda) = \lambda(\mathcal{C})F(q^{-1}u^{-1},\bar{\lambda})$. Since $F(u,\lambda) = q^{(g-1)s}L(s,\lambda) = \Lambda(s,\lambda)$, the functional equation we have proven for $F(u,\lambda)$ translates into the functional equation for $\Lambda(s,\lambda)$ given in the statement of the Proposition.

It remains to prove that $L(s,\lambda)$ is a polynomial in q^{-s} of degree $2g-2$. Rewriting the functional equation for $f(u,\lambda)$ we derive

$$u^{-(2g-2)}f(u,\lambda) = \lambda(\mathcal{C})q^{g-1}f(q^{-1}u^{-1},\bar{\lambda}) \ .$$

The constant term of $f(u,\bar{\lambda})$ is 1 (this is immediate from the definition). Thus, the right-hand side tends to $\lambda(\mathcal{C})q^{g-1}$ as $u \to \infty$. It follows that $f(u,\lambda)$ has degree $2g-2$ and also that the coefficient of the leading term is $\lambda(\mathcal{C})q^{g-1}$. Translating back to "s" language shows that $L(s,\lambda)$ is a polynomial in q^{-s} of degree $2g-2$, as asserted.

Exercises

1. Let L/K be a Galois extension of function fields with Galois group G. Let $\mathcal{D}_{L/K}$ be the different divisor. Show that $\sigma \mathcal{D}_{L/K} = \mathcal{D}_{L/K}$ for all $\sigma \in G$.

2. Let L/K be a Galois extension of function fields with Galois group G. Suppose there is a prime P of K which is inert in L; i.e., there is a prime $\mathfrak{P}$ in L, lying above P, such that $f(\mathfrak{P}/P) = [L : K]$. Show that G is cyclic.

3. (Continuation). Conversely, if G is cyclic, show that there exist infinitely many primes P in K which are inert in L. What is the density of this set of primes?

4. Suppose that E/K is a geometric and separable extension of function fields. Let L be the smallest Galois extension of K containing L. Show by example that the constant field of L may be larger than the constant field of K.

5. Let E_1 and E_2 be two finite Galois extensions of a function field K. Suppose that there is a prime P of K which is totally ramified in E_1 and unramified in E_2. Show that $E_1 E_2/E_2$ is Galois with group isomorphic to $\mathrm{Gal}(E_1/K)$ and that every prime in E_2 lying above P is totally ramified in $E_1 E_2$.

6. Let L/K be a Galois extension of function fields with Galois group G. Let N be a normal subgroup of G and L' the fixed field of N. Let $\mathfrak{P}$ be a prime of L and P the prime of K lying below $\mathfrak{P}$. If $I(\mathfrak{P}/P) \subseteq N$, show that P is unramified in L'. If $Z(\mathfrak{P}/P) \subseteq N$, show that P splits completely in L'.

7. Suppose L/K is a Galois extension of function fields and that $\mathfrak{P}$ is a prime of L. If $a \in L^*$, and $\sigma \in \mathrm{Gal}(L/K)$, show $\mathrm{ord}_{\sigma\mathfrak{P}}(a) = \mathrm{ord}_{\mathfrak{P}}(\sigma^{-1}a)$. In particular, if $\mathfrak{P}$ is fixed by $\mathrm{Gal}(L/K)$, then for any $a \in L^*$, all the conjugates of a have the same order at $\mathfrak{P}$.

8. Let L/K be a Galois extension of function fields with Galois group G. Let $\mathfrak{P}$ be a prime of L and P the prime of K lying below it. We assume that $G = Z(\mathfrak{P}/P)$ (if this isn't true, simply replace K by the fixed field of $Z(\mathfrak{P}/P)$). Define subsets of G as follows: $G_m = \{\sigma \in G \mid \mathrm{ord}_{\mathfrak{P}}(\sigma a - a) \geq m+1, \forall a \in O_{\mathfrak{P}}\}$. Show that these sets are normal subgroups of G. Note that $G_{-1} = G = Z(\mathfrak{P}/P)$ and $G_0 = I(\mathfrak{P}/P)$.

9. (Continuation) We wish to study the structure of $G_0 = I(\mathfrak{P}/P)$. We can replace K with the fixed field of $I(\mathfrak{P}/P)$. Once this is done, we can assume $\mathfrak{P}$ is totally ramified over P. If Π is a uniformizing parameter at $\mathfrak{P}$, it can be shown that $O_{\mathfrak{P}}$ is free as a module over O_P

with a basis $\{1, \Pi, \Pi^2, \ldots, \Pi^{e-1}\}$. Using this, show for each $m > 0$ that $G_m = \{\sigma \in G_0 \mid \text{ord}_{\mathfrak{P}}(\sigma\Pi - \Pi) \geq m + 1\}$. Also show that $G_m = (e)$ for all sufficiently large integers m.

10. (Continuation) Define $U^{(m)} = \{u \in O_{\mathfrak{P}}^* \mid \text{ord}_{\mathfrak{P}}(u - 1) \geq m\}$. For $m \geq 0$ we define maps $\rho_m : G_m/G_{m+1} \to U^{(m)}/U^{(m+1)}$ by sending $\sigma \in G_m$ to the residue class of $\sigma\Pi/\Pi$ in $U^{(m)}/U^{(m+1)}$. Show that ρ_m is independent of the choice of uniformizing parameter. Show that ρ_m is a homomorphism and that it is one to one.

11. (Continuation) Let $F_{\mathfrak{P}} = O_{\mathfrak{P}}/\mathfrak{P}$. Show that there is a monomorphism from $U/U^{(1)}$ to $F_{\mathfrak{P}}^*$ and monomorphisms from $U^{(m)}/U^{(m+1)}$ to $F_{\mathfrak{P}}$ for all $m \geq 1$. Deduce that G_0/G_1 is cyclic of order prime to p=char $\mathbb{F}$, and that G_1 is a p-group, in fact, the unique p-Sylow subgroup of $I(\mathfrak{P}/P)$.

12. Let L be a function field over a constant field F and suppose that σ is an automorphism of L which is the identity on F. Let $\mathfrak{P}$ be a prime of L. Then, σ induces a continuous map from $O_{\sigma^{-1}\mathfrak{P}} \to O_{\mathfrak{P}}$ which extends to an isomorphism (which we continue to call σ) from $\hat{L}_{\sigma^{-1}\mathfrak{P}} \to \hat{L}_{\mathfrak{P}}$. Define $\tilde{\sigma} : A_L \to A_L$ by $\tilde{\sigma}(a_{\mathfrak{P}}) = (b_{\mathfrak{P}})$ where $b_{\mathfrak{P}} = \sigma a_{\sigma^{-1}\mathfrak{P}}$ for all primes $\mathfrak{P}$ of L. Show that $\tilde{\sigma}$ is a ring automorphism of A_L and that its restriction to L is σ.

13. Let L/K be a finite separable extension of function fields. In Chapter 7, we defined a trace map $tr_{L/K} : A_L \to A_K$. We now define a map in the other direction, $i_{L/K} : A_K \to A_L$ sending the adele (a_P) to the adele $(b_{\mathfrak{P}})$ whose $\mathfrak{P}$-th coordinate is a_P for every prime $\mathfrak{P}$ lying over P. Show that $i_{L/K}$ is a one-to-one ring homomorphism which sends K to L.

14. (Continuation) Show that $tr_{L/K} \circ i_{L/K}$ is multiplication by $n = [L : K]$. If n is not divisible by the characteristic of L, conclude that $tr_{L/K}$ is onto.

15. (Continuation) Suppose L/K is a Galois extension of function fields with Galois group G. Show that the adeles of L which are fixed by G, A_L^G, are equal to $i_{L/K}A_K$.

16. Let L/K be a Galois extension of function fields and σ an automorphism of L which leaves the constant field fixed. If $\omega \in \Omega_L$ is a differential, define $\sigma\omega$ by $\sigma\omega(a) = \omega(\sigma^{-1}a)$ for all $a \in A_L$. If ω vanishes on $A_L(D)$ for a divisor D, show that $\sigma\omega$ vanishes on $A_L(\sigma D)$. Use this to prove that $\sigma\omega$ is a differential. If (ω) is the divisor of ω show that the divisor of $\sigma\omega$ is $\sigma(\omega)$.

17. (Continuation) Let O denote the zero divisor and $\Omega_K(O)$ the space of holomorphic differentials. Show σ maps $\Omega_L(O)$ into itself. Assume

that $g \geq 2$, where g is the genus of L. Also assume that σ is non-trivial and of finite order prime to the characteristic of L. Show that the action of σ on $\Omega_L(O)$ is non-trivial. (Hint: Let K be the fixed field of σ. If σ acts trivially on $\Omega_L(O)$, show the map $\omega \to \omega^* = \omega \circ tr_{L/K}$ is an isomorphism between $\Omega_K(O)$ and $\Omega_L(O)$. Conclude that $g_L = g_K$. This contradicts the Riemann-Hurwitz formula). The proof outlined here is due to R. Accola.

18. Let $k = \mathbb{F}(T)$ be the rational function field and $A = \mathbb{F}[T]$ be the ring of polynomials. Let $m \in A$, $m \notin \mathbb{F}^*$, and suppose $m = \alpha P_1^{a_1} P_2^{a_2} \ldots P_t^{a_t}$ is its prime decomposition. Each P_i corresponds to a prime $\mathfrak{P}_i$ of k. Let $M = \sum a_i \mathfrak{P}_i$ be the effective divisor of k corresponding to m. We set $M_\infty = M + \infty$. Show that $Cl^0_{M_\infty} \cong (A/mA)^*$ and deduce that a ray class character modulo M_∞, restricted to the divisors of degree zero, is the same as a Dirichlet character modulo m.

10
Artin's Primitive Root Conjecture

By now we have developed a lot of foundational material about the arithmetic of function fields. In this chapter we will put this material to work and give the beautiful proof, due to H. Bilharz, of E. Artin's conjecture about primitive roots in function fields.

The work we will describe is the PhD thesis of Bilharz, who wrote the thesis under the direction of H. Hasse. His paper appeared in 1937 (see Bilharz [1]).

Bilharz dates the origin of the conjecture very precisely. He claims Artin made his conjecture in a private conversation with Hasse which took place on September 12, 1927. Artin considered an integer $a \in \mathbb{Z}$ which is not in the set $\{0, 1, -1\}$. Let $\mathcal{M}_a$ be the set of primes, not dividing a, for which a is a primitive root. Does this set have a Dirichlet density and if so can a formula be found for it? On heuristic grounds, Artin conjectured that the density was

$$\delta(\mathcal{M}_a) = \prod_{l \notin S_a} \left(1 - \frac{1}{l(l-1)}\right) \prod_{l \in S_a} \left(1 - \frac{1}{l-1}\right),$$

where the first product is over all primes for which a is not an l-th power in $\mathbb{Q}$ and the second over the finitely many primes (maybe the empty set) for which a is an l-th power. The first product is convergent and, since all the terms are non-zero, so is the product. The second term is zero if and only if $2 \in S_a$, i.e. a is a square in $\mathbb{Q}$. Thus, assuming this formula is correct, it follows that if a is not $0, \pm 1$, and not a square, then there are infinitely many primes for which a is a primitive root.

Artin's formula is not correct as it stands, but it may be modified slightly to give what is believed to be the correct result . The qualitative consequence described above remains unaffected by this. It is this latter statement that is known as Artin's conjecture on primitive roots. It remains open to this day. In 1967, C. Hooley gave a conditional proof of the conjecture, with the correct formula for the Dirichtlet density, by assuming the truth of the generalized Riemann hypothesis for a certain set of algebraic number fields (Hooley [1]). Even without the Riemann hypothesis, great progress has been made in recent years by R. Gupta, M. Ram Murty, and D.R. Heath-Brown (see the survey article of M. Ram Murty [1]).

The thesis problem of Bilharz was to formulate the primitive root conjecture in global function fields and give a proof of it in this context. He did this brilliantly except that his proof was conditional on the truth of the Riemann hypothesis for global function fields. In 1948, Weil published his proof of this result and one consequence was that the Artin conjecture on primitive roots was no longer a conjecture, but a theorem, in the function field context.

Let us fix a global function field K with constant field $\mathbb{F}$ having q elements. Let $a \in K^*$ and $P \in S_K$, a prime of K which is prime to a. We say that a is a primitive root modulo P if its residue class in $(O_P/P)^*$ has order $NP - 1$, i.e. it is a cyclic generator of $(O_P/P)^*$. If $a \in \mathbb{F}^*$ its order divides $q - 1$ and thus a can be a primitive root only for the finitely many primes of degree 1. We assume from now on that $a \in K^*$, but not in $\mathbb{F}^*$. The following simple lemma is crucial to what follows.

Lemma 10.1. *Let P be a prime of K not containing $a \in K^*$. Then, a is a primitive root modulo P if and only if there is no prime $l \in \mathbb{Z}$ satisfying both of the following conditions:*

$$i)\ \ NP \equiv 1 \pmod{l} \quad \text{and} \quad ii)\ \ a^{\frac{NP-1}{l}} \equiv 1 \pmod{P}\ .$$

Proof. If there is a prime l satisfying both conditions, then the order of a modulo P divides $(NP - 1)/l$, so that a cannot be a primitive root. So, if a is a primitive root, there is no prime l for which both conditions hold.

Now, suppose there is no prime l for which both conditions hold and let h be the order of a modulo P. We claim that $h = NP - 1$. If not, there is a prime l dividing $(NP - 1)/h$. In this case, h divides $(NP - 1)/l$ and so both conditions of the lemma are satisfied, which is a contradiction. Thus, $h = NP - 1$ and a is a primitive root modulo P.

We can assume from now on that $l \neq p$ since condition $i)$ of the lemma never holds for the characteristic p of $\mathbb{F}$. For each prime $l \neq p$ in $\mathbb{Z}$ let ζ_l be a primitive l-th root of unity in a fixed algebraic closure of K. We define an extension K_l of K to be the field obtained by adjoining ζ_l and any l-th root of a, $\sqrt[l]{a}$, to K, i.e., $K_l = K(\zeta_l, \sqrt[l]{a})$. Since K_l is the splitting field of the separable polynomial $X^l - a$ over K, K_l is a Galois extension of K.

We will show in a while that both conditions of Lemma 9.1 are satisfied if and only if P splits completely in K_l. This will tie our discussion in with some of the material developed in the last chapter. Before doing so we have to take a short detour to discuss cyclotomic and Kummer extensions in function fields.

Proposition 10.2. *Let $L = K(\zeta_l)$. Then $[L : K] = f(l)$, the smallest positive integer f such that $q^f \equiv 1 \pmod{l}$. A prime $P \in S_K$ splits completely in L if and only if $NP \equiv 1 \pmod{l}$.*

Proof. Since $L = K\mathbb{F}(\zeta_l)$ it is a constant field extension and $[L : K] = [\mathbb{F}(\zeta_l) : \mathbb{F}]$ by Proposition 8.1. Now, $\mathrm{Gal}(\mathbb{F}(\zeta_l) : \mathbb{F})$ is generated by ϕ_q the automorphism that takes an element to its $q-th$ power. Thus, $\phi^f(\zeta_l) = \zeta_l^{q^f}$. It follows that ϕ_l^f fixes ζ_l if and only if $q^f \equiv 1 \pmod{l}$. It follows that the order of ϕ_l is $f(l)$ since ζ_l generates $\mathbb{F}(\zeta_l)$ over $\mathbb{F}$. This proves $[L : K] = f(l)$.

Since L/K is a constant field extension, every prime of K is unramified in L by Proposition 8.5. If $\mathfrak{P}$ is a prime of L above P, then by definition $(P, L/K)\zeta_l \equiv \zeta_l^{NP} \pmod{\mathfrak{P}}$. Both sides of this congruence are constants, so we must have equality. Thus,

$$(P, L/K)(\zeta_l) = \zeta_l^{NP} \, ,$$

and it follows that $(P, L/K)$ is the identity on $\mathbb{F}(\zeta_l)$ if and only if $NP \equiv 1 \pmod{l}$. This proves the second assertion of the proposition.

A more elementary proof of the second part of the proposition can be obtained by using the corollary to Proposition 8.13.

Our next task is to investigate extensions of the type $K(\sqrt[l]{a})/K$. These are called Kummer extensions. We want to know which primes split completely, which primes ramify, and also a formula for the genus of $K(\sqrt[l]{a})$ if we know the genus of K. The answers to these questions are given in the next three propositions.

Proposition 10.3. *Let K be a function field over a constant field F of characteristic p. Let l be a prime number not equal to p and $a \in K^*$, not an l-th power in K^*. Let α be a root of $X^l - a = 0$ and $L = K(\alpha)$. A prime P of K is ramified in L if and only if l does not divide $\mathrm{ord}_P(a)$. If P is ramified, it is totally ramified, i.e., there is only one prime $\mathfrak{P}$ above it in L and $e(\mathfrak{P}/P) = l$.*

Proof. Note to begin with that since $l \neq p$, $X^l - a$ is a separable polynomial and so L is a separable extension of K. Also, since a is not an l-th power, $x^l - a$ is irreducible (see Lang [4]) and so $[L : K] = l$.

Suppose first that $l \mid \mathrm{ord}_P(a)$. Let π be a uniformizing parameter in the valuation ring of P, O_P. Then, $a = \pi^{lh}u$, where $h \in \mathbb{Z}$ and u is a unit in O_P. Thus, $\alpha/\pi^h = \mu$ is an $l - th$ root of u and $L = K(\alpha) = K(\mu)$. Let R_P be the integral closure of O_P in L. We claim $\{1, \mu, \mu^2, \cdots, \mu^{l-1}\}$ is a basis of R_P over O_P. It is certainly a basis of L/K since $X^l - u$ is irreducible. Suppose $\beta \in R_P$. Then

$$\beta = \sum_{i=0}^{l-1} c_i \mu^i \quad \text{with} \quad c_i \in K .$$

By the usual argument, the coefficients c_i will be in O_P if we can show that the determinant of the matrix $(\mathrm{tr}_{L/K}(\mu^i \mu^j))$ is a unit in O_P. We claim $\mathrm{tr}(\mu^h) = 0$ unless $l | h$, in which case the answer is $l u^{h/l}$. This is because μ^h is a root of $X^l - u^h$, which is irreducible when u^h is not an l-th power, i.e., when h is not divisible by l. If $l | h$, then $\mu^h = u^{h/l}$ and the result is clear.

We now have enough information to show easily

$$\det(\mathrm{tr}_{L/K}(\mu^i \mu^j)) = \pm l^l u^{l-1} .$$

This is a unit in O_P since $l \neq p$ and u is a unit in O_P. This shows, simultaneously, that $\{1, \mu, \cdots, \mu^{l-1}\}$ is an integral basis for R_P/O_P and that the discriminant $\mathfrak{d}_{R_P/O_P} = O_P$. From Proposition 7.9, we conclude that P is unramified in L.

Now suppose that l does not divide $\mathrm{ord}_P(a)$. Let $\mathfrak{P}$ be a prime above P in L. Since $\alpha^l = a$, we have

$$l \, \mathrm{ord}_{\mathfrak{P}}(\alpha) = \mathrm{ord}_{\mathfrak{P}}(a) = e(\mathfrak{P}/P) \mathrm{ord}_P(a) .$$

This implies that $l | e(\mathfrak{P}/P)$. By Proposition 7.1, $e(\mathfrak{P}/P) \leq l$. This shows that $e(\mathfrak{P}/P) = l$ and so P is totally ramified as claimed.

Before stating the next proposition we pause to give a somewhat technical definition which will be useful here and later. Let K/F be a function field with the constant field of characteristic p, possibly zero. An element $a \in F^*$ is said to be geometric at a prime $l \neq p$ if $K(\sqrt[l]{a})$ is a geometric field extension of K; i.e., the constant field of $K(\sqrt[l]{a})$ is F. Here $\sqrt[l]{a}$ is some root of $X^l - a = 0$ in an extension of K. The definition does not depend on which root is chosen (in a given algebraic closure of K). It is an exercise to show that a is geometric at l unless it has the form μb^l where $\mu \in F^* - F^{*l}$ and $b \in K^*$. One way is clear. If a has this form, then $K(\sqrt[l]{a}) = K(\sqrt[l]{\mu})$ which is certainly a constant field extension.

Proposition 10.4. *Let K/F be a function field, with constant field F of characteristic p (possibly, $p = 0$). Let $l \neq p$ be a prime and $L = K(\alpha)$ where $\alpha^l = a \in K^*$. Assume that a is geometric at l and that a is not an l-th power in K^*. Then, $2g_L - 2 = l(2g_K - 2) + R_a(l - 1)$ where R_a is the sum of the degrees of the finitely many primes P of K where $\mathrm{ord}_P(a)$ is not divisible by l.*

Proof. This is an application of the Riemann-Hurwitz Theorem, Theorem 7.16, which asserts that

$$2g_L - 2 = [L : K](2g_K - 2) + \deg_L \mathfrak{D}_{L/K} .$$

Here, L/K is presumed to be a finite, separable, geometric extension of function fields and $\mathfrak{D}_{L/K}$ is the different divisor of the extension. These hypotheses apply to $L = K(\alpha)$, given our assumptions. If $\mathfrak{P}$ is a ramified prime of L lying above P in K, by Proposition 10.3 we must have $e(\mathfrak{P}/P) = l$, $f(\mathfrak{P}/P) = 1$, and $g(\mathfrak{P}/P) = 1$.

We have to figure out the quantities on the right-hand side of the equation. We already know $[L : K] = l$ since a is not an l-th power in K. Since $p \neq l$, by Corollary 2 to Lemma 7.10, the coefficient of a ramified prime $\mathfrak{P}$ in the different is $e(\mathfrak{P}/P) - 1 = l - 1$. Since $\mathfrak{P}$ is totally ramified, $f(\mathfrak{P}/P) = 1$ and so $\deg_L \mathfrak{P} = \deg_K P$. Thus, the degree of the different is just $(l - 1)$ times the sum of the K-degrees of the primes P of K with $\mathrm{ord}_P(a)$ not divisible by l (again using Proposition 10.3). This sum is R_a, by definition, so the proof is complete.

It is worthwhile noticing that if one fixes the base field K and an element $a \in K^*$ satisfying the hypotheses of Proposition 10.4, then, as l varies over the prime numbers, the genus of $K(\sqrt[l]{a})$ is a linear function of l. This observation will be of use later.

An interesting special case is to take $K = F(T)$, the rational functional field, $f(T) \in F[T]$ a square-free polynomial of degree N, and $L = K(\sqrt{f(T)})$ (assuming that char $(F) \neq 2$). A calculation, using the proposition (and not forgetting the prime at infinity) yields the fact that the genus of L is $(N - 1)/2$ if N is odd and $(N/2) - 1$ if N is even.

Proposition 10.5. *Let K be a global function field over a constant field $\mathbb{F}$ with q elements. Let l be a prime different from the characteristic of $\mathbb{F}$. Let $a \in K^*$. Assume that K contains a primitive l-th root of unity, ζ_l. Suppose that P is a prime of K and that $\mathrm{ord}_P(a) = 0$. Then, P splits completely in $L = K(\sqrt[l]{a})$ if and only if*

$$a^{\frac{NP-1}{l}} \equiv 1 \pmod{P} .$$

Proof. Since $\zeta_l \in K$, the extension L/K is a cyclic Galois extension. Also, we must have $\zeta_l \in \mathbb{F}^*$, which implies $q \equiv 1 \pmod{l}$ and so $NP \equiv 1 \pmod{l}$ for all primes P of K.

P is unramified in L if $\mathrm{ord}_P(a) = 0$ by Proposition 10.3. Thus, the Artin automorphism $(P, L/K)$ is defined. The order of this automorphism is $f(\mathfrak{P}/P)$ where $\mathfrak{P}$ is any prime of L lying over P. Thus, P splits completely if and only if $(P, L/K)$ is the identity automorphism.

Any two roots of $X^l - a = 0$ differ by an l-th root of unity. Let α be any root of this equation. Then $(P, L/K)\alpha$ is another root. Thus $(P, L/K)\alpha/\alpha$ is an l-th root of unity which is easily seen to depend only on a and not on α. The usual notation for this l-root of unity is $(a/P)_l$, the l-th power residue symbol. We have

$$(a/P)_l\alpha = (P, L/K)\alpha \equiv \alpha^{NP} \pmod{\mathfrak{P}} ,$$

where $\mathfrak{P}$ is any prime of L lying above P. By hypothesis $a \notin P$, so we can divide this congruence by α to obtain

$$(a/P)_l \equiv \alpha^{NP-1} \equiv a^{\frac{NP-1}{l}} \pmod{\mathfrak{P}} .$$

Since $\zeta_l \in K$ we can conclude that

$$(a/P)_l \equiv a^{\frac{NP-1}{l}} \pmod{P} .$$

Thus, if P splits completely, $(P, L/K)$ is the identity, which implies $(a/P)_l = 1$ and the above congruence shows $a^{NP-1/l} \equiv 1 \pmod{P}$. Conversely, if $a^{NP-1/l} \equiv 1 \pmod{P}$, then $(a/P)_l \equiv 1 \pmod{P}$. Since both sides are constants, they must be equal; i.e., $(a/P)_l = 1$. This implies $(P, L/K)$ is the identity (since α generates L) and so, P splits completely.

Notice that the conclusion is true even if a is an l-th power in K^*. Of course, everything is trivial in this case.

Proposition 10.6. *Let K be a global function field with constant field $\mathbb{F}$. Let l be a prime different from the characteristic of $\mathbb{F}$. Let $a \in K^*$. Let $E_l = K(\zeta_l, \sqrt[l]{a})$. Let P be a prime of K such that $\mathrm{ord}_P(a) = 0$. Then, P splits completely in L if and only if*

$$NP \equiv 1 \pmod{l} \quad and \quad a^{\frac{NP-1}{l}} \equiv 1 \pmod{P} .$$

Proof. Consider the tower of fields $K \subseteq K(\zeta_l) \subseteq E_l$. A prime of K splits completely in L if and only if it splits completely in $K(\zeta_l)$ and every prime above it in $K(\zeta_l)$ splits completely in E_l.

By Proposition 10.2, P splits in $K(\zeta_l)$ iff $NP \equiv 1 \pmod{l}$. Let $\mathfrak{P}$ be a prime of $K(\zeta_l)$ lying above P. We apply Proposition 10.5 to $E_l/K(\zeta_l)$; i.e., in that proposition we replace K by $K(\zeta_l)$ and L by E_l. It follows that $\mathfrak{P}$ splits completely in E_l if and only if

$$a^{\frac{N\mathfrak{P}-1}{l}} \equiv 1 \pmod{\mathfrak{P}} .$$

Since both sides of this congruence are in O_P, we may replace the modulus with P. Also, if P splits in $K(\zeta_l)$, then $N\mathfrak{P} = NP$; so the condition is

$$a^{\frac{NP-1}{l}} \equiv 1 \pmod{P} .$$

This completes the proof.

We now return to the original problem about Artin's primitive root conjecture in the function field case. By combining Propositions 10.1 and 10.6 we see that a is a primitive roots modulo a prime P if and only if $\mathrm{ord}_P(a) = 0$ and P does not split completely in any of the fields $E_l = K(\zeta_l, \sqrt[l]{a})$, where l runs through all the primes different from the characteristic of $\mathbb{F}$. Let $\mathcal{M}_a$

denote the set of primes in K which do not split completely in any of the fields E_l. $\mathcal{M}_a$ differs from the set of primes for which a is a primitive root by at most the finitely many primes P with $\text{ord}_P(a) \neq 0$. In particular, if one of the sets has Dirichlet density, both do, and the Dirichlet densities are equal. Bilharz's proof proceeds by showing that the Dirichlet density of $\mathcal{M}_a$, $\delta(\mathcal{M}_a)$, exists and is non-zero with the exception of some very special circumstances for which a cannot be a primitive root for infinitely many primes. We will make the simplifying assumption that a is geometric. We remind the reader that this means that for all primes $l \neq p$, the field extension $K(\sqrt[l]{a})/K$ is geometric. If this assumption is not made, there is a gap in Bilharz's proof of his theorem. This observation was also made by J. Yu, who has filled in the gap in a paper which is to appear, Yu [2]. By assuming a is geometric we avoid these difficulties and lose nothing essential about the original proof.

The outline of the proof is clear and elegant, but the details are somewhat complicated. We will begin by sketching the outline of the proof and, afterwards, go back and fill in the details.

For any field extension L/K, recall that $\{L\}$ denotes the set of primes of K which split completely in L. Note that $\{K\} = S_K$ in our previous notation.

We will need the following key result.

Proposition 10.7. *Assume a is geometric and not an l-th power in K. Let m be a square-free integer prime to q and E_m the compositum of the fields E_l for all $l|m$. Let $f(m)$ denote the order of q modulo m. Then, $[E_m : K] = m_a f(m)$, where m_a is the product of the primes l dividing m for which a is not an l-th power. The Dirichlet density of the set of primes which do not split completely in any of the fields E_l with $l|m$ is given by*

$$\sum_{d|m} \frac{\mu(d)}{d_a f(d)} \cdot$$

Proof. The field E_m contains ζ_m, a primitive m-th root of unity.

Let $l|m$. We claim a is an l-th power in K if and only if it is an l-th power in $K(\zeta_m)$. One way is obvious, so suppose a is an l-th power in $K(\zeta_m)$. This implies that $K(\sqrt[l]{a}) \subseteq K(\zeta_m)$. A subfield of a constant field extension is a constant field extension. Since, by assumption, a is geometric, this can only happen if $K(\sqrt[l]{a}) \subseteq K$, i.e., when a is an l-th power in K.

Clearly, E_m is the compositum of the field extensions $K(\zeta_m, \sqrt[l]{a})/K(\zeta_m)$ as l runs through those primes dividing m for which a is not an l-th power in $K(\zeta_m)$. By the last paragraph, these primes are the same as those for which a is not an l-th power in K. Since all these field extensions are cyclic of prime order, for distinct primes, we conclude that $[E_m : K(\zeta_m)] = m_a$.

To finish the proof that $[E_m : K] = m_a f(m)$ it remains to show that $[K(\zeta_m) : K] = f(m)$. Since $[K(\zeta_m) : K] = [\mathbb{F}(\zeta_m) : \mathbb{F}]$, this follows in the usual way by computing the order of the finite field Frobenius automorphism ϕ_q.

To compute the density of the primes $\mathcal{M}_{(m)}$ which do not split completely in E_m, we resort to the inclusion-exclusion principle of set theory. Let $\{l_1, l_2, \cdots, l_t\}$ be the primes dividing m. Applied to our situation, the inclusion-exclusion principle yields the following expression for the set in question:

$$\mathcal{M}_{(m)} = \{K\} - \bigcup_i \{E_{l_i}\} + \bigcup_{i,j} \{E_{l_i}\} \cap \{E_{l_j}\} - \cdots \text{ etc.}$$

For any subset $T \subset \{K\}$ define

$$\delta(T, s) = \frac{\sum_{P \in T} NP^{-s}}{\log \zeta_K(s)} .$$

It is clear that $\delta(T, s)$ is well defined for $s > 1$. From our discussion of Dirichlet density in Chapter 9 (immediately after Proposition 9.11) we see that $\lim_{s \to 1+} \delta(T, s) = \delta(T)$.

The above set theoretic-expression for $\mathcal{M}_{(m)}$ yields the following identity on the level of functions:

$$\delta(\mathcal{M}_{(m)}, s) = \delta(\{K\}, s) - \sum_i \delta(\{E_{l_i}\}, s) + \sum_{i,j} \delta(\{E_{l_i}\} \cap \{E_{l_j}\}, s) - \cdots \text{ etc.}$$

Taking the limit as $s \to 1^+$, we see

$$\delta(\mathcal{M}_{(m)}) = 1 - \sum_i \delta(\{E_{l_i}\}) + \sum_{i,j} \delta(\{E_{l_i}\} \cap \{E_{l_j}\}) - \cdots \text{ etc.}$$

In a finite set of Galois extensions the set of primes which split completely in the compositum is the intersection of the sets of primes which split completely in the individual extensions. Using this, the above expression, Theorem 9.13, and the computation of $[E_m : K]$ in the first part of the proof yields

$$\delta(\mathcal{M}_{(m)}) = \sum_{d|m} \frac{\mu(d)}{d_a f(d)} ,$$

as asserted.

Let $l_1, l_2, l_3, \cdots$ be an enumeration of the primes different from p, the characteristic of $\mathbb{F}$. Let $m_n = l_1 l_2 \cdots l_n$ and let E_{m_n} be the compositum of the fields E_l with $l | m_n$, i.e. $l \in \{l_1, l_2, \cdots, l_n\}$. Let $\mathcal{M}_n$ be the set of primes which do not split completely in any of the field E_{l_i} with $1 \le i \le n$ (in the notation of the above proof, $\mathcal{M}_n = \mathcal{M}_{(m_n)}$). Note that $\mathcal{M}_n \supset \mathcal{M}_{n+1} \supset \mathcal{M}_a$ and that $\bigcap_n \mathcal{M}_n = \mathcal{M}_a$. We have just shown that

$$\delta(\mathcal{M}_n) = \sum_{d|m_n} \frac{\mu(d)}{d_a f(d)} .$$

It is tempting to just pass to the limit as $n \to \infty$ to get Bilharz's Theorem (Theorem 10.19 below). In fact, that is how the theorem is proved, but the passage to the limit requires a rather elaborate justification. An important part of the justification is played by the following theorem, whose proof we postpone.

Theorem 10.8. (Romanoff) *Let $q > 1$ be an integer and for any integer m relatively prime to q, let $f(m)$ be the order of q modula m. Then, the series*

$$\sum_m \frac{1}{mf(m)}$$

converges. The sum is over all square-free integers m relatively prime to q.

We can now state the principal result of this chapter.

Theorem 10.9. (Bilharz) *Assume a is geometric element of K^*. Then, with the above notations, the Dirichlet density of $\mathcal{M}_a$ exists and is given by*

$$\delta(\mathcal{M}_a) = \sum_{\substack{m=1 \\ (m,p)=1}}^{\infty} \frac{\mu(m)}{m_a f(m)} \, .$$

The sum is easily seen to be absolutely convergent using Romanoff's theorem.

As we have already pointed out, Bilharz does not make the restriction that a be geometric. His proof seems to contain a small error involving the computation of the degree $[E_m : K]$. This problem has recently been corrected by J. Yu, but we will be content with proving the theorem as stated.

Before going on to the proof, we discuss the consequences of Bilharz's theorem. The main difficulty is to determine when $\delta(\mathcal{M}_a)$ is zero and when it is not. The above expression for $\delta(\mathcal{M}_a)$ as sum does not immediately resolve this problem. One needs the following special case of a result of H. Heilbronn, whose proof we also postpone.

Proposition 10.10. *With the same notations as Theorem 10.9,*

$$\sum_{m=1}^{\infty} \frac{\mu(m)}{m_a f(m)} \geq \prod_{l \neq p} \left(1 - \frac{1}{l_a f(l)}\right) \, .$$

Let S denote the set of primes $l \neq p$ for which a is an l-th power. S is a finite set, possibly empty. We can rewrite the right-hand side of the equation in Proposition 10.10 as follows:

$$\prod_{l \notin S} \left(1 - \frac{1}{l f(l)}\right) \prod_{l \in S} \left(1 - \frac{1}{f(l)}\right) \, .$$

The first (infinite) product converges and is not zero since the individual factors are non-zero and the product converges, since $\sum_{l\notin S} 1/lf(l)$ converges by Romanoff's theorem. The second (finite) product is zero if and only if $f(l) = 1$ for some prime $l \in S$. In other words, the whole product is non-zero unless there is a prime l dividing $q - 1$ for which a is an l-th power. This leads to the following theorem.

Theorem 10.11. (Bilharz) *Let a be a geometric element of K^*. Then there are infinitely many primes $P \in \mathcal{S}_K$ for which a is a primitive root provided that there is no prime divisor l of $q - 1$ for which a is an l-th power. If there is such a prime divisor, then a is not a primitive root of any prime $P \in \mathcal{S}_K$.*

Proof. We have just shown on the basis of Theorem 10.9 and Proposition 10.10 that if a is not an l-th power for some prime $l|\,q-1$, then $\delta(\mathcal{M}_a) \neq 0$. Since we pointed out earlier that $\mathcal{M}_a$ differs from the set of primes for which a is a primitive root by a finite set, it follows that the latter set has non-zero Dirichlet density and so must be infinite.

If $l|\,q - 1$ and $a = b^l$ for some $b \in K^*$, then for any prime P of K not dividing a we have (using $l|\,q - 1 \mid NP - 1$)

$$a^{\frac{NP-1}{l}} \equiv b^{NP-1} \equiv 1 \pmod{P}.$$

Thus, a is not a primitive root for any prime P of K.

We now turn to the proof of Theorem 10.9. We first show how to reduce the proof of the theorem to the proof that a certain infinite sum of functions converges uniformly. We then prove that assertion. This is the hardest part of the proof. After that we give a proof of Romanoff's Theorem. Finally, we prove Heilbronn's result, Proposition 10.10.

Suppose $a \in K^*$ is a fixed geometric element. Recall that $\mathcal{M}_n$ is the set of primes in K which do not split in any field E_{l_i} for $1 \leq i \leq n$. We note again that

$$i)\ \mathcal{M}_n \supset \mathcal{M}_{n+1} \supset \mathcal{M}_a \quad ii)\ \bigcap_{n=1}^{\infty} \mathcal{M}_n = \mathcal{M}_a.$$

To these two properties we add a third:

$$iii)\ \mathcal{M}_n - \mathcal{M}_a \subseteq \bigcup_{i=n+1}^{\infty} \{E_{l_i}\}.$$

This follows since any prime in $\mathcal{M}_n$ which is not in $\mathcal{M}_a$ must split completely in E_{l_i} for some $i > n$.

From property $i)$ we see that for any s with $s > 1$ we have

$$\delta(\mathcal{M}_n, s) \geq \delta(\mathcal{M}_{n+1}, s) \geq \delta(\mathcal{M}_a, s).$$

Thus, $\lim_{n\to\infty} \delta(\mathcal{M}_n, s)$ exists and is $\geq \delta(\mathcal{M}_a, s)$.

We want to estimate $\delta(\mathcal{M}_n, s) - \delta(\mathcal{M}_a, s)$. To do this we first observe that by property *iii* we have

$$\delta(\mathcal{M}_n, s) - \delta(\mathcal{M}_a, s) \leq \sum_{i=n+1}^{\infty} \delta(\{E_{l_i}\}, s) \ .$$

To go further we need an observation and a key lemma. The observation is that for any Galois extension L of K and s which is real and greater than 1 we have

$$\delta(\{L\}, s) < \frac{1}{[L:K]} \frac{\log \zeta_L(s)}{\log \zeta_K(s)} \ .$$

To justify this, one can go back to the proof of Theorem 9.13 and note that the proof shows $\log \zeta_L(s) = [L:K] \sum_{P\in\{L\}} NP^{-s} + R(s)$ where $R(s)$ is positive when s is real and bigger than 1. Dividing both sides of this equation by $[L:K] \log \zeta_K(s)$ proves what we want.

Putting the last two inequalities together yields

$$\delta(\mathcal{M}_n, s) - \delta(\mathcal{M}_a, s) \leq \sum_{i=n+1}^{\infty} \frac{1}{[E_{l_i}:k]} \frac{\log \zeta_{E_{l_i}}(s)}{\log \zeta_K(s)} \ . \tag{1}$$

We can now state the main lemma

Lemma 10.12. *There is a real number $s_1 > 1$ such that*

$$\sum_{i=1}^{\infty} \frac{1}{[E_{l_i}:K]} \frac{\log \zeta_{E_{l_i}}(s)}{\log \zeta_K(s)}$$

converges uniformly on the interval $(1, s_1)$.

Assuming this lemma we will conclude the proof of Theorem 10.9. We will then give the proof of the lemma.

From Lemma 10.12 and Equation 1 we see that $\lim_{n\to\infty} \delta(\mathcal{M}_n, s) = \delta(\mathcal{M}_a, s)$ on $(1, s_1)$ and that the convergence is uniform. We can use the following standard fact from real analysis.

Fact. Let $\{f_n(s)\}$ be a sequence of functions on the interval (s_0, s_1) which converges uniformly to a function $f(s)$. For each n suppose $\lim_{s\to s_0} f_n(s) = A_n$ exists and that $\lim_{n\to\infty} A_n = A$ exists. Then, $\lim_{s\to s_0} f(s) = A$. In other words,

$$\lim_{s\to s_0} \lim_{n\to\infty} f_n(s) = \lim_{n\to\infty} \lim_{s\to s_0} f_n(s) \ .$$

We now apply this fact to the sequence of functions $\delta(\mathcal{M}_n, s)$.

$$\lim_{s \to 1+} \delta(\mathcal{M}_a, s) = \lim_{s \to 1+} \lim_{n \to \infty} \delta(\mathcal{M}_n, s) = \lim_{n \to \infty} \lim_{s \to 1+} \delta(\mathcal{M}_n, s) =$$

$$\lim_{n \to \infty} \sum_{d \mid m_n} \frac{\mu(d)}{d_a f(a)} = \sum_{m=1,\ (m,p)=1}^{\infty} \frac{\mu(m)}{m_a f(m)} .$$

In the next to the last equality we have used Proposition 10.7. The last equality uses Theorem 10.8, Romanoff's Theorem.

This sequence of equalities gives the proof of Bilharz's Theorem, Theorem 10.9, once we have proved Theorem 10.8, Proposition 10.10, and Lemma 10.12.

Still assuming the truth of Romanoff's Theorem, we next tackle the proof of Lemma 10.12.

Since a is an l-th power for only finitely many primes, to prove Lemma 10.12 it suffices to prove that the sum

$$\sideset{}{'}\sum_{l \neq p} \frac{1}{l f(l)} \frac{\log \zeta_{E_l}(s)}{\log \zeta_K(s)} \tag{2}$$

is uniformly convergent on some interval $(1, s_1)$, where the sum is over all primes $l \neq p$ for which a is not an l-th power.

Let $R = \mathbb{F}(T)$ denote the rational function field over $\mathbb{F}$ and $R_l = \mathbb{F}(\zeta_l)(T)$ denote the rational function field with $\mathbb{F}(\zeta_l)$ as constant field. It follows from Theorem 5.9 that

$$\zeta_{E_l}(s) = L_{E_l}(s) \zeta_{R_l}(s),$$

where $L_{E_l}(s)$ is a polynomial in $q^{-f(l)s}$ of degree $2g_l$, where g_l is the genus of E_l. (Remember that $f(l) = [\mathbb{F}(\zeta_l) : \mathbb{F}]$.) Taking the logarithm of both sides of this relation and substituting into Equation 2 gives

$$\sideset{}{'}\sum_{l \neq p} \frac{1}{l f(l)} \frac{\log \zeta_{E_l}(s)}{\log \zeta_K(s)} = \sideset{}{'}\sum_{l \neq p} \frac{1}{l f(l)} \frac{\log L_{E_l}(s)}{\log \zeta_K(s)} + \sideset{}{'}\sum_{l \neq p} \frac{1}{l f(l)} \frac{\log \zeta_{R_l}(s)}{\log \zeta_K(s)} . \tag{3}$$

It thus suffices to prove that these two sums are uniformly convergent on some interval $(1, s_1)$. We shall first prove this for the second sum.

For $s > 1$ we note that

$$\zeta_{R_l}(s) = \frac{1}{(1 - q^{-f(l)s})(1 - q^{f(l)(1-s)})} \leq \frac{1}{(1 - q^{-s})(1 - q^{1-s})} = \zeta_R(s) .$$

Thus,

$$\sideset{}{'}\sum_{l \neq p} \frac{1}{l f(l)} \frac{\log \zeta_{R_l}(s)}{\log \zeta_K(s)} \leq \sideset{}{'}\sum_{l \neq P} \frac{1}{l f(l)} \frac{\log \zeta_R(s)}{\log \zeta_K(s)} . \tag{4}$$

Since both $\zeta_R(s)$ and $\zeta_K(s)$ have a simple pole at $s = 1$, it follows easily that the ratio $\log \zeta_R(s) / \log \zeta_K(s) \to 1$ as $s \to 1^+$. Thus, there is an interval $(1, s_1)$ such that this ratio is less than 2 for $s \in (1, s_1)$. It follows that the right-hand sum in Equation 4 is dominated by the convergent sum $\sum'_{l \neq p} \frac{2}{lf(l)}$. This establishes the uniform convergence of the second sum in Equation 3 on the interval $(1, s_1)$. We now turn our attention to the first sum.

From Theorem 5.10, we deduce the following:

$$L_{E_l}(s) = \prod_{j=1}^{2g_l} \left(1 - \pi_j q^{-f(l)s}\right) ,$$

where each π_j has absolute value $q^{f(l)/2}$ (it is here that the Riemann hypothesis for function fields is used). Assuming that s is real and bigger than 1, we obtain the following inequalities:

$$\left(1 - q^{-\frac{f(l)}{2}}\right)^{2g_l} \leq L_{E_l}(s) \leq \left(1 + q^{-\frac{f(l)}{2}}\right)^{2g_l} .$$

By Proposition 10.4 and the remark following it, we see that there is a constant r, independent of l, such that $2g_l \leq rl$ for all l. Substituting this into the last equation and taking logarithms of the terms of the resulting inequalities yields

$$rl \log\left(1 - q^{-\frac{f(l)}{2}}\right) \leq \log L_{E_l}(s) \leq rl \log\left(1 + q^{-\frac{f(l)}{2}}\right) . \tag{5}$$

If $0 < x$, then $\log(1 + x) < x$, so

$$\log L_{E_l}(s) < rl q^{-\frac{f(l)}{2}} .$$

To deal with the left-hand side of Equation 5, note that for $0 < x < 1$

$$-\log(1 - x) = \sum_{k=1}^{\infty} \frac{x^k}{k} < \sum_{k=1}^{\infty} x^k = \frac{x}{1-x} .$$

Substitute $x = q^{-f(l)/2}$ into this and also use the fact that $(1 - q^{-f(l)/2})^{-1} \leq (1 - q^{-1/2})^{-1}$. We obtain

$$-rl \frac{\sqrt{q}}{\sqrt{q} - 1} q^{-\frac{f(l)}{2}} < \log L_{E_l}(s) .$$

Altogether, we have established that

$$|\log L_{E_l}(s)| < rl \frac{\sqrt{q}}{\sqrt{q} - 1} q^{-\frac{f(l)}{2}} .$$

Since $\log \zeta_K(s) \to \infty$ as $s \to 1^+$, we see that $1/|\log \zeta_K(s)|$ is bounded by some constant, say, C, on the interval $(1, s_1)$. Thus, for $s \in (1, s_1)$ we have

$$\sum'_{l \neq p} \frac{1}{lf(l)} \frac{|\log L_{E_l}(s)|}{|\log \zeta_K(s)|} < rC \frac{\sqrt{q}}{\sqrt{q} - 1} \sum'_{l \neq p} \frac{1}{f(l) q^{\frac{f(l)}{2}}} .$$

We will have established that the first sum on the right-hand side of Equation 3 is uniformly convergent on the interval $(1, s_1)$ once we prove the following lemma. This will also finish the proof of Lemma 10.12.

Lemma 10.13. *The sum*

$$\sideset{}{'}\sum_{l \neq p} \frac{1}{f(l)q^{\frac{f(l)}{2}}}$$

is convergent.

Proof. We will break the sum up into two subsums, the first over all primes l such that $l \leq q^{f(l)/2}$ and the second over all primes such that $l > q^{f(l)/2}$.

For the first subsum we have

$$\sum_{\substack{l \neq p \\ l \leq q^{f(l)/2}}} \frac{1}{f(l)q^{\frac{f(l)}{2}}} \leq \sum_{l \neq p} \frac{1}{f(l)l} \, .$$

The latter sum converges by Romanoff's Theorem.

To analyze the second subsum, we first try to figure out how many primes l there are such that $f(l)$ takes on a fixed value f. Such primes must divide $q^f - 1$. Let $l_1, l_2, \cdots, l_{t_f}$ be the set of such primes. We have

$$l_1 l_2 \cdots l_{t_f} \mid q^f - 1 \, .$$

We are now considering primes l such that $q^{f(l)/2} < l$, so it follows that

$$q^{\frac{t_f f}{2}} < q^f \, ,$$

and it follows that $t_f \leq 1$. Thus,

$$\sum_{\substack{l \neq p \\ q^{\frac{f(l)}{2}} < l}} \frac{1}{f(l)q^{\frac{f(l)}{2}}} \leq \sum_{f=1}^{\infty} \frac{1}{f q^{\frac{f}{2}}} \, ,$$

which is a convergent series. In fact, its sum is $-\log(1 - q^{-\frac{1}{2}})$.

Since both subsums converge, the proof is complete.

The two remaining things which need proof are the results of Romanoff and Heilbronn. Both these proofs belong to elementary number theory and not to the arithmetic of function fields. Nevertheless, for the sake of completeness we will sketch these proofs.

The proof of Romanoff's Theorem uses the following lemma.

Lemma 10.14. *Let γ denote Euler's constant.*

$$\sum_{d \mid n} \frac{\mu(d)^2}{d} \leq \frac{6}{\pi^2} e^{\gamma} \log \log(n) + O(1) \, .$$

Proof. First of all, note that

$$\sum_{d|n} \frac{\mu(d)^2}{d} = \prod_{p|n} \left(1 + \frac{1}{p}\right) .$$

The product is over all prime divisors p of n. We break this product up into two parts - first the product over prime divisors $> \log(n)$ and secondly over prime divisors $< \log(n)$.

If $p_1, p_2, \cdots, p_{g(n)}$ are the prime divisors of n which are greater than $\log(n)$ we find, using $\prod p_i \le n$, that

$$\log(n)^{g(n)} \le n ,$$

and, consequently, $g(n) \le \log(n)/\log\log(n)$. It follows that

$$\prod_{\substack{p|n \\ \log(n) < p}} \left(1 + \frac{1}{p}\right) \le \left(1 + \frac{1}{\log(n)}\right)^{g(n)} = 1 + O\left(\frac{1}{\log\log(n)}\right) .$$

We now consider the product over prime divisors less than $\log(n)$.

$$\prod_{\substack{p|n \\ p < \log(n)}} \left(1 + \frac{1}{p}\right) = \prod_{\substack{p|n \\ p < \log(n)}} \left(1 - \frac{1}{p^2}\right) \prod_{\substack{p|n \\ p < \log(n)}} \left(1 - \frac{1}{p}\right)^{-1} .$$

The first product on the right-hand side is $1/\zeta(2) + O(1/\log(n)) = 6/\pi^2 + O(1/\log(n))$. The second product is $\le e^{\gamma} \log\log(n) + O(1)$ by Merten's Theorem (see Hardy and Wright [1], Theorem 429).

Putting all these estimates together gives the result.

We now have everything we need to give the proof of Theorem 10.8. We begin by rewriting the sum in question as follows (all sums are over square-free m prime to q):

$$\sum_m \frac{1}{mf(m)} = \sum_{n=1}^{\infty} \frac{d(n)}{n} ,$$

where $d(n) = \sum_{m, f(m)=n} 1/m$.

Define $D(n) = \sum_{k=1}^{n} d(k)$. A moment's reflection shows that

$$D(n) = \sum_{m, f(m) \le n} \frac{1}{m} .$$

Any m entering the definition of $D(n)$ must be a square-free divisor of $A(n) = (q^n - 1)(q^{n-1} - 1)\cdots(q - 1)$. Clearly, $A(n) < q^{n^2}$. Thus, using Lemma 10.14 we find

$$D(n) \le \sum_{m|A(n)} \frac{\mu(m)^2}{m} \le \frac{6}{\pi^2} e^{\gamma} \log\log A(n) \le M \log(n) ,$$

for an appropriate positive constant M.

Let's define $D(0) = 0$ and calculate

$$\sum_{n=1}^{N} \frac{d(n)}{n} = \sum_{n=1}^{N} \frac{D(n) - D(n-1)}{n} = \sum_{n=1}^{N-1} \frac{D(n)}{n(n+1)} + \frac{D(N)}{N} \, .$$

Since $D(n) \leq M \log(n)$, we see that for large N the right-hand side of this equation is less than

$$1 + M \sum_{n=1}^{\infty} \frac{\log(n)}{n(n+1)} \, .$$

This is finite, and this implies $\sum_{n=1}^{\infty} d(n)/n$ converges. The proof is complete.

This simple proof is due to M. Ram Murty. The same argument can be made to give much stronger and more general results of the same nature. For these improvements and generalizations see Murty-Rosen-Silverman [1].

Our final task in his chapter is to prove the inequality due to Heilbronn, Proposition 10.10. This will be seen to follow from a more general inequality belonging to the elementary theory of numbers (see Heilbronn [1]).

For any subset $S \subseteq \mathbb{Z}^+$, the positive integers, we define its natural density to be

$$d(S) = \lim_{X \to \infty} X^{-1} \#\{n \in S \mid n \leq X\} \, ,$$

provided that the limit exists.

If S has a natural density, then it is not too hard to show it has a Dirichlet density as well, and that the two are equal. On the other hand, there exist sets with Dirichlet density, but not natural density.

Natural density has a number of simple and easily proven properties.

1. $d(S) = 0$ if S is a finite set.

2. $d(\mathbb{Z}^+) = 1$, where $\mathbb{Z}^+$ is the set of all positive integers.

3.   If $S_1 \subseteq S_2$, then $d(S_1) \leq d(S_2)$, provided that both densities exist.

4.   If $S_1 \cap S_2 = \phi$, then $d(S_1 \cup S_2) = d(S_1) + d(S_2)$, provided that $d(S_1)$ and $d(S_2)$ both exist.

5.   Let h be a positive integer and define $hS = \{hs \mid s \in S\}$. Then, $d(hS) = \frac{1}{h} d(S)$, provided that $d(S)$ exists.

It follows that the natural density of the set of integers divisible by a positive integer a is exactly $1/a$ and, consequently, the natural density of the set of integers not divisible by a is $1 - 1/a$. Let $\{a_1, a_2, \ldots, a_n\}$ be a finite set of positive integers and T_n the set of positive integers not divisible by any a_i with $1 \leq i \leq n$. It is an exercise to show, using the inclusion-exclusion principle, that T_n has a natural density given by

$$d(T_n) = 1 - \sum_{i=1}^{n} \frac{1}{a_i} + \sum_{1 \leq i < j \leq n} \frac{1}{[a_i, a_j]} - \cdots + (-1)^n \frac{1}{[a_1, a_2, \ldots, a_n]} .$$

In this equation, the square brackets denote least common multiple.

Heilbronn gives the following lower bound for this density. It is worth pointing out that the inequality becomes an equality if the integers a_i are pairwise relatively prime.

Lemma 10.15.

$$d(T_n) \geq \prod_{i=1}^{n} \left(1 - \frac{1}{a_i} \right) .$$

Proof. The proof is by induction on n. If $n = 1$ the assertion reads $d(T_1) = 1 - 1/a_1$ which we have already noted. So, we assume the result is true for n and prove it for $n + 1$.

Note that $T_n = T_{n+1} \cup S$ where S is the set of positive integers divisible by a_{n+1} and by none of the integers a_i with $1 \leq i \leq n$. This is a disjoint union. A moments reflection shows that $S \subseteq a_{n+1} T_n$. Thus, $d(S) \leq \frac{1}{a_{n+1}} d(T_n)$. Consequently,

$$
\begin{aligned}
d(T_{n+1}) \;\; &= \;\; d(T_n) - d(S) \geq d(T_n) - \frac{1}{a_{n+1}} d(T_n) \\
&= \;\; \left(1 - \frac{1}{a_{n+1}} \right) d(T_n) \geq \prod_{i=1}^{n+1} \left(1 - \frac{1}{a_i} \right) .
\end{aligned}
$$

The last inequality follows from the induction assumption.

To go from this elementary lemma to the inequality of Proposition 10.10 we need to first give a mild generalization which is proven, as we shall see, by an amusing geometric argument.

Lemma 10.16. *Let $x_1, x_2, \ldots, x_n$ be real numbers with $0 \leq x_i \leq 1$ for each i. As above, let $\{a_1, a_2, \ldots, a_n\}$ be a finite set of positive integers. Then*

$$1 - \sum_{i=1}^{n} \frac{x_i}{a_i} + \sum_{1 \leq i < j \leq n} \frac{x_i x_j}{[a_i, a_j]} - \cdots + (-1)^n \frac{x_1 x_2 \cdots x_n}{[a_1, a_2, \ldots, a_n]} \geq \prod_{i=1}^{n} \left(1 - \frac{x_i}{a_i} \right) .$$

Proof. (Sketch) Let $F(x_1, x_2, \ldots, x_n)$ denote the difference between the left-hand side and the right-hand side of the inequality given in the statement of the Lemma. We think of F as a function on the unit cube and prove that it is non-negative on this domain. That will prove the lemma.

If we fix the values of all the variables except x_i the resulting function is an inhomogeneous linear function of x_i. Consequently, on the interval $[0, 1]$

it takes its minimum value at one of the endpoints. Using this fact, and a simple induction on n, we find that on the unit cube, F takes its minimum at a vertex. A vertex has coordinates $(\epsilon_1, \epsilon_2, \ldots, \epsilon_n)$, where each ϵ_i is either 0 or 1. At such a point, the value of F is non-negative by Lemma 10.15. Thus, F is non-negative on the unit cube. This completes the proof.

We can now prove Proposition 10.10. Let $l_1 < l_2 < l_3 < \cdots$ be an enumeration of the positive prime numbers different from p. Recall that $f(m)$ is the order of q modulo m and that, for square-free m, $f(m)$ is the least common multiple of $\{f(l) \mid l|m\}$

Define $m_n = l_1 l_2 \cdots l_n$. Then

$$\sum_{d|m_n} \frac{\mu(d)}{df(d)} = 1 - \sum_{i=1}^{n} \frac{1}{l_i f(l_i)} + \sum_{1 \le i < j \le n} \frac{1}{l_i l_j [f(l_i), f(l_j)]}$$

$$- \cdots + (-1)^{n+1} \frac{1}{l_1 l_2 \cdots l_n [f(l_1), f(l_2), \cdots, f(l_n)]} \ .$$

By Lemma 10.16, setting $a_i = f(l_i)$ and $x_i = 1/l_i$, we obtain

$$\sum_{l|m_n} \frac{\mu(d)}{df(d)} \ge \prod_{l|m_n} \left(1 - \frac{1}{lf(l)}\right) \ge \prod_{l \ne p} \left(1 - \frac{1}{lf(l)}\right) \ .$$

Using Romanoff's theorem one more time, it is easy to see that the left-hand side of this inequality tends to $\sum_{m,(m,q)=1} \mu(m)/mf(m)$ as n tends to ∞.

We have now proven Theorem 10.10 in the case where a is not an l-th power for any prime $l \ne p$. In the general case one proceeds the same way as above except that one sets $x_i = 1/l_i$ if a is not an l_i-th power and $x_i = 1$ if a is an l_i-th power. We leave it to the reader to check that this procedure leads to the correct result.

Exercises

1. Let $K/\mathbb{F}$ be a function field and suppose K contains a primitive l-th root of unity, where l is a prime unequal to the characteristic of K. If $a \in K^*$ is not geometric at l, show there is a $\mu \in \mathbb{F}^*$ and a $b \in K^*$ such that $a = \mu b^l$.

2. In the course of the proof of Proposition 10.5 the l-th power residue symbol, $(a/P)_l$, was defined. Show that it is a good generalization of the Legendre symbol of elementary number theory, by proving that it has the following three properties: (i) $(a/P)_l = (b/P)_l$ if $a \equiv b \pmod{P}$, (ii) $(ab/P)_l = (a/P)_l(b/P)_l$, and $X^l \equiv a \pmod{P}$ is solvable (for $a \ne 0$) if and only if $(a/P)_l = 1$.

3. Prove that $[\mathbb{F}(\zeta_m) : \mathbb{F}] = f(m)$ is the order of q modulo m, provided that $(q, m) = 1$.

4. Prove Property 5 of natural density. Namely, if $S \subseteq \mathbb{Z}^+$ and $h \in \mathbb{Z}^+$, then $\delta(hS) = h^{-1}\delta(S)$ provided that either density exists.

5. In Lemma 10.5, show the inequality is an equality if the a_i are pairwise coprime. What happens if all the a_i are all equal?

6. Suppose $K/\mathbb{F}$ is a function field of characteristic p. If $a \in K^*$ is not an l-th power for any prime l, then it follows from the text that the Dirichlet density of the set of primes for which a is a primitive root exceeds $\prod_{l \neq p}(1 - l^{-1}f(l)^{-1}) = c_p$, which does not depend on a. If $p > 2$ show $c_p < .5$. Is $c_2 > .5$? (Recall that $f(l)$ is the order of q modulo l, where $q = |\mathbb{F}|$.)

7. Let $K/\mathbb{F}$ be a function field and suppose $l \mid q - 1$ where $q = |\mathbb{F}|$. If a and b are geometric at l, show the constant field of $K(\sqrt[l]{a}, \sqrt[l]{b})$ is $\mathbb{F}$ unless $ab^i \in \mathbb{F}^* K^{*l}$ for some i with $1 \leq i < l$.

8. Let $K/\mathbb{F}$ be a function field and let l be a prime different from the characteristic. Two elements of K^*, a and b, are said to be l-independent if for all integers m and n, $a^m b^n \in \mathbb{F}^* K^{*l}$ if and only if $l|m$ and $l|n$. Assume $a, b \in K^*$ are geometric at l and l-independent. Define $K_l = K(\zeta_l, \sqrt[l]{a}, \sqrt[l]{b})$. Prove that $[K_l : K] = f(l)l^2$.

9. With the same notation as the previous problem, suppose $a, b \in K^*$ are l-independent for all primes l different from the characteristic of K. Show there is a constant c depending only on a and b such that the genus of K_l is bounded by cl^2 for all l.

10. (Continuation) Let $a, b \in K^*$ be geometric and l-independent for all primes l different from p, the characteristic of K. Define $\mathcal{M}_{a,b}$ to be the set of primes P of K such that $(O_P/P)^*$ is generated by the residues of a and b modulo P. Calculate $\delta(\mathcal{M}_{a,b})$. Use Heilbronn's Theorem to show that $\delta(\mathcal{M}_{a,b}) \geq 6/\pi^2(1 - p^{-2})^{-1}$. (Hint: Imitate the proof of Bilharz's Theorem. Because many of the sums involved converge for trivial reasons, one need not use Romanoff's Theorem.)

11. The reader may wonder if there are elements which satisfy the hypotheses of the previous problem. In fact, they exist in abundance. Suppose S is a finite set of primes with more than three elements. Let $E(S) = \{a \in K^* \mid \mathrm{ord}_P(a) = 0, \forall P \notin S\}$, the group of S-units of K^*. In Chapter 14 we will show that $E(S)/\mathbb{F}^*$ is a free group on $|S| - 1$ generators. Let $a, b \in E(S)$ map onto elements of a basis of $E(S)/\mathbb{F}^*$. Show that a and b are geometric and that they are l-independent for all primes l.

11

The Behavior of the Class Group in Constant Field Extensions

In Chapter 8, we discussed constant field extensions and, toward the end of the chapter, we gave particular attention to the case when the base field has a finite field of constants. We begin by recalling some notation.

Let K be an algebraic function field over a finite field $\mathbb{F}$ with q elements. Fix an algebraic closure $\bar{\mathbb{F}}$ of $\mathbb{F}$, and let $\mathbb{F}_n$ be the unique subfield of $\bar{\mathbb{F}}$ such that $[\mathbb{F}_n : \mathbb{F}] = n$. Let $K_n = K\mathbb{F}_n$ be the constant field extension of K by $\mathbb{F}_n$ and $h(K_n)$ the class number of K_n. By definition, $h(K_n)$ is the number of elements in the group of divisor classes of degree zero of the field K_n. Formerly we denoted this group by $Cl^0_{K_n}$. We simplify the notation by writing $Cl^0_{K_n} = J(\mathbb{F}_n)$. The choice of the letter "J" is not completely arbitrary. If one regards K as the field of rational functions on a smooth curve defined over $\mathbb{F}$, then J is closely related to the Jacobian variety of that curve.

To begin with we will discuss the question of how the class numbers $h(K_n)$ vary with n. The main tool will be the formula

$$h(K_n) = \prod_{i=1}^{2g}(1 - \pi_i^n),$$

given in the corollary to Proposition 8.16. We will also use some elementary facts from algebraic number theory and the theory of l-adic numbers.

Afterwards we will deal with the more difficult question of how the finite abelian groups $J(\mathbb{F}_n)$ vary with n. Once again we will need to deal with some elementary results from the theory of l-adic numbers. We will also use some results from the theory of cohomology of groups. Readers who lack this background may want to just skim these proofs and move on. Finally, we will need some basic facts which can be stated entirely in the language of function fields, but whose proof requires a substantial amount of algebraic geometry. We will state these facts as clearly as possible and use them to derive a number of interesting results on the behavior of the class group under constant field extension. For the proof of the basic facts we will have to refer the reader to other sources.

The chapter will conclude with a brief discussion of how the material presented in this chapter was imported into algebraic number theory by K. Iwasawa. The resulting theory has been a fundamental tool in many of the most important developments in number theory over the past 50 years.

Let $u = q^{-s}$. Then the zeta function of K written as a function of u has the form

$$Z_K(u) = \frac{L_K(u)}{(1-u)(1-qu)} \, ,$$

where $L_K(u)$ is a polynomial of degree twice the genus with integer coefficients and constant term 1. If we factor $L_K(u)$ over the algebraic closure of $\mathbb{Q}$, we obtain

$$L_K(u) = \prod_{i=1}^{2g}(1 - \pi_i u) \, ,$$

where the π_i are algebraic integers. It was pointed out after Theorem 5.9 that the functional equation for $Z_K(u)$ is equivalent to $\pi \to q/\pi$ being a permutation of the roots of $L_K(u)$. Write $L_K(u) = \sum_{k=0}^{2g} a_k u^k$. It is then easy to see that another way of stating the functional equation is

$$q^{g-k} a_k = a_{2g-k} \quad \text{for} \quad 0 \le k \le g \, .$$

We will now use these facts to prove several things about how class numbers behave in constant field extensions.

Proposition 11.1. Let l be a prime which divides $h(K)$. If $l \mid n\frac{q^n-1}{q-1}$, then $l \mid h(K_n)/h(K)$.

Proof. Let E be the number field obtain by adjoining all the elements π_i to the rational numbers $\mathbb{Q}$, and let $\mathcal{L}$ be a prime ideal of E lying over l. Then, $l \mid h(K)$ implies

$$\prod_{i=1}^{2g}(1 - \pi_i) \equiv 0 \pmod{\mathcal{L}} \, .$$

It follows that there is an i such that $\pi_i \equiv 1 \pmod{\mathcal{L}}$. Let j be such that $\pi_i \pi_j = q$. Then, $\pi_j \equiv q \pmod{\mathcal{L}}$. Now write

$$h(K_n)/h(K) \equiv \prod_{i=1}^{2g}(1 + \pi_i + \pi_i^2 + \cdots + \pi_i^{n-1}) \pmod{\mathcal{L}} \, .$$

The i-th term is congruent to n and the j-th term is congruent to $q^n - 1/q - 1$ modulo $\mathcal{L}$. The Proposition follows from this.

Corollary 1. *If $l \mid h(K)$ and $l \mid n$, then $l \mid h(K_n)/h(K)$.*

Corollary 2. *If $l \mid h(K)$ and $n = l^t$, then $l^t \mid h(K_n)/h(K)$.*

Proof. Just use Corollary 1 and induction on t.

Corollary 3. *If the genus of K is 1 and $l \mid h(K)$, then $l \mid h(K_n)/h(K)$ if and only if $l \mid n(q^n - 1)/(q - 1)$.*

Proposition 11.1 is about divisibility. The next result will be about non-divisibility. First, a definition. If l is a prime different from p, the characteristic of $\mathbb{F}$, let $d(l)$ be the least common multiple of the numbers $l^k - 1$ for $1 \le k \le 2g$. Let $d(p)$ be the least common multiple of the numbers $p^k - 1$ for $1 \le k \le g$.

Lemma 11.2. *As above, let $\mathcal{L}$ be a prime of E lying above l. If $\pi_i \notin \mathcal{L}$, then $\pi_i^{d(l)} \equiv 1 \pmod{\mathcal{L}}$.*

Proof. Let $L_K^*(u) = u^{2g} L_K(1/u)$. Then, $L_K^*(u)$ is a monic polynomial with integer coefficients whose roots are the numbers π_i. Suppose $\pi_i \notin \mathcal{L}$ and let $\bar{\pi}_i$ be the residue of π_i modulo $\mathcal{L}$. Then, $\bar{\pi}_i$ is a root of $L_K^*(u)$ modulo l, i.e. of $\overline{L_K^*}(u) \in \mathbb{Z}/l\mathbb{Z}[u]$. Thus, it satisfies an irreducible polynomial over $\mathbb{Z}/l\mathbb{Z}$ of degree k where $1 \le k \le 2g$, if $l \ne p$, and $1 \le k \le g$ if $l = p$. The last restriction holds because if $l = p$, the first g coefficients of $\overline{L_K^*}(u)$ are zero (recall that $q^{g-k} a_k = a_{2g-k}$ for $0 \le k \le g$). Since $[\mathbb{Z}/l\mathbb{Z}[\bar{\pi}_i] : \mathbb{Z}/l\mathbb{Z}] = k$, $\bar{\pi}_i$ is a non-zero element of a finite field with l^k elements. This implies $\bar{\pi}_i^{l^k - 1} = \bar{1}$, which in turn implies that $\pi_i^{l^k - 1} \equiv 1 \pmod{\mathcal{L}}$, and so $\pi_i^{d(l)} \equiv 1 \pmod{\mathcal{L}}$, as asserted.

Since for every π_i there is a π_j with $\pi_i \pi_j = q$, it follows that every prime dividing π_i in E must lie above p. Thus, the hypothesis $\pi_i \notin \mathcal{L}$ is only necessary when $l = p$.

Proposition 11.3. *Suppose l does not divide $h(K)$ and that n is relatively prime to $d(l)$. Then, l does not divide $h(K_n)$.*

Proof. Suppose $l \mid h(K_n)$. Then, for some k, $\pi_k^n - 1 \equiv 0 \pmod{\mathcal{L}}$. By the above lemma, $\pi_k^{d(l)} \equiv 1 \pmod{\mathcal{L}}$. Since $(n, d(l)) = 1$ it follows that $\pi_k \equiv 1 \pmod{\mathcal{L}}$, but this implies $l \mid h(K)$ contrary to the hypothesis. Thus, l does not divide $h(K_n)$.

Corollary. *If $n = l^t$ and l does not divide $h(K)$, then l does not divide $h(K_n)$.*

Proof. From the definition of $d(l)$, it is clear that l does not divide $d(l)$ and so $(l^t, d(l)) = 1$. Now apply the proposition.

Proposition 11.4. *Let n be he smallest integer such that $l \mid h(K_n)$. Then, $n \mid d(l)$.*

Proof. There must be an index k such that $\pi_k^n \equiv 1 \pmod{\mathcal{L}}$. From the definition of n it follows that n is the order of π_k modulo $\mathcal{L}$. By Lemma 11.2, $\pi_k^{d(l)} \equiv 1 \pmod{\mathcal{L}}$. Thus, $n \mid d(l)$, as asserted.

We are next going to consider the behavior of the class group in the l-tower. By this we mean the ascending sequence of fields $K \subset K_l \subset K_{l^2} \subset \cdots \subset K_{l^n} \subset \cdots$. To begin with we will be interested in the l-primary part of the class groups. We denote these by $J(\mathbb{F}_{l^n})(l)$. The orders of these groups are given by

$$l^{e_n} \quad \text{where} \quad e_n = \operatorname{ord}_l h(K_{l^n}) .$$

Of course, the numbers e_n depend on the prime l, but we have decided not to complicate the notation overmuch by making this dependence explicit. The main result is as follows.

Theorem 11.5. *There are constants λ_l, ν_l, and a positive integer n_0 such that for all $n \geq n_0$, $e_n = \lambda_l \, n + \nu_l$; i.e., the numbers e_n grow linearly with n.*

Proof. We will again make use of the formula $h(K_n) = \prod_{i=1}^{2g}(1 - \pi_i^n)$, but it will be convenient to reformulate it as an l-adic formula. Consider $\mathbb{Q}$ as a subset of its l-adic completion $\mathbb{Q}_l$. Let $\bar{\mathbb{Q}}$ and $\bar{\mathbb{Q}}_l$ be the algebraic closures of $\mathbb{Q}$ and $\mathbb{Q}_l$, respectively. Finally, let ρ be an embedding of $\bar{\mathbb{Q}}$ into $\bar{\mathbb{Q}}_l$. Applying ρ to the above formula yields $h(K_n) = \prod_{i=1}^{2g}(1 - \rho(\pi_i)^n)$. Having done this, we now simply rename $\rho(\pi_i)$ as π_i and assume that our original formula takes place inside $\bar{\mathbb{Q}}_l$.

In the usual way, the additive valuation ord_l from $\mathbb{Q}_l$ to $\mathbb{Z} \cup \infty$ extends to an additive valuation from $\bar{\mathbb{Q}}_l$ to $\mathbb{Q} \cup \infty$. Namely, if $\alpha \in \bar{\mathbb{Q}}_l$, let a be the norm of α from $\mathbb{Q}_l(\alpha)$ down to $\mathbb{Q}_l$. Then, $\operatorname{ord}_l(\alpha) = [\mathbb{Q}_l(\alpha) : \mathbb{Q}_l]^{-1}\operatorname{ord}_l(a)$. If $\zeta \in \bar{\mathbb{Q}}_l$ is any primitive l^n-th root of unity, it is a standard fact that $\operatorname{ord}_l(\zeta - 1) = 1/\phi(l^n) = 1/l^{n-1}(l-1)$. Thus, as n increases, these valuations tend to zero.

Let $\pi \in \bar{\mathbb{Q}}_l$ be integral over $\mathbb{Z}_l$. If $\operatorname{ord}_l(\pi - 1) = 0$, we claim that $\operatorname{ord}_l(\pi^{l^n} - 1) = 0$ for all positive integers n. To see this, note

$$\operatorname{ord}_l(\pi^{l^n} - 1) = \sum_{j=0}^{l^n - 1} \operatorname{ord}_l(\pi - \zeta_j) = \operatorname{ord}_l(\pi - 1) + \sum_{j=1}^{l^n - 1} \operatorname{ord}_l(\pi - \zeta_j) .$$

Here, the ζ_j run through all the l^n-th roots of unity and $\zeta_0 = 1$. We are assuming $\operatorname{ord}_l(\pi - 1) = 0$. It follows that for $j > 0$, $\operatorname{ord}_l(\pi - \zeta_j) = \operatorname{ord}_l(\pi - 1 + 1 - \zeta_j) = \operatorname{ord}_l(\pi - 1) = 0$. The next to the last equality follows since $\operatorname{ord}_l(\zeta_j - 1) > 0 = \operatorname{ord}_l(\pi - 1)$. Thus, all the terms on the right-hand side of the equation displayed above are zero and we have proven our claim.

Now assume that $\mathrm{ord}_l(\pi - 1) > 0$. Consider all l^n-th roots of unity for all n. Among all these we claim $\mathrm{ord}_l(\pi - \zeta) = \mathrm{ord}_l(1 - \zeta)$ with only finitely many exceptions. This is because

$$\mathrm{ord}_l(\pi - \zeta) = \mathrm{ord}_l(\pi - 1 + 1 - \zeta) = \mathrm{ord}_l(1 - \zeta) \, ,$$

as soon as ζ is a primitive l^n-th root of unity with n so large that $\mathrm{ord}_l(1 - \zeta) = 1/\phi(l^n) < \mathrm{ord}_l(\pi - 1)$. Let S be the set of l-power roots of unity where this fails to happen. Choosing n sufficiently large we find

$$\mathrm{ord}_l(\pi^{l^n} - 1) = \sum_{j=0}^{l^n-1} \mathrm{ord}_l(\pi - \zeta_j) = \sum_{\zeta_j \in S} \mathrm{ord}_l(\pi - \zeta_j) + \sum_{j=1, \, \zeta_j \notin S}^{l^n-1} \mathrm{ord}_l(\pi - \zeta_j) \, .$$

In the last sum all the terms can be replaced by $\mathrm{ord}_l(1 - \zeta_j)$. Adding in and simultaneously subtracting the remaining terms yields

$$\sum_{\zeta_j \in S} (\mathrm{ord}_l(\pi - \zeta_j) - \mathrm{ord}_l(1 - \zeta_j)) + \sum_{j=1}^{l^n-1} \mathrm{ord}_l(1 - \zeta_j) \, .$$

Call the first sum c. The second sum is actually equal to n as we see by differentiating the equation $x^{l^n} - 1 = \prod_{j=0}^{l^n-1}(x - \zeta_j)$, setting $x = 1$ in the result, and then taking ord_l of both sides.

Summarizing, we have shown that if $\mathrm{ord}_l(\pi - 1) > 0$, then for all sufficiently large n there is a constant c such that $\mathrm{ord}_l(\pi^{l^n} - 1) = n + c$. Of course, c depends on π.

To return to our original situation, among the numbers $\{\pi_i \mid 1 \le i \le 2g\}$ label them in such a way that for $1 \le i \le \lambda_l$ we have $\mathrm{ord}_l(\pi_i - 1) > 0$ and for $\lambda_l < i \le 2g$ we have $\mathrm{ord}_l(\pi_i - 1) = 0$. For the first range, let c_i be the constant associated to π_i by the above considerations. Then, for all n sufficiently large,

$$e_n = \mathrm{ord}_l(h(K_{l^n})) = \sum_{i=1}^{2g} \mathrm{ord}_l(1 - \pi_i^{l^n}) =$$

$$\sum_{i=1}^{\lambda_l} \mathrm{ord}_l(1 - \pi_i^{l^n}) = \sum_{i=1}^{\lambda_l}(n + c_i) = \lambda_l \, n + \nu_l \, ,$$

where $\nu_l = \sum_{i=1}^{\lambda_l} c_i$. The proof is complete!

Let l' run through the primes other than l. What can be said about the behavior of the l'-primary components of the class group of K_{l^n}? Surprisingly, these behave in an entirely different manner than the l-primary component.

Theorem 11.6. *The numbers* $\mathrm{ord}_{l'}(h(K_{l^n}))$ *are bounded from above. In other words, they increase up to some point n_0 and then stay the same for all $n \geq n_0$.*

Proof. Once again, we start from the formula $h(K_n) = \prod_{i=1}^{2g}(1 - \pi_i^n)$ taking place in $\bar{\mathbb{Q}}$, but now we reinterpret it to hold in $\bar{\mathbb{Q}}_{l'}$ (see the first paragraph in the proof of Theorem 11.5).

If $\zeta \neq 1$ is an l^n-th root of unity, then $\mathrm{ord}_{l'}(\zeta - 1) = 0$. This follows from the identity

$$0 = \mathrm{ord}_{l'}(l^n) = \sum_{j=1}^{l^n - 1} \mathrm{ord}_{l'}(1 - \zeta_j) ,$$

where the sum is over all l^n-th roots of unity except 1.

Suppose $\pi \in \bar{\mathbb{Q}}_{l'}$ is integral over $\mathbb{Z}_{l'}$ and $\mathrm{ord}_{l'}(\pi - 1) > 0$. If $\zeta \neq 1$ is any l^n-th root of unity, we must have $\mathrm{ord}_{l'}(\pi - \zeta) = \mathrm{ord}_{l'}(\pi - 1 + 1 - \zeta) = \mathrm{ord}_{l'}(1 - \zeta) = 0$. Thus, $\mathrm{ord}_{l'}(\pi^{l^n} - 1) = \sum_{j=0}^{l^n - 1} \mathrm{ord}_{l'}(\pi - \zeta_j) = \mathrm{ord}_{l'}(\pi - 1)$.

Now, suppose $\mathrm{ord}_{l'}(\pi - 1) = 0$. There are two possibilities. Either $\mathrm{ord}_{l'}(\pi^{l^n} - 1) = 0$ for all n, or there is a t such that $\mathrm{ord}_{l'}(\pi^{l^t} - 1) > 0$. In the latter case we must have $\mathrm{ord}_{l'}(\pi^{l^n} - 1) = \mathrm{ord}_{l'}(\pi^{l^t} - 1)$ for all $n \geq t$ by the considerations of the last paragraph.

For the set of elements $\{\pi_i \mid 1 \leq i \leq 2g\}$ label the indices in such a way that for each i with $1 \leq i \leq d$ there is a $t_i \geq 0$ such that $\mathrm{ord}_{l'}(\pi_i^{l^{t_i}} - 1) > 0$ and for $i > d$, $\mathrm{ord}_{l'}(\pi_i^{l^n} - 1) = 0$ for all $n > 0$. Then, if n is bigger than $\max_{1 \leq i \leq d}(t_i)$ we have

$$\mathrm{ord}_{l'}(h(K_{l^n})) = \sum_{i=1}^{2g} \mathrm{ord}_{l'}(\pi_i^{l^n} - 1) = \sum_{i=1}^{d} \mathrm{ord}_{l'}(\pi_i^{l^{t_i}} - 1) .$$

Since this last sum does not depend on n, the result follows.

Corollary. *Let $S(l^n)$ be the set of primes which divide $h(K_{l^n})$. Then, $\#S(l^n) \to \infty$ as $n \to \infty$.*

Proof. By Proposition 5.11, there is a constant C such that $h(K_{l^n}) \geq Cq^{gl^n}$. If $S(l^n)$ remained constant from some point on, then combining Theorems 11.5 and 11.6, it would follow that $h(K_{l^n})$ would be equal to a constant times $l^{\lambda_l n}$ for large n. Clearly, this is incompatible with the lower bound for the growth of $h(K_{l^n})$ just given.

Up to now, we have not paid too much attention to the fact that the prime p, the characteristic of $\mathbb{F}$, behaves differently from the other primes. Let's pay a little more attention to this now.

As before, let E be the field obtained by adjoining all the elements π_i to the rational numbers $\mathbb{Q}$. Let $\mathcal{P}$ be a prime in E lying above p. Given an index i let j be determined by $\pi_i\pi_j = q$. Then either π_i or π_j or both must lie in $\mathcal{P}$. It follows that at most g of the π_i do not belong to $\mathcal{P}$. Let λ be the exact number that do not belong to $\mathcal{P}$. We have $0 \leq \lambda \leq g$. Relabel the indices, if necessary, so that $\pi_i \notin \mathcal{P}$ for $0 \leq i \leq \lambda$ and $\pi_i \in \mathcal{P}$ for $\lambda < i \leq 2g$.

Lemma 11.7. *Let $\bar{L}_K(u)$ be the reduction of $L_K(u)$ modulo p. Then λ is equal to the degree of $\bar{L}_K(u)$.*

Proof. $L_K(u) = \sum_{m=0}^{2g} a_m u^m = \prod_{i=1}^{2g}(1 - \pi_i u)$. Thus, each a_m is the m-th elementary symmetric function of the π_i. If $m > \lambda$, each term of the m-th elementary symmetric function contains a factor π_j with j in the range $\lambda < j \leq 2g$. Thus, $a_m \in \mathcal{P}$, which implies $p | a_m$. Now consider a_λ, the λ-th elementary symmetric function of the π_i. One of the terms is $\pi_1 \pi_2 \ldots \pi_\lambda$, which is not in $\mathcal{P}$. All the other terms are in $\mathcal{P}$. Thus, $a_\lambda \notin \mathcal{P}$. This implies that p does not divide a_λ, and so the degree of $\bar{L}_K(u)$ is λ, as asserted.

Later we will give a nice algebraic interpretation to the number λ, which is sometimes called the p-invariant of K. For the moment, we consider in more detail the case $\lambda = 0$, i.e., the situation when all the π_i are in $\mathcal{P}$. By the lemma, this is equivalent to $\bar{L}_K(u)$'s having degree zero. It also yields the following interesting congruences:

$$h(K_n) = \prod_{i=1}^{2g}(1 - \pi^n) \equiv 1 \pmod{\mathcal{P}}.$$

It follows from this that none of the class groups $J(\mathbb{F}_n)$ contain an element of order p. We make this property into a definition.

Definition. Let $K/\mathbb{F}$ be a function field over a finite field of constants $\mathbb{F}$. Let p be the characteristic of $\mathbb{F}$. We say that K is super-singular if for all integers $n > 0$, p does not divide $h(K_n)$.

This definition can be given in even greater generality. Assume F has characteristic $p > 0$ (but is not necessarily finite) and that K/F is a function field with F as its field of constants. One defines K to be super-singular if the class group of every constant field extension of K contains no element of order p.

The next proposition gives several equivalent conditions for a global function field to be supersingular. Before we state it, recall that for each integer $m \geq 0$, $b_m(K)$ denotes the number of effective divisors of K of degree m.

Proposition 11.8. *The following conditions are equivalent:*

  a) K is supersingular.

  b) All the π_i are in $\mathcal{P}$.

  c) The degree of $\bar{L}_K(u)$ is zero.

d) $h(K_n) \equiv 1 \pmod{\mathcal{P}}$ *for all* $n > 0$.

e) $b_m(K) \equiv 1 \pmod{p}$ *for* $1 \leq m \leq g$.

Proof. We have already seen that b implies a. To show the reverse, suppose that some $\pi_i \notin \mathcal{P}$. Then, for some $n > 0$ we have $\pi_i^n \equiv 1 \pmod{\mathcal{P}}$. This implies $p|h(K_n)$ contrary to a. Thus, a implies b so these two conditions are equivalent.

Lemma 11.7 shows the equivalence of b and c.

We have already seen that b implies d. If d holds then a obviously follows, and we have shown d implies b. Thus, the first four conditions are all equivalent.

Recall that $Z_K(u) = \sum_{n=0}^{\infty} b_n(K)u^n$ (see Chapter 5). From consideration of the equation $(1 - u)(1 - qu)Z_K(u) = L_K(u)$, we obtain the following congruence:

$$(1 - u) \sum_{n=0}^{\infty} b_n(K)u^n \equiv \sum_{m=0}^{2g} a_m u^m \pmod{p} .$$

It follows easily that $b_m(K) \equiv \sum_{k=0}^{m} a_k \pmod{p}$ for all $m \geq 0$ (we define $a_m = 0$ if $m > 2g$).

Now, assuming that c holds, it follows from the last paragraph that e is true. Assuming e, it follows again from the last paragraph, that $p|a_m$ for $1 \leq m \leq g$. Since $a_{2g-m} = q^g a_m$ for m in the same range, we have $p|a_m$ for $1 \leq m \leq 2g$. This is the same as c. All equivalences have been demonstrated.

Before giving two corollaries, we note that $h(K) \equiv b_g(K) \pmod{p}$. This is because $h(K) = L_K(1) = \sum_{m=0}^{2g} a_m \equiv \sum_{m=0}^{g} a_m \equiv b_g(K) \pmod{p}$ by what has been proven above.

Corollary 1. *Suppose K has genus 1. Then, K is supersingular if and only if $h(K) \equiv 1 \pmod{p}$.*

Corollary 2. *Suppose K has genus 2. Then, K is supersingular if and only if the number of primes of degree 1 and the class number, $h(K)$, are both congruent to 1 modulo p.*

Both corollaries are consequences of condition e and the above remark.

We are now going to consider not just the size of the class groups $J(\mathbb{F}_n)$, i.e., the class numbers $h(K_n)$, but the actual structure of these groups. To do this it will be necessary to use more algebraic tools than we have previously, and also, as already has been said, a number of fairly difficult results whose only known proofs require a substantial amount of algebraic geometry. So we will only sketch these developments and not attempt to provide proofs for everything. Nevertheless, we hope to tell a coherent and interesting mathematical story with enough detail so that the interested reader can, perhaps with some additional work, fill in the gaps.

We begin with some generalities about constant field extensions. For a while we only require that the constant field F of the function field K be a perfect field. With this assumption, we can use all the results about constant field extensions proven in Chapter 8.

The first result we will need is even more general.

Proposition 11.9. *Let L be a finite, unramified, Galois extension of the function field K. Let $G = \mathrm{Gal}(L/K)$. Then, $\mathcal{D}_L^G = i_{L/K}\mathcal{D}_K$. Here, $\mathcal{D}_K$ and $\mathcal{D}_L$ denote the divisor groups of K and L, respectively, and $\mathcal{D}_L^G$ denotes the divisors of L left fixed by the elements of G.*

Proof. If P is a prime of K, then, by definition, $i_{L/K}P = \sum_{\mathfrak{P}|P} e(\mathfrak{P}/P)\mathfrak{P}$. Since we are assuming that L/K is unramified, this becomes $i_{L/K}P = \sum_{\mathfrak{P}|P} \mathfrak{P}$.

Suppose $D = \sum_{\mathfrak{P}} a(\mathfrak{P})\mathfrak{P}$ is a divisor of L fixed by all the elements $\sigma \in G$. Applying σ to both sides, using $\sigma D = D$, and equating coefficients, yields the result that $a(\sigma\mathfrak{P}) = a(\mathfrak{P})$ for all primes $\mathfrak{P}$ of L. By Proposition 9.2, G acts transitively on the set of primes of L lying over a fixed prime P of K. It follows that D is a $\mathbb{Z}$-linear combination of divisors of the form $i_{L/K}P$. This shows that $D \in i_{L/K}\mathcal{D}_K$. Thus, $\mathcal{D}_L^G \subseteq i_{L/K}\mathcal{D}_K$. The converse is obvious.

From the definition, it is clear that $i_{L/K}$ gives a one-to-one homomorphism from $\mathcal{D}_K$ to $\mathcal{D}_L$. By the corollary to Proposition 7.8, $i_{L/K}$ induces a homomorphism from Cl_K to Cl_L. On this level it need not be one to one in general. However, in the case of separable constant field extensions it is one to one. Even more is true when the constant field extension is Galois.

Proposition 11.10. *Let K be a function field with constant field F. Let E be a finite, Galois extension of F and set $L = KE$. Let $G = \mathrm{Gal}(E/F) = \mathrm{Gal}(L/K)$. Then $i_{L/K} : Cl_K \to Cl_L$ is one to one, and $Cl_L^G = i_{L/K}Cl_K$.*

Proof. The proof will use some facts from cohomology of groups.

Let $\mathcal{P}_L$ and $\mathcal{P}_K$ be the principal divisors of L and K, respectively. Consider the exact sequence:

$$(0) \to E^* \to L^* \to \mathcal{P}_L \to (0) .$$

Passing to the long exact sequence and using the fact that $H^1(G, E^*) = (0)$ (Hilbert's Theorem 90), we find that the following sequence is exact

$$(0) \to F^* \to K^* \to \mathcal{P}_L^G \to (0) .$$

This shows that $i_{L/K}\mathcal{P}_K = \mathcal{P}_L^G$.

Using the same associated long exact sequence, but starting at the term $H^1(G, L^*)$, which is zero, again by Hilbert's Theorem 90, we find

$$(0) \to H^1(G, \mathcal{P}_L) \to H^2(G, E^*) \to H^2(G, L^*)$$

is exact. The third arrow can be shown to be one to one by a localization argument. It follows that $H^1(G, \mathcal{P}_L) = (0)$. In the case of finite constant fields, which is our principal interest, the result is even easier. In this case $H^2(G, E^*)$ is isomorphic to $F^*/N_{E/F}E^*$ since G is cyclic. For finite fields the norm map is onto. Thus, in this case, $H^2(G, E^*) = (0)$ and so $H^1(G, \mathcal{P}_L) = (0)$ as well.

Next, consider a second exact sequence:

$$(0) \to \mathcal{P}_L \to \mathcal{D}_L \to Cl_L \to (0) \ .$$

Passing to the associated long exact sequence, we find

$$(0) \to \mathcal{P}_L^G \to \mathcal{D}_L^G \to Cl_L^G \to H^1(G, \mathcal{P}_L)$$

is exact. By what we have already proven, the first term is $i_{L/K}\mathcal{P}_K$, the second is $i_{L/K}\mathcal{D}_K$ (by Proposition 11.9, since constant field extensions are unramified), and the fourth term is zero. Thus,

$$(0) \to \mathcal{P}_K \to \mathcal{D}_K \to Cl_L^G \to (0)$$

is exact if we define the third arrow to be the map $i_{L/K}$. This is equivalent to the assertions of the proposition, so the proof is complete.

Corollary. *With the same hypotheses as the proposition, $i_{L/K}$ restricted to Cl_K^o is one to one and $i_{L/K}Cl_K^o = (Cl_L^o)^G$.*

Proof. The restriction of a one-to-one map is still one to one, so the first assertion is obvious.

To prove the second assertion, suppose $\bar{\mathfrak{D}} \in (Cl_L^o)^G$. By the proposition, there is a class $\bar{D} \in Cl_K$ such that $i_{L/K}\bar{D} = \bar{\mathfrak{D}}$. By Proposition 7.7, $\deg_L i_{L/K}\bar{D} = \deg_K \bar{D}$. Since $\deg_L \bar{\mathfrak{D}} = 0$, it follows that $\deg_K \bar{D} = 0$, which completes the proof.

We now attempt to package all the class groups of constant field extensions of K into one big group. To do this we will need the notion of an infinite Galois extension of fields and the associated Galois group. Let L/K be an algebraic extension of fields, of finite or infinite degree. Let $\text{Aut}(L/K)$ be the group of all field automorphisms of L which leave K fixed. If K is the fixed field of $\text{Aut}(L/K)$, we say that L/K is a Galois extension and define $\text{Gal}(L/K) = \text{Aut}(L/K)$ to be its Galois group. It is easy to check that L/K is a Galois extension if and only if it is the union of all finite Galois extensions of K contained in L. The fundamental theorem of Galois theory relating subgroups of $\text{Gal}(L/K)$ with intermediate fields does not hold in the general case, but it can be reestablished by defining a topology on the Galois group. One proclaims a neighborhood basis for the identity element to be the set of subgroups of finite index. This leads to a unique topology on the Galois group which makes it into a topological group. This topology is called the Krull topology. It then turns out that the usual procedure now yields a one to one correspondence between closed subgroups of the Galois group and intermediate fields. All the standard properties of this correspondence continue to hold. One simply has to be careful that all subgroups under consideration are closed.

We continue to assume that F is perfect. Let $\bar{F}$ denote an algebraic closure of F. Then, one sees easily that $\bar{F}/F$ is an infinite Galois extension. We set $G_F = \mathrm{Gal}(\bar{F}/F)$. If $F \subseteq E \subseteq \bar{F}$ is an intermediate field we easily see that $\bar{F}/E$ is a Galois extension and we write $G_E = \mathrm{Gal}(\bar{F}/E)$. G_E is a closed subgroup of G_F. It is a normal subgroup if and only if E is a Galois extension of F. In this case $\mathrm{Gal}(E/F) \cong G_F/G_E$.

We now return to the situation where K is a function field over a perfect constant field F. For every field E between F and $\bar{F}$, we define $J(E)$ to be the group Cl^o_{KE}. Proposition 11.10 and its corollary which we proved for finite constant field extensions continue to hold for infinite constant field extensions. Thus, there are one-to-one maps from all the groups $J(E)$ into $J(\bar{F})$ given by extension of divisors. We will identify $J(E)$ with its image in $J(\bar{F})$. In this way all the groups $J(E)$ are subgroups of $J(\bar{F})$ and one can use the corollary to Proposition 11.10 to show that $J(\bar{F})^{G_E} = J(E)$. Of course, the same corollary shows that if E/F is Galois, then $J(E)^{\mathrm{Gal}(E/F)} = J(F)$.

Another useful property is that $J(\bar{F})$ is the union of the groups $\{J(E) \mid [E : F] < \infty\}$. The idea of the proof is to use Propositions 8.10 and 8.11 to show that every prime $\mathfrak{P}$ of $K\bar{F}$ is the extension of a prime coming from a finite level. If P is a prime of K lying below $\mathfrak{P}$, then Proposition 8.11 shows there is a finite extension E/F such that P splits into a product of primes of degree 1 in KE. Let $\mathfrak{p}$ be the prime in KE lying below $\mathfrak{P}$. Then Proposition 8.10 can be used to show that $\mathfrak{P}$ is the extension of $\mathfrak{p}$ to $K\bar{F}$.

It will simplify the notation to simply call $J(\bar{F}) = J$. The advantage of considering the group J is that algebraic geometry gives a method for determining a great deal about its structure. We then attempt to use this information to investigate the structure of the groups $J(E)$ where E is a finite extension of F.

Theorem 11.12. *Let K/M be a function field over an algebraically closed field of constants M. Set $J = Cl^o_K$. Then, J is a divisible group (for all integers n, the map $x \to nx$ is onto). Denote by g the genus of K. If l is a prime different from the characteristic of M, then the l-primary subgroup of J, $J(l)$, is the direct sum of $2g$ copies of $\mathbb{Q}_l/\mathbb{Z}_l$. If the characteristic of M is $p > 0$, then $J(p)$ is the direct sum of γ copies of $\mathbb{Q}_p/\mathbb{Z}_p$ where $0 \le \gamma \le g$.*

In the case where $M = \mathbb{C}$, the complex numbers, it is a classical theorem (Abel-Jacobi) that J is isomorphic to the direct sum of g copies of $\mathbb{C}$ modulo a lattice Λ of maximal rank, i.e., $J \cong \mathbb{C}^g/\Lambda$. Λ must be a free $\mathbb{Z}$ module of rank $2g$. It follows that the elements of J of order dividing l^n constitute a group isomorphic to $\frac{1}{l^n}\Lambda/\Lambda \cong \bigoplus_1^{2g} \frac{1}{l^n}\mathbb{Z}/\mathbb{Z}$. Passing to the limit as $n \to \infty$ gives the theorem.

The first proof of the theorem in the abstract case was due to Weil [2]. The reader can also consult Mumford [1].

The group $\mathbb{Q}_l/\mathbb{Z}_l$ has a very simple structure. It is the union of the subgroups $\frac{1}{l^n}\mathbb{Z}_l/\mathbb{Z}_l$ which are isomorphic to $\mathbb{Z}_l/l^n\mathbb{Z}_l \cong \mathbb{Z}/l^n\mathbb{Z}$, a cyclic group of order l^n. Thus, there is a one-to-one correspondence between proper subgroups of $\mathbb{Q}_l/\mathbb{Z}_l$ and non-negative powers of l. This allows us to deduce a simple corollary from the theorem.

Corollary. *Define $J[N]$ to be the subgroup of J consisting of elements whose order divides N. If N is not divisible by the characteristic of M, then*

$$J[N] \cong \bigoplus_1^{2g} \mathbb{Z}/N\mathbb{Z} .$$

Proof. By the Chinese Remainder Theorem, it is enough to check the result when $N = l^n$ is a power of a prime different from the characteristic. In this case the corollary is immediate from the theorem and the above remark.

Let $Y \subset J$ be a finite subgroup of J which is invariant under G_F, i.e., $\sigma y \in Y$ for all $\sigma \in G_F$ and all $y \in Y$. For example, $Y = J[N]$ is such a subgroup since $Ny = 0$ implies $0 = \sigma(Ny) = N(\sigma y)$. It is immediate that each $\sigma \in G_F$ induces a group automorphism of Y. Thus, we get a map $G_F \to \text{Aut}(Y)$. This is easily seen to be a group homomorphism. The kernel is a normal subgroup H of G_F of finite index. By definition, such a group is closed and thus it corresponds uniquely to its fixed field which we will call $F(Y)$. Thus, $H = G_{F(Y)}$. Since $G_{F(Y)}$ fixes Y we see that $Y \subseteq J^{G_{F(Y)}} = J(F(Y))$. $F(Y)$ is called the field of rationality of Y. It is the smallest extension E/F such that $Y \subseteq J(E)$.

The exact sequence

$$(0) \to G_{F(Y)} \to G_F \to \text{Aut}(Y),$$

gives rise to a monomorphism from $\text{Gal}(F(Y)/F) \cong G_F/G_{F(Y)} \to \text{Aut}(Y)$.

Suppose N is a positive integer prime to the characteristic of F. By the Corollary to Proposition 11.12 we have

$$J[N] \cong \bigoplus_1^{2g} \mathbb{Z}/N\mathbb{Z} .$$

It follows that $\text{Aut}(J[N]) \cong GL_{2g}(\mathbb{Z}/N\mathbb{Z})$. Putting it all together we get a monomorphism

$$\rho_N : \text{Gal}(F(J[N])/F) \to GL_{2g}(\mathbb{Z}/N\mathbb{Z}) . \tag{1}$$

This "Galois representation" plays a big role in the more advanced arithmetic theory of curves. Here we will leave off this general development and return to the special case where the constant field is finite. So, once again, let us suppose that $F = \mathbb{F}$, a finite field with q elements. Let $\phi_q \in G_{\mathbb{F}} = \mathrm{Gal}(\bar{\mathbb{F}}/\mathbb{F})$ be the automorphism defined by $\phi_q(\alpha) = \alpha^q$ for all $\alpha \in \bar{\mathbb{F}}$. For every positive integer m, let $\mathbb{F}_m \subseteq \bar{\mathbb{F}}$ be the unique intermediate field with $[\mathbb{F}_m : \mathbb{F}] = m$. The restriction of ϕ_q to $\mathbb{F}_m$ generates the Galois group $\mathrm{Gal}(\mathbb{F}_m/\mathbb{F})$. The image of ϕ_q in this group has order m. It follows that the subgroup of $G_{\mathbb{F}}$ corresponding to the field $\mathbb{F}_m$ is the closure of the cyclic group $< \phi_q^m >$ generated by ϕ_q^m in $G_{\mathbb{F}}$. We note the fact that $G_{\mathbb{F}}$ is isomorphic to the inverse limit of the groups $\mathbb{Z}/m\mathbb{Z}$, i.e. the group $\hat{\mathbb{Z}}$, the completion of $\mathbb{Z}$ with respect to all subgroups of finite index.

Proposition 11.13. *Let $J[N] \subset J$ be the subgroup of J consisting of elements whose order divides N. Then, $[\mathbb{F}(J[N]) : \mathbb{F}]$ is equal to the order of the matrix $\rho_N(\phi_q)$ (see Equation 1 above).*

Proof. Since ρ_N is a monomorphism, the order of $\rho_N(\phi_q)$ is the smallest power of ϕ_q which is the identity on $\mathbb{F}(J[N])$. By the Galois theory of finite fields, this number is the dimension $[\mathbb{F}(J[N]) : \mathbb{F}]$.

Proposition 11.14. *For all but finitely many primes l, the dimension $[\mathbb{F}(J[l]) : \mathbb{F}]$ is prime to l.*

Proof. We will need to use one of those basic facts about J whose proof involves a substantial amount of algebraic geometry. It concerns the characteristic polynomial of the matrix $\rho_l(\phi_q)$. We will assume $l \neq p$, the characteristic of $\mathbb{F}$.

Let $L_K(u)$ be numerator of the zeta function $Z_K(u)$. Define $L^*(u) = u^{2g}L_K(1/u)$. $L^*(u)$ is a monic polynomial of degree $2g$ with coefficients in $\mathbb{Z}$. Let $D \in \mathbb{Z}$ denote the discriminant of this polynomial. The fact we need is that the characteristic polynomial of $\rho_l(\phi_q)$ is the reduction of $L^*(u)$ modulo l. This is implied by the characterization, due to Weil, of $L^*(u)$ as the "characteristic polynomial of Frobenius" acting on the Tate module at l. See Milne [1], Mumford [1], or, for the original formulation, Weil [2]. It follows from this that if l does not divide D, then the characteristic polynomial of $\rho_l(\phi_q)$ does not have repeated factors in $\mathbb{Z}/l\mathbb{Z}[u]$. By linear algebra $\rho(\phi_q)$ is diagonalizable over the algebraic closure of $\mathbb{Z}/l\mathbb{Z}$. The eigenvalues , which are non-zero since $l \neq p$, have order prime to l. It follows that the matrix $\rho_l(\phi_q)$ has order prime to l. The proposition is now a consequence of Proposition 11.13.

Because of these propositions, among other reasons, it is of interest to investigate the structure of the matrix groups $GL_r(\mathbb{Z}/N\mathbb{Z})$ where r and N are positive integers. We will sketch some of the main facts about them.

Suppose $N = l_1^{m_1} l_2^{m_2} \ldots l_t^{m_t}$ is the prime decomposition of N. Then using the Chinese Remainder Theorem, it can be seen that

$$GL_r(\mathbb{Z}/N\mathbb{Z}) \cong GL_r(\mathbb{Z}/l_1^{m_1}\mathbb{Z}) \times GL_r(\mathbb{Z}/l_2^{m_2}\mathbb{Z}) \times \ldots GL_r(\mathbb{Z}/l_t^{m_t}\mathbb{Z}) \ .$$

This reduces the problem to the structure of the groups $GL_r(\mathbb{Z}/l^m\mathbb{Z})$, where l is a prime number. If $m > 1$, reduction modulo l gives rise to an exact sequence:

$$(I) \to I + lM_r(\mathbb{Z}/l^m\mathbb{Z}) \to GL_r(\mathbb{Z}/l^m\mathbb{Z}) \to GL_r(\mathbb{Z}/l\mathbb{Z}) \to (I) \ .$$

Here, I denotes the identity $r \times r$ matrix and $M_r(\mathbb{Z}/l^m\mathbb{Z})$ the ring of $r \times r$ matrices with coefficients in $\mathbb{Z}/l^m\mathbb{Z}$. The order of the group $I + lM_r(\mathbb{Z}/l^m\mathbb{Z})$ is $l^{r^2(m-1)}$. It is an l-group whose structure can be investigated more closely. However, we will not enter into further details about this here.

The size of the group $GL_r(\mathbb{Z}/l\mathbb{Z})$ is given by

$$\begin{aligned} |GL_r(Z/l\mathbb{Z})| &= (l^r - 1)(l^r - l) \ldots (l^r - l^{r-1}) \\ &= l^{r(r-1)/2}(l^r - 1)(l^{r-1} - 1) \ldots (l - 1) \ . \end{aligned}$$

The proof of this is obtained by noticing that an $r \times r$ matrix with coefficients in a field is invertible if and only if its rows are linearly independent. Thus, the first row must be non-zero. It can be any of $l^r - 1$ row vectors. The second row must be linearly independent from the first, so it cannot be a multiple of the first row. Thus, there are $l^r - l$ choices for the second row. There are l^2 vectors in the linear span of the first two rows, so there are $l^r - l^2$ choices for the third row. The general pattern is now clear.

We summarize a portion of this discussion as follows.

Proposition 11.15. *The group $GL_r(\mathbb{Z}/l^m\mathbb{Z})$ has order equal to*

$$l^{f(r,m)}(l^r - 1)(l^{r-1} - 1) \ldots (l - 1)$$

where $f(r, m) = (m - 1)r^2 + r(r - 1)/2$.

Note, in particular, that $GL_r(\mathbb{Z}/l^m\mathbb{Z})$ has a large l-Sylow subgroup whose order depends on m, whereas the "prime to l" part of the group has order which is independent of m. This will be of importance later.

We have developed most of the tools we will need for the next task, which is to give a structural, algebraic interpretation to Theorems 11.5 and 11.6. To recall the situation, fix a prime l and consider the tower of fields $K \subset K_l \subset K_{l^2} \subset \ldots$. The two theorems in question were about the way the class numbers $h(K_{l^n})$ behave as a function of n. We will now look at the more general question of how the groups $J(\mathbb{F}_{l^n})$ behave as a function of n.

It will be convenient to define $\mathbb{F}_{l^\infty}$ to be the union of the fields $\mathbb{F}_{l^n}$ where n varies over the positive integers. It is not hard to check that for every finite extension $\mathbb{E}$ of $\mathbb{F}_{l\infty}$ we have $[\mathbb{E} : \mathbb{F}_{l\infty}]$ is prime to l. Thus, $F_{l\infty}$ is characterized as the maximal l-extension of $\mathbb{F}$ in $\bar{\mathbb{F}}$. Note that the group $J(\mathbb{F}_{l\infty})$ is the union of the groups $J(\mathbb{F}_{l^n})$ over all n.

We first consider the l-primary components. Recall

$$J(\mathbb{F}_{l\infty})(l) \subseteq J(l) \cong \bigoplus_1^{2g} \mathbb{Q}_l/\mathbb{Z}_l . \qquad (2)$$

Proposition 11.16. *We have the following group isomorphism:*

$$J(\mathbb{F}_{l\infty})(l) \cong \bigoplus_1^{r_l} \mathbb{Q}_l/\mathbb{Z}_l,$$

where r_l is the dimension over $\mathbb{Z}/l\mathbb{Z}$ of $J[l] \cap J(\mathbb{F}_{l\infty})$.

Proof. (Sketch) We first show that $J(\mathbb{F}_{l\infty})(l)$ is a divisible group. Consider the exact sequence of $G_{\mathbb{F}}$ modules

$$(0) \to J[l] \to J(l) \to J(l) \to (0) ,$$

where the third arrow is the map "multiplication by l." This is onto since J is a divisible group and this easily implies that the l-primary component of J is divisible. Let $H \subset G_{\mathbb{F}}$ be the subgroup corresponding to $\mathbb{F}_{l\infty}$. Passing to the associated long exact sequence we find

$$J(\mathbb{F}_{l\infty})(l) \to J(\mathbb{F}_{l\infty})(l) \to H^1(H, J[l])$$

is exact. The first arrow is, again, multiplication by l. Every finite quotient of H is prime to l. It follows that $H^1(H, J[l]) = (0)$. Thus, multiplication by l is onto and $J(\mathbb{F}_{l\infty})(l)$ is divisible.

From Equation 2 it now follows that

$$J(\mathbb{F}_{l\infty})(l) \cong \bigoplus_1^{r_l} \mathbb{Q}_l/\mathbb{Z}_l ,$$

where $1 \leq r_l \leq 2g$. If one considers the subgroups of both sides of this isomorphism consisting of the elements of order dividing l we obtain the characterization of r_l given in the proposition.

The following group theoretic lemma and its corollaries are the key to understanding the behavior of the groups $J(\mathbb{F}_{l^n})(l)$.

Lemma 11.17. *Suppose l is an odd prime, and that A is an abelian group isomorphic to $\bigoplus_1^r \mathbb{Q}_l/\mathbb{Z}_l$. Let $\phi : A \to A$ be an endomorphism. Define $A_0 = \{x \in A \mid \phi(x) = x\}$ and $A_1 =: \{x \in A \mid \phi^l(x) = x\}$. Suppose that A_0 contains $A[l] = \{a \in A \mid la = 0\}$. Then, we have*

$$A_1 = \{x \in A \mid lx \in A_0\} .$$

If $l = 2$, the same result holds provided we assume A_0 contains $A[4]$.

Proof. We assume l is odd and begin by showing there is an endomorphism ψ of A such that $\phi = I + l\psi$ where I is the identity endomorphism.

Since A is divisible, given $x \in A$, there is a $y \in A$ such that $ly = x$. Set $\psi(x) = \phi(y) - y$. This is well defined, because if $ly' = x$, then $y - y' \in A[l]$. Since $\phi - I$ vanishes on $A[l]$ by hypothesis, we must have $\phi(y) - y = \phi(y') - y'$. It is easily verified that ψ is an endomorphism. Finally, $l\psi(x) = \phi(ly) - ly = \phi(x) - x$, so $\phi = I + l\psi$ as asserted.

Thus,

$$\phi^l = (I + l\psi)^l = I + \binom{l}{1}l\psi + \binom{l}{2}l^2\psi^2 + \cdots = I + l^2\psi(I + l\mu) \;,$$

where μ is an endomorphism of A which commutes with ψ.

The endomorphism $I + l\mu$ is invertible because A being a torsion group implies that the formal inverse $(I + l\mu)^{-1} = I - l\mu + l^2\mu^2 - \cdots$ gives an actual inverse.

We find that $\phi^l(x) = x$ if and only if $l^2\psi(1 + l\mu)(x) = 0$ if and only if $l^2\psi(x) = 0$ (since ψ and $1 + l\mu$ commute). This last condition can be rewritten as $l\psi(lx) = 0$. Adding lx to both sides, we see this condition is equivalent to $lx + l\psi(lx) = lx$, or, what is the same $\phi(lx) = lx$. We have shown that x is fixed by ϕ^l if and only if lx is fixed by ϕ. This is equivalent to the statement of the lemma.

If $l = 2$ and $A[4] \subseteq A_0$, the proof is entirely similar. We leave the details to the reader.

Corollary 1. *In addition to the hypotheses of the lemma, assume that A_0 is finite. Then,*

$$A_0 \cong \bigoplus_{i=1}^{r} l^{-\nu_i}\mathbb{Z}_l/\mathbb{Z}_l \quad \text{and} \quad A_1 \cong \bigoplus_{i=1}^{r} l^{-\nu_i-1}\mathbb{Z}_l/\mathbb{Z}_l \;.$$

Proof. Since $A[l] \subset A_0$ and A_0 is finite, it follows that A_0 is the direct sum of r cyclic groups each of l-power order. This gives the first assertion.

Rephrasing what has been shown so far, A_0 has a set of generators $\{e_1, e_2, \ldots, e_r\}$ such that $\sum_{i=1}^{r} n_i e_i = 0$ if and only if $l^{\nu_i} | n_i$ for $1 \le i \le r$. Since A is divisible, for each i there is an element $e_i^1 \in A$ with $le_i^1 = e_i$. Let A_1' be the group generated by the set $\{e_1^1, e_2^1, \ldots, e_r^1\}$. Clearly, $A_1' \subseteq A_1$. They both have the same number of elements since multiplication by l maps both sets onto A_0 and the kernel in both cases is $A[l]$. Thus, $A_1' = A_1$. It is now a simple matter to show $\sum_{i=1}^{r} n_i e_i^1 = 0$ if and only if $l^{\nu_i+1} | n_i$ for $1 \le i \le r$. This is equivalent to the second isomorphism in the statement of the Corollary.

Corollary 2. *With the notation of Corollary 1, define $A_n = \{x \in A \mid \phi^{l^n}(x) = x\}$. Then,*

$$A_n \cong \bigoplus_{i=1}^{r} l^{-\nu_i-n}\mathbb{Z}_l/\mathbb{Z}_l \;.$$

Proof. The proof is by induction on n. Corollary 1 shows the result is true for $n = 1$. Now assume it is true for $n - 1$. Then apply Corollary 1 with A_0 replaced by A_{n-1} and ϕ replaced by $\phi^{l^{n-1}}$. The result follows.

We remark that the proofs may be given using properties of modules over $\mathbb{Z}_l$. The Pontyagin dual of A, $\hat{A}$, is a free module of rank r over $\mathbb{Z}_l$ and ϕ induces an endomorphism of $\hat{A}$. One can then prove the dual statements to those in the lemma and its corollaries by using properties of modules over $\mathbb{Z}_l$. Then dualizing once again we get what we want. It seemed more straightforward to work directly with A.

The next result is, perhaps, the main result of this chapter.

Theorem 11.18. *Let $K/\mathbb{F}$ be a function field of genus g over a finite field $\mathbb{F}$ with q elements. Let $J = Cl^o(K\bar{\mathbb{F}})$ and define r_l to be the dimension over $\mathbb{Z}/l\mathbb{Z}$ of $J[l] \cap J(\mathbb{F}_{l^\infty})$. There is an integer $n_0 > 0$ and integers ν_i with $1 \le i \le r_l$ such that for all $n \ge n_0$ we have*

$$J(\mathbb{F}_{l^n})(l) \cong \bigoplus_1^{r_l} l^{-\nu_i + n_0 - n} \mathbb{Z}_l / \mathbb{Z}_l .$$

Proof. Define n_0 by the equation $\mathbb{F}_{l^\infty} \cap \mathbb{F}(J[l]) = \mathbb{F}_{l^{n_0}}$. It is not hard to see that every element in $J[l] \cap J(\mathbb{F}_{l^\infty})$ is rational over this field. Invoking Proposition 11.16, we find

$$J(\mathbb{F}_{l^\infty})(l) \cong \bigoplus_1^{r_l} \mathbb{Q}_l / \mathbb{Z}_l .$$

Also, there exist integers ν_i for $1 \le i \le r_l$ such that for each i, $\nu_i > 0$, and

$$J(\mathbb{F}_{l^{n_0}})(l) \cong \bigoplus_1^{r_l} l^{-\nu_i} \mathbb{Z}_l / \mathbb{Z}_l .$$

Now, define $A = J(\mathbb{F}_{l^\infty})(l)$, $A_0 = J(\mathbb{F}_{l^{n_0}})(l)$, and $\phi = \phi_q^{l^{n_0}}$. One sees that ϕ is the Frobenius automorphism for the extension $\bar{\mathbb{F}}/\mathbb{F}_{n_0}$ and that the triple A, A_0, and ϕ satisfy the hypotheses of Lemma 11.17. Invoking Corollary 2 to that lemma, we see that for all $m \ge 0$ we have

$$J(\mathbb{F}_{l^{n_0+m}})(l) \cong \bigoplus_1^{r_l} l^{-\nu_i - m} \mathbb{Z}_l / \mathbb{Z}_l .$$

If we simply set $n = n_0 + m$, the theorem follows.

Corollary. *For all $n \ge n_0$, we have*

$$\operatorname{ord}_l h(K_{l^n}) = r_l n + \nu_l ,$$

where $\nu_l = \operatorname{ord}_l h(K_{l^{n_0}}) - r_l n_0$.

Proof. We recall that $h(K_{l^n})$ is the order of $J(\mathbb{F}_{l^n})$. From the isomorphism given in the theorem, we see

$$\text{ord}_l \; h(K_{l^n}) = \sum_{i=1}^{r_l}(\nu_i + n - n_0) = r_l n + \sum_{i=1}^{r_l}(\nu_i - n_0) \; .$$

Define $\nu_l = \sum_{i=1}^{r_l}(\nu_i - n_0)$ (and ignore the fact that the notation here is somewhat ambiguous). If we set $n = n_0$ in the above formula, we get the characterization of ν_l given in the corollary.

The reader will not fail to notice that the corollary is a sharpened version of Proposition 11.5. The qualitative content is exactly the same, but now we have given group theoretic interpretations of the constants r_l and ν_l. Moreover, even n_0 is now made more precise.

The situation becomes much simpler when $\mathbb{F}_{l^\infty} \cap \mathbb{F}(J[l]) = \mathbb{F}$. By Proposition 11.14, this happens for all but finitely many primes l. In this case, $n_0 = 0$. Thus, r_l is just the number of cyclic groups whose sum is $J(\mathbb{F})(l)$ and the regular behavior of the l primary parts of the class group begins at the first step. Also, in this case $\nu_l = \text{ord}_l \; h(K)$. Everything is really simple!

We can now give a group theoretic interpretation of Proposition 11.6 as well. For notational convenience we will use k for the second prime instead of l'.

Theorem 11.19. *Let l and k be primes with $l \neq k$. Assume $k \neq p$. Define n_0 as follows*

$$n_0 = \text{ord}_l \; \prod_{i=1}^{2g}(k^i - 1) \; .$$

Then, $J(\mathbb{F}_{l^\infty}) \cap J(k) \subseteq J(\mathbb{F}_{l^{n_0}})$.

Proof. Let $P \in J(\mathbb{F}_{l^\infty})$ have order k^m. Then, $P \in J[k^m]$ and so P is rational over both $\mathbb{F}_{l^\infty}$ and $\mathbb{F}(J[k^m])$, and is thus rational over their intersection.

By Equation 1 (just before Proposition 11.13) we have a monomorphism

$$\text{Gal}(\mathbb{F}(J[k^m])/\mathbb{F}) \to GL_{2g}(\mathbb{Z}/k^m\mathbb{Z}) \; .$$

By Proposition 11.15, the later group has an order which is a power of k times $\prod_{i=1}^{2g}(k^i - 1)$. Thus, the highest power of l which divides $[\mathbb{F}(J[k^m]) : \mathbb{F}]$ is less than or equal to n_0. It follows that

$$\mathbb{F}_{l^\infty} \cap \mathbb{F}(J[k^m]) \subseteq \mathbb{F}_{l^{n_0}} \; .$$

This containment holds for all $m \geq 0$. Thus,

$$\mathbb{F}_{l^\infty} \cap \mathbb{F}(J(k)) \subseteq \mathbb{F}_{l^{n_0}} \; ,$$

which is equivalent to the theorem.

It is clear that Theorem 11.19 implies Theorem 11.6 together with the improvement that we get an upper bound on n_0. It also gives some added insight into why the result is true.

The restriction in the statement of the theorem that $k \neq p$ is not essential. One simply has to modify the definition of n_0 slightly and then the result holds when $k = p$ as well.

The final theorem of this chapter concerns the reduction of the polynomial $L_K(u)$ modulo p. In Proposition 11.7 we showed this reduced polynomial has degree λ where $0 \leq \lambda \leq g$. In Proposition 11.8, we showed that K is supersingular if and only if $\lambda = 0$. This can be restated as, $J[p] = (0)$ if and only if $\lambda = 0$. Using the tools developed to this point we can give a far-reaching generalization of this.

Proposition 11.20. *With previous notations, let γ be the dimension over $\mathbb{Z}/p\mathbb{Z}$ of $J[p]$. Then, γ is equal to λ, the degree of the polynomial $L_K(u)$ reduced modulo p.*

Proof. Let $L_K(u) = \prod_{i=1}^{2g}(1 - \pi_i u)$ be the factorization of $L_K(u)$ over the algebraic closure of the p-adic numbers. Let E be the subfield of $\bar{\mathbb{Q}}_p$ obtained by adjoining the elements π_i to $\mathbb{Q}_p$. Finally, let $\mathcal{P}$ be the maximal ideal of the integral closure of $\mathbb{Z}_p$ in E.

By the above remarks we can assume $\lambda \geq 1$. By using the proof of Proposition 11.7 we can assume $\mathrm{ord}_p \pi_i = 0$ for $1 \leq i \leq \lambda$ and $\mathrm{ord}_p \pi_i > 0$ for $\lambda < i \leq 2g$. By passing to a constant field extension $K_n = K\mathbb{F}_n$, the elements π_i are replaced by π_i^n. Using the proof of Lemma 11.7, we easily see that the degrees of $L_K(u)$ modulo p and $L_{K_n}(u)$ modulo p are the same. By an appropriate choice of n we can insure $\pi_i^n \equiv 1 \pmod{\mathcal{P}}$ for $1 \leq i \leq \lambda$. By passing to a further constant field extension $K_{nm} = K\mathbb{F}_{nm}$, we can insure $J[p] \subseteq J(\mathbb{F}_{nm})$. This does not affect the congruences already established, so we can assume from the beginning that $\mathrm{ord}_p(\pi_i - 1) > 0$ for $1 \leq i \leq \lambda$ and $J[p] \subseteq J(\mathbb{F})$.

Since $J[p] \subseteq J(\mathbb{F})$ it follows from the proof of Theorem 11.18 and its Corollary that $\mathrm{ord}_p h(K_{p^n}) = \gamma n + \nu$ for all n sufficiently large. We will now establish a similar formula with γ replaced with λ.

Recall that

$$h(K_{p^n}) = \prod_{i=1}^{2g}(1 - \pi_i^{p^n}) .$$

The terms in the product with $\lambda < i \leq 2g$ are units since in this range $\mathrm{ord}_p \pi_i > 0$. Thus,

$$\mathrm{ord}_p h(K_{p^n}) = \sum_{i=1}^{\lambda} \mathrm{ord}_p(1 - \pi_i^{p^n}) . \tag{3}$$

In the proof of Proposition 11.5, we showed that if $\mathrm{ord}_p(1 - \pi_i) > 0$, then for all sufficiently large n, $\mathrm{ord}_p(1 - \pi_i^{p^n}) = n + c_i$. It now follows from

Equation 3 that $\operatorname{ord}_p h(K_{p^n}) = \lambda n + c$ for all sufficiently large n, where $c = \sum_{i=1}^{\lambda} c_i$.

We have now established that $\operatorname{ord}_p h(K_{p^n}) = \gamma n + \nu = \lambda n + c$ for all sufficiently large n. This can only happen if $\lambda = \gamma$.

This theorem is well known, but the usual proof uses much more sophisticated methods, e.g., crystalline cohomology. The above proof is due to H. Stichtenoth.

In the late 1950s, K. Iwasawa began publishing papers which carry over some of the theory developed in this chapter into algebraic number theory. The key idea is that constants in function fields are like roots of unity in number fields. To see why this is so, let K/F be a function field over a field of constants F. Let $C_P > 1$ be a constant and define for $x \in K$ and P a prime divisor of K,

$$|x|_P = C_P^{-\operatorname{ord}_P(x)} .$$

It is easily checked that $|x|_P$ is a non-archimedean valuation of K; i.e., it satisfies

$$(a) \ |0|_P = 0 \quad (b) \ |1|_P = 1$$
$$(c) \ |xy|_P = |x|_P|y|_P \quad (d) \ |x+y|_P \leq \max(|x|_P, |y|_P) .$$

Up to a standard equivalence relation this set of valuations is a complete set of valuations of K. By Proposition 5.1, an element $x \in K^*$ is a constant if and only if $\operatorname{ord}_P(x) = 0$ for all $P \in \mathcal{S}_K$. This is equivalent to saying that $x \in K^*$ is a constant if and only if $|x|_P = 1$ for all $P \in \mathcal{S}_K$.

If K is an algebraic number field, let O_K be the ring of integers. If P is a maximal ideal in O_K and $x \in K$, define

$$|x|_P = NP^{-\operatorname{ord}_P(x)} .$$

This is a non-archimedean valuation for each maximal ideal $P \subset O_K$. The equivalence classes of these valuations are called the finite primes of K. In addition to the non-archimedean valuations there are finitely many archimedean valuations. These are obtained by imbedding K into the complex numbers and then applying the usual absolute value. By definition, these archimedean valuations correspond to primes at infinity. It is a well known theorem, due to Kronecker, that $x \in K^*$ is a root of unity if and only if $|x|_P = 1$ for all primes of K, both finite and infinite.

From this discussion, it is clear that constants and roots of unity are analogous concepts, so it makes sense to think of cyclotomic extensions of a number field as analogous to constant field extensions of a function field. This is exactly what Iwasawa did.

What is the analogue of the cyclic l-towers of constant field extensions we have considered in this chapter? Consider the cyclotomic field $\mathbb{Q}(\zeta_{l^{n+1}})$

where $\zeta_{l^{n+1}}$ is a primitive l^{n+1}-th root of unity in a fixed algebraic closure $\bar{\mathbb{Q}}$ of $\mathbb{Q}$. The dimension of this field over $\mathbb{Q}$ is $l^n(l-1)$ and there is a unique subfield B_{l^n} such that $[B_{l^n} : \mathbb{Q}] = l^n$. The tower $\mathbb{Q} \subset B_l \subset B_{l^2} \subset \dots$ is called the basic $\mathbb{Z}_l$ extension of $\mathbb{Q}$. Note that $\mathrm{Gal}(B_{l^n}/\mathbb{Q})$ is a cyclic group of order l^n. Let $B_{l^\infty} = \cup_{n=0}^\infty B_{l^n}$. The Galois group of B_{l^∞} over $\mathbb{Q}$ is isomorphic to $\mathbb{Z}_l$, which is why this tower of fields is called a $\mathbb{Z}_l$-tower.

If K is an arbitrary number field we define K_{l^∞} to be the compositum $K B_{l^\infty}$. The Galois group of K_{l^∞} over K is isomorphic to $\mathrm{Gal}(B_{l^\infty}/K \cap B_{l^\infty})$ which is isomorphic to a subgroup of $\mathbb{Z}_l$ of finite index and is thus isomorphic to $\mathbb{Z}_l$. It follows that K_{l^∞} is a $\mathbb{Z}_l$ extension of K and there is a tower of extensions $K \subset K_l \subset K_{l^2} \subset \dots K_{l^\infty}$ with K_{l^n} being a cyclic extension of K of degree l^n. The field K_{l^∞} is taken to be the analogue of the constant field extension $K\mathbb{F}_{l^\infty}$. It is called the cyclotomic $\mathbb{Z}_l$ extension of K.

One can now ask the question if there is an analogue of Proposition 11.5? Is there an asymptotic formula for $\mathrm{ord}_l\, h(K_{l^n})$ in the number field case? It is a remarkable fact that the answer is yes. The zeta function of a number field is a much more complicated and mysterious object than that of a function field. Also, there is no obvious appeal to algebraic geometry as in the case of function fields. Nevertheless, Iwasawa was able to prove the following result (see Iwasawa [1] and [3], Lang [6], and Washington [1]).

Theorem 11.21. (K. Iwasawa) *Let K be an algebraic number field, l a prime number, and K_{l^∞} the cyclotomic $\mathbb{Z}_l$ extension of K. Let $h(K_{l^n})$ be the class number of K_{l^n}. Then there are integers λ_l, μ_l, ν_l, and n_0, such that for all $n \geq n_0$,*

$$\mathrm{ord}_l\, h(K_{l^n}) = \lambda_l n + \mu_l l^n + \nu_l .$$

Actually, Iwasawa was able to prove this for any $\mathbb{Z}_l$ extension of K. An infinite number field L is said to be a $\mathbb{Z}_l$ extension of K if it is a Galois extension and $\mathrm{Gal}(L/K) \cong \mathbb{Z}_l$. In general, there exist many such extensions in addition to the cyclotomic $\mathbb{Z}_l$ extension.

The formula given in the theorem is of precisely the same type as that given in Proposition 11.5 and the analogy is even more precise if $\mu_l = 0$. Iwasawa conjectured that $\mu_l = 0$ for the cyclotomic $\mathbb{Z}_l$ extension. For the case of general number fields this is still an open question. However, if K is a Galois extension of $\mathbb{Q}$ with $\mathrm{Gal}(K/\mathbb{Q})$ abelian, then Ferraro and L. Washington were able to show $\mu_l = 0$ (see Washington [1], Chapter 7), so in these cases the analogy is perfect.

It is worth pointing out that μ_l is not always zero. Iwasawa gave examples of non-cyclotomic $\mathbb{Z}_l$ extensions where $\mu_l \neq 0$ and even showed that μ_l can be made to be as large as you like (see Iwasawa [2]).

Washington was also able to show that the analogue of Proposition 11.6 is true for cyclotomic $\mathbb{Z}_l$ of number fields (see Washington [1], Chapter 16).

There are many interesting open questions remaining in Iwasawa's theory of number fields. There are also generalizations of these concepts to arithmetic-geometric contexts. For example, if A is an abelian variety over a number field K and L is a $\mathbb{Z}_l$ extension of K, is there an asymptotic formula for the Mordell-Weil rank of $A(K_{l^n})$ as a function of n? This is the subject of a fascinating paper of B. Mazur [1]. However, we will have to be content with pointing out these directions and pass on to other matters.

Exercises

1. Let $K/\mathbb{F}$ be a function field over a finite field with q elements. Let $L_K(u) = \sum_{i=0}^{2g} a_k u^k$ be the numerator of the zeta-function of K. Show that the functional equation (see Chapter 5) implies that $q^{g-k} a_k = a_{2g-k}$ for $0 \le k \le g$.

2. Suppose $K/\mathbb{F}$ has genus 1 and that l is an odd prime dividing $q - 1$. Show that $l \mid h_n$ for some n dividing $(l^2 - 1)/2$. (Hint: If $L_K(u) = (1 - \pi_1 u)(1 - \pi_2 u)$, show that $\pi_1^l \equiv \pi_1$ or $\pi_2 \pmod{\mathcal{L}}$, in the notation of Proposition 11.1). This fact is due to J. Leitzel.

3. Let $K/\mathbb{F}$ be a function field of genus 1 over a finite field with q elements. Write $L_K(u) = 1 - au + qu^2$. Suppose l is a prime such that $(a^2 - 4q \; / \; l) = 1$ (the Legendre symbol). Show $l \mid h_n$ for some $n \mid l - 1$.

4. Let $K/\mathbb{F}$ be a function field of genus 2 and characteristic p. Let h be the class number of K and N_1 the number of primes of degree 1. Suppose that $h \equiv N_1 \pmod{p}$ and that $N_1 \not\equiv 1 \pmod{p}$. Show that $p \mid h_n$ for some integer n dividing $p - 1$. (Hint: Make use of the proof of Proposition 11.8.)

5. Prove the case $l = 2$ of Lemma 11.17.

6. Let $K/\mathbb{F}$ be a function field over a finite field and let J be the associated divisor class group over the algebraic closure of $\mathbb{F}$. Suppose $l \ne 2$ is a regular prime in this situation, i.e., $[\mathbb{F}(J[l]) : \mathbb{F}]$ is prime to l. Use Lemma 11.17 to show directly, i.e., not as a corollary to Theorem 11.18, that $\mathrm{ord}_l \, h(K_{l^n}) = r_l n + \mathrm{ord}_l \, h(K)$, where r_l is the dimension over $\mathbb{Z}/l\mathbb{Z}$ of $J(\mathbb{F})[l]$.

7. Let $\mathbb{F}$ be a field with q elements and $K = \mathbb{F}(x, y)$ be a function field with generators x and y which satisfy $Y^2 = f(X)$, where $f(X) \in \mathbb{F}[X]$ is a cubic polynomial without repeated roots. Assume that q is odd. Prove that 2 is an irregular prime if and only if $f(X)$ is the product of a linear factor and an irreducible quadratic. (One needs some elementary facts about points of order 2 on an elliptic curve).

8. Suppose $\mathbb{F} = \mathbb{Z}/5\mathbb{Z}$ and $K = \mathbb{F}(x, y)$, where x and y satisfy $Y^2 = X^3 - 3X$. This is a curve of genus 1. The primes of degree 1 of K are in one-to-one correspondence with the solutions of this equation over $\mathbb{F}$ together with one prime at infinity. Use this to show $N_1(K) = h(K) = 2$ and deduce $L_K(u) = 1 - 4u + 5u^2$. Deduce from this that (we drop K from the notation): $N_1 = 2$, $N_2 = 2^2 5$, $N_4 = 2^7 5$, and $N_8 = 2^9 3^2 5 \cdot 17$.

9. (Continuation) One can show that all the points of order dividing 4 are in $J(\mathbb{F}_4)$. Use this and the previous exercise to show

$$J(\mathbb{F}_{2^j})(2) \cong 2^{-j-1}\mathbb{Z}/\mathbb{Z} \oplus 2^{-j-2}\mathbb{Z}/\mathbb{Z} ,$$

for all $j \geq 2$.

10. (Continuation) Show $[\mathbb{F}(J[5]) : \mathbb{F}] = 2$ and deduce that $J(\mathbb{F}_{5^j})(5) = (0)$ for $j \geq 0$, $J(\mathbb{F}_{2 \cdot 5^j})(5) \cong \mathbb{Z}/5^j\mathbb{Z}$ for $j \geq 0$, and $J(\mathbb{F}_{2^j})(5) \cong \mathbb{Z}/5\mathbb{Z}$ for $j \geq 1$.

12
Cyclotomic Function Fields

In the last chapter we explored the arithmetic of constant field extensions and noted (as was pointed out by Iwasawa) that these extensions can be thought of as function field analogues of cyclotomic extensions of number fields. This analogy led to various conjectures about the behavior of class groups in number fields which have proved very fruitful for the development of algebraic number theory and arithmetic geometry. There is another function field analogy to cyclotomic number fields which was first discovered by L. Carlitz [3] in the late 1930s. This ingenious analogy was not well known until D. Hayes, in 1973, published an exposition of Carlitz's idea and showed that it provided an explicit class field theory for the rational function field (see Hayes [1]). Later developments, due independently to Hayes and V. Drinfeld, showed that Carlitz's ideas can be generalized to provide an explicit class field theory for any global function field, i.e., an explicit construction of all abelian extensions of such a field (see Drinfeld [1] and Hayes [2]). This is a complete solution to Hilbert's 9-th problem in the function field case. Nothing remotely so satisfying is known for number fields except for the field of rational numbers (cyclotomic theory) and imaginary quadratic number fields (the theory of complex multiplication).

It is interesting to note that this discussion shows how the power of the number field-function field analogy is useful in both directions. The theory of constant field extensions in function fields gave rise to Iwasawa theory in number fields. The extensive theory of cyclotomic number fields gave rise to the work of Carlitz, Drinfeld, and Hayes which provided a way explicitly to construct all abelian extensions of a global function field.

The impetus to function field arithmetic given by these developments has led to many new ideas and developments. One direction, which we will not be able to discuss, is the invention of characteristic p-valued L-functions by D. Goss. These functions share many properties of their classical analogues. In particular, their special values can be related to the arithmetic of fields generated by adding torsion points on Drinfeld modules. On the other hand, they do not seem to possess functional equations. For a survey of these developments, see Goss [2]. Another interesting reference is Thakur [2]. For a more comprehensive treatment, one should consult the treatise by Goss [4].

In this chapter we will deal almost exclusively with Carlitz's construction of what we will call cyclotomic function fields. This is a special case of far more general constructions, but it contains most of the features of the general theory and has the advantage of being very down to earth and very close to our initial theme in this book; the analogy between the rational integers $\mathbb{Z}$ and the ring of polynomials over a finite field $A = \mathbb{F}[T]$.

We begin by recalling, mostly without proof, several features of the theory of cyclotomic number fields. Let $m > 2$ be a positive integer and $\zeta_m \in \mathbb{C}$ a primitive m-th root of unity. Then, the field $K_m := \mathbb{Q}(\zeta_m)$ is called the m-th cyclotomic number field. It is the splitting field of $x^m - 1 \in \mathbb{Q}[x]$, so it is a Galois extension of $\mathbb{Q}$. If $\sigma \in \mathrm{Gal}(K_m/\mathbb{Q})$, then $\sigma(\zeta_m) = \zeta_m^a$, where a is relatively prime to m and is only determined modulo m. This gives rise to a monomorphism $\mathrm{Gal}(K_m/\mathbb{Q}) \to (\mathbb{Z}/m\mathbb{Z})^*$. This map can be shown to be onto (the irreducibility of the m-th cyclotomic polynomial). Thus, $\mathrm{Gal}(K_m/\mathbb{Q}) \cong (\mathbb{Z}/m\mathbb{Z})^*$. It follows that $K_m/\mathbb{Q}$ is an abelian extension of degree $\phi(m)$, where ϕ is the Euler ϕ-function.

If $a \in (\mathbb{Z}/m\mathbb{Z})^*$, we denote by σ_a the corresponding automorphism. It is characterized by

$$\sigma_a(\zeta_m) = \zeta_m^a \ .$$

There are two important consequences of this. The first is immediate. Namely, σ_{-1} is complex conjugation on K_m. Indeed, $\sigma_{-1}(\zeta_m) = \zeta_m^{-1} = \bar{\zeta}_m$, and ζ_m, by definition, generates $\mathbb{Q}(\zeta_m)$ over $\mathbb{Q}$. The second consequence is that if $p > 0$ is a prime not dividing m, then σ_p is the Artin automorphism for the prime ideal $p\mathbb{Z}$. To see this, we must first investigate ramification in K_m and learn something about the ring of integers $\mathcal{O}_m$ of K_m.

Consider first the case when $m = p^e$ is a prime power. Set $\zeta_{p^e} = \zeta$. Since ζ satisfies the polynomial $f(x) = x^{p^e} - 1$ and $f'(\zeta) = p^e \zeta^{p^e - 1}$ one can deduce that $K_{p^e}/\mathbb{Q}$ is unramified at all primes different from p. We claim it is totally ramified at p and that the prime ideal lying above $p\mathbb{Z}$ in $\mathcal{O}_m$ is just $(\zeta - 1)$. Here is a sketch of the proof. Let $a \in \mathbb{Z}$ be relatively prime to p. There is a $b \in \mathbb{Z}$ such that $ab \equiv 1 \pmod{p^e}$. One has

$$\frac{\zeta^a - 1}{\zeta - 1} = \zeta^{a-1} + \zeta^{a-2} + \cdots + \zeta + 1,$$

and

$$\frac{\zeta - 1}{\zeta^a - 1} = \frac{\zeta^{ab} - 1}{\zeta^a - 1} = (\zeta^a)^{b-1} + (\zeta^a)^{b-2} + \cdots + \zeta^a + 1 .$$

It follows that $\zeta^a - 1/\zeta - 1$ is a unit in $\mathcal{O}_m$. Now, the irreducible polynomial in $\mathbb{Q}[x]$ for ζ is

$$g(x) = \frac{x^{p^e} - 1}{x^{p^{e-1}} - 1} = (x^{p^{e-1}})^{p-1} + (x^{p^{e-1}})^{p-2} + \cdots + x^{p^{e-1}} + 1 .$$

The irreducible polynomial for $\zeta - 1$ is thus $g(x + 1)$, which has constant term p. The other roots of $g(x + 1)$ are $\zeta^a - 1$, where $1 \le a < p^e$ and $(a, p) = 1$. Thus,

$$p = \prod_{\substack{a=1 \\ (a,p)=1}}^{p^e} (\zeta^a - 1) = (\zeta - 1)^{\phi(p^e)} \times \text{unit} .$$

Passing to ideals in $\mathcal{O}_m$, we find $p\mathcal{O}_m = (\zeta-1)^{\phi(p^e)}$. Since $[K_m : \mathbb{Q}] = \phi(m)$, this can only happen if $(\zeta - 1)$ is a prime ideal in $\mathcal{O}_m$, which shows that $p\mathbb{Z}$ is totally ramified, as asserted.

We continue to assume that $m = p^e$ and set $\zeta_{p^e} = \zeta$. Under these circumstances we claim $\mathcal{O}_m = \mathbb{Z}[\zeta]$. To this end we note that the discriminant of the ring $\mathbb{Z}[\zeta]$ over $\mathbb{Z}$ is a power of p. This is a calculation (see Lang [5]). Note that $\mathbb{Z}[\zeta] = \mathbb{Z}[\zeta - 1]$. Let $\omega \in \mathcal{O}_{p^e}$ and write

$$\omega = \sum_{i=0}^{\phi(p^e)-1} a_i(\zeta - 1)^i, \quad a_i \in \mathbb{Q}.$$

From the usual deduction via discriminants and Cramer's rule we find that the rational numbers a_i have denominators a power of p. We want to show the denominators are ± 1 so that the $a_i \in \mathbb{Z}$. Let the least common multiple of the denominators be p^n with $n \ge 0$. Then $a_i = b_i/p^n$ with the $b_i \in \mathbb{Z}$ and not all the b_i divisible by p. We have

$$p^n \omega = \sum_{i=0}^{\phi(p^e)-1} b_i(\zeta - 1)^i .$$

Extend ord_p to K_{p^e} by writing $\text{ord}_p(\alpha) = \phi(p^e)^{-1}\text{ord}_{(\zeta-1)}(\alpha)$ for all $\alpha \in K_{p^e}$. Let i_0 be the smallest integer such that $\text{ord}_p(b_i) = 0$. Then, as is easily seen, ord_p of the right-hand side of the above equation is $i_0/\phi(p^e) < 1$. On the other hand, ord_p of the left-hand side is $\ge n$. This shows $n = 0$ and it follows that all the a_i are integers, as required.

In the general case, write out the prime decomposition of m, $m = p_1^{e_1} p_2^{e_2} \cdots p_t^{e_t}$. We require that m not be twice an odd integer. This is not a big restriction since if m_0 is odd, $K_{2m_0} = K_{m_0}$. Then, K_m is the compositum of the fields $K_{p_i^{e_i}}$. It follows that all the p_i ramify in K_m and all other primes are unramified in K_m. Moreover, using what we have shown in the prime power case it follows that $\mathcal{O}_m = \mathbb{Z}[\zeta_m]$ (see Lang [5]).

We now can prove what was promised earlier. Namely, if $p > 0$ does not divide m, then the Artin automorphism of the prime ideal $P = p\mathbb{Z}$ for the extension $K_m/\mathbb{Q}$ is precisely σ_p provided $p \nmid m$. Let $\mathfrak{P}$ be a prime ideal in $\mathcal{O}_m$ lying over P. The Artin automorphism for P is characterized by the congruence

$$(P, K_m/\mathbb{Q})\, \omega \equiv \omega^{NP} \pmod{\mathfrak{P}} \quad \forall \omega \in \mathcal{O}_m .$$

If $p > 0$ is prime and $P = p\mathbb{Z}$, then $NP = p$. Since $\mathcal{O}_m = \mathbb{Z}[\zeta_m]$ we can check the congruence on elements of the form $\sum a_i \zeta_m^i$ where the $a_i \in \mathbb{Z}$ and the sum is from 0 to $\phi(m) - 1$. We calculate

$$\sigma_p\Big(\sum a_i \zeta_m^i\Big) = \sum a_i \zeta_m^{pi} \equiv \Big(\sum a_i \zeta_m^i\Big)^p \pmod{\mathfrak{P}} .$$

This shows that $(P, K_m/\mathbb{Q}) = \sigma_p$ as asserted.

From this fact about σ_p one can calculate the way primes in $\mathbb{Z}$ split in K_m. If $P = p\mathbb{Z}$ is unramified, then P splits into $\phi(m)/f$ primes of degree f in K_m, where f is the order of $(P, K_m/\mathbb{Q})$ in $\mathrm{Gal}(K_m/\mathbb{Q})$ (see Proposition 9.10). Since we have shown $\mathrm{Gal}(K_m/\mathbb{Q}) \cong (\mathbb{Z}/m\mathbb{Z})^*$, the order of σ_p in $\mathrm{Gal}(K_m/\mathbb{Q})$ is the smallest integer f such that $p^f \equiv 1 \pmod{m}$.

We summarize a portion of this discussion in a theorem.

Theorem 12.1. *Let $m > 0$ be an integer not equal to twice an odd number. Let $\zeta_m \in \mathbb{C}$ be a primitive m-th root of unity and $K_m = \mathbb{Q}(\zeta_m)$. Then $K_m/\mathbb{Q}$ is an abelian extension of degree $\phi(m)$. The Galois group is isomorphic to $(\mathbb{Z}/m\mathbb{Z})^*$. A rational prime p is ramified in K_m if and only if $p|m$. If $p > 0$ does not divide m, the Artin automorphism corresponding to the prime ideal $P = p\mathbb{Z}$ takes ζ_m to ζ_m^p. Let f be the smallest positive integer such that $p^f \equiv 1 \pmod{m}$. Then, $P = p\mathbb{Z}$ splits into $\phi(m)/f$ primes of degree f in K_m. Finally, if $\mathcal{O}_m$ denotes the ring of integers in K_m, then $\mathcal{O}_m = \mathbb{Z}[\zeta_m]$.*

The last thing about cyclotomic fields which we wish to discuss at this point is the behavior of the prime at infinity. The field of rational numbers $\mathbb{Q}$ has only one archimedean prime given by the usual absolute value. The field K_m is such that every embedding into $\mathbb{C}$ is complex since the only roots of unity in the real numbers $\mathbb{R}$ are ± 1. Consider the subfield $K_m^+ = \mathbb{Q}(\zeta_m + \zeta_m^{-1})$. This field is real and so is every embedding of it into the complex numbers. Moreover, it is of index 2 in K_m since ζ_m satisfied the quadratic equation $x^2 - (\zeta_m + \zeta_m^{-1})x + 1 = 0$. Thus, the prime at infinity in $\mathbb{Q}$ splits into $\phi(m)/2$ real primes in K_m^+ and each of these ramifies to a complex prime in K_m. It is clear that the Galois group of K_m/K_m^+ is generated by σ_{-1} which is complex conjugation. Thus, σ_{-1} can be thought of as generating the inertia group of the primes at infinity in K_m.

Having now reviewed the cyclotomic theory in the number field case we will next consider how to construct an analogous theory in the function field case. Considering roots of unity will yield only constant field extensions which, as we have seen, are everywhere unramified. How can we generate abelian extensions which are geometric? The answer is not at all obvious. To provide the necessary background, we begin by exploring the notion of an additive polynomial over a field.

Let k be a field. A polynomial $f(x) \in k[x]$ is said to be additive if inside the polynomial ring in two variables $k[x,y]$ we have $f(x+y) = f(x)+f(y)$. For any element $a \in k$, $f(x) = ax$ is such a polynomial. We shall see that in characteristic zero this collection of homogeneous linear polynomials constitutes all additive polynomials. In characteristic $p > 0$ the polynomial $\tau(x) = x^p$ is additive, as is easily seen using the binomial theorem. It is easy to check that the set of additive polynomials is closed under addition, subtraction, multiplication by elements of k, and composition. The last is seen from the calculation

$$
\begin{aligned}
(f \circ g)(x+y) &= f(g(x+y)) = f(g(x)+g(y)) \\
&= f(g(x)) + f(g(y)) = (f \circ g)(x) + f \circ g)(y) \ .
\end{aligned}
$$

This leads us to additive polynomials of the form $a_0 x + a_1 x^p + \cdots + a_r x^{p^r}$. With these we have exhausted the collection of additive polynomials as we now show.

Proposition 12.2. *Let k be a field and $f(x) \in k[x]$ an additive polynomial. If the characteristic of k is zero then $f(x) = ax$ for some $a \in k$. If the characteristic of k is $p > 0$, then there are elements $a_i \in k$ with $0 \leq i \leq r$ such that $f(x) = a_0 x + a_1 x^p + \cdots + a_r x^{p^r}$.*

Proof. By definition, if $f(x)$ is additive, $f(x+y) = f(x) + f(y)$. Take the formal partial derivative with respect to x. Then, $\partial_x f(x+y) = \partial_x f(x)$. Setting $x = 0$ we see that the formal derivative of f is a constant. If $f(x) = \sum b_i x^i$, then $f'(x) = \sum i b_i x^{i-1}$. In characteristic zero this shows that $f'(x)$ is a constant if and only if $f(x) = b_0 + b_1 x$, a linear polynomial. However, $f(x + y) = f(x) + f(y)$ implies $f(0) = 0$. Thus, in this case $f(x) = b_1 x$.

Now, if the characteristic of k is $p > 0$, then $f'(x)$ is a constant if and only if $b_i = 0$ for all $i > 1$ with i not divisible by p. We may write

$$
f(x) = b_1 x + \sum_{j \geq 1} b_{pj} x^{pj} = b_1 x + g(x)^p \ ,
$$

where g(x) has coefficients in the field k_1 obtained from k by adjoining the p-th roots of the coefficients b_{pj}. It is a simple matter to check that $g(x)$ is additive in $k_1[x]$. By induction on the degree of $f(x)$ we can assume that $g(x) = \sum_{h \geq 0} c_h x^{p^h}$. Thus,

$$f(x) = b_1 x + \sum_{h \geq 0} c_h^p x^{p^{h+1}} \ ,$$

which is a polynomial of the required form since $c_h^p \in k$ for all h.

From now on, we assume that we are working in characteristic $p > 0$. Suppose that k is a field of characteristic p and let $\mathcal{A}(k)$ denote the set of additive polynomials with coefficients in k. $\mathcal{A}(k)$ is easily seen to be a ring with addition being given by the standard addition of polynomials and multiplication being given by composition (as is easily seen, $\mathcal{A}(k)$ is not closed under ordinary multiplication). We will reformulate the structure of $\mathcal{A}(k)$ in a more convenient manner by associating with every additive polynomial

$$f(x) = \sum_{i=0}^{r} a_i \, x^{p^i}$$

the polynomial in τ (recall, $\tau(x) = x^p$)

$$g(\tau) = \sum_{i=0}^{r} a_i \tau^i \ .$$

Clearly, $f(x) = g(\tau)(x)$ and the map $f(x) \to g(\tau)$ sets up a bijection between $\mathcal{A}(k)$ and $k < \tau >$, the ring of polynomials in τ with "twisted" multiplication. This means that for all $a \in k$ we have

$$\tau a = a^p \tau \ . \tag{1}$$

This follows from the calculation,

$$(\tau a)(x) = \tau(ax) = (ax)^p = a^p x^p = (a^p \tau)(x) \ .$$

Thus, multiplication in $k < \tau >$ is just like that in a polynomial ring except that when multiplying an element of k by a power of τ one must use the Relation 1. For obvious reasons, $k < \tau >$ is often referred to as a twisted polynomial ring. It is now easy to check that the ring of additive polynomials with coefficients in k is isomorphic to $k < \tau >$ under the map $f(x) \to g(\tau)$ given above. We will work primarily with $k < \tau >$.

It is possible, and desirable, to give a group scheme interpretation to this ring. Let G_a/k be the additive group scheme over k. Among other things, G_a/k assigns to every commutative k-algebra B the underlying additive group B_+. Every additive polynomial gives rise to an endomorphism of B_+ in the obvious way. If $u \in B$ and $\sum a_i \tau^i \in k < \tau >$, then

$$\left(\sum a_i \tau^i \right)(u) = \sum a_i u^{p^i} \ .$$

From these considerations it is not hard to show $\text{End}(G_a/k) \cong k < \tau >$. In what follows, we will identify these rings. We will not need to invoke any facts from the theory of group schemes, but this point of view is illuminating. All of the theory which we will develop is made possible by the fact that in characteristic p, $\text{End}(G_a/k)$ is a "big" ring.

We need to make one modification in these definitions before returning to function fields. Let $\mathbb{F}$ be a finite field with $q = p^s$ elements. We want to work only with $\mathbb{F}$-algebras and we want our endomorphisms to respect the $\mathbb{F}$-algebra structure. So we assume $\mathbb{F} \subseteq k$ and only look at additive polynomials $f(x)$ which commute with the elements of $\mathbb{F}$. This requirement is that $f(\alpha x) = \alpha f(x)$ for all $\alpha \in \mathbb{F}$. If $f = \sum a_i \tau^i$ this requirement is easily seen to be equivalent to

$$\alpha^{p^i} = \alpha \quad \forall \alpha \in \mathbb{F} \quad \text{whenever} \quad a_i \neq 0 \ .$$

From the theory of finite fields, we see that these conditions hold if and only if $s|i$ for all i such that $a_i \neq 0$. Another way of saying this is that $f \in k < \tau^s >$. Note that $\tau^s(x) = x^q$. Since $\mathbb{F}$ will be fixed in our further considerations, we will redefine τ to be the mapping which raises to the q-th power and write

$$\text{End}_\mathbb{F}(G_a/k) = k < \tau > \ ,$$

where now the fundamental commutation Relation 1 will be replaced by

$$\tau a = a^q \tau \quad \forall a \in k \ . \tag{2}$$

As usual, set $A = \mathbb{F}[T]$ and $k = \mathbb{F}(T)$.

Definition. A Drinfeld module for A defined over k will be an $\mathbb{F}$-algebra homomorphism $\rho : A \to k < \tau >$ such that for all $a \in A$ the constant term of ρ_a is a and, moreover, for at least one $a \in A$, $\rho_a \notin k$.

The notion of a Drinfeld module is much more general, but for the purposes of this chapter, this definition will suffice. The idea behind the definition is that given a Drinfeld module ρ every commutative k-algebra B can be made into an A algebra in a new way. B is already an A-algebra since A is a subset of k and B is a k-algebra. However, given ρ we can define a new multiplication by

$$a \cdot u = \rho_a(u) \quad \forall a \in A \quad \text{and} \quad \forall u \in B \ .$$

The condition that $\rho_a \notin k$ for at least one $a \in A$ is to insure that this action is indeed different from the standard action of A on B. We will call B with this new A-module structure, B_ρ. The k-algebra which will receive the most attention is the algebraic closure of k, $\bar{k}$.

We have said nothing yet about the existence of Drinfeld modules. In the general case (which we have yet to define) this is a delicate question.

Here, however, it is a trivial matter. A is generated freely as an algebra over $\mathbb{F}$ by one element T. Thus, for any element $h \in k < \tau >$ there is a unique homomorphism from A to $k < \tau >$ which takes T to h. We must only make sure that the constant term of h is T and that $h \notin k$ to get a Drinfeld module. Perhaps the simplest choice for h is $T + \tau$. The resulting Drinfeld module is called the Carlitz module, C. Thus, $C_T = T + \tau$, $C_{T^2} = T^2 + (T + T^q)\tau + \tau^2$, etc. Carlitz discovered and exploited this module decades before anyone else had any idea of the value of this construction. The reader may wish to consult the papers by Carlitz [1, 2]. In these papers Carlitz actually works with the module C' defined by the relation $C'(T) = T - \tau$. In almost all modern treatment the plus sign is chosen. We shall stay with this convention.

We will discuss the properties of the Carlitz module in some detail, but for a while we will continue to develop the theory more generally.

Suppose ρ is a Drinfeld module and

$$\rho_T = T + c_1 \tau + c_2 \tau^2 + \cdots + c_r \tau^r ,$$

where the $c_i \in k$ and $c_r \neq 0$. Using $\rho_{T^2} = \rho_T \rho_T$, we see that the constant term of ρ_{T^2} is T^2 and that the highest power of τ that occurs is $2r$ and the leading coefficient is c_r raised to the power $1 + q^r$. Continuing this way we find that the constant term of ρ_{T^n} is T^n and the highest power of τ that occurs is nr and the leading coefficient is c_r raised to the power $1 + q^r + q^{2r} + \cdots + q^{(n-1)r}$. Using these comments and the fact that ρ is an $\mathbb{F}$-algebra homomorphism we find for $a \in A$ that the constant term of ρ_a is a and the degree in τ of ρ_a is $r \deg(a)$. It is important to note that the degree of the polynomial $\rho_a(x)$ is $q^{r \deg(a)}$. Under these conditions, we say that the Drinfeld module ρ has rank r. We shall now see how the rank plays an important role in the theory of the A-module $\bar{k}_\rho$.

Let's consider the A-module $\bar{k}_\rho$ and its torsion submodule:

$$\Lambda_\rho = \{\lambda \in \bar{k} \mid \rho_a(\lambda) = 0 \text{ for some } a \in A, a \neq 0\} .$$

For any $a \in A$, $a \neq 0$, we define the submodule $\Lambda_\rho[a] \subset \Lambda_\rho$ as follows:

$$\Lambda_\rho[a] = \{\lambda \in \bar{k} \mid \rho_a(\lambda) = 0\} .$$

It is possible to identify the A-module structure of these modules with some precision. The following abstract lemma is the key.

Lemma 12.3. *Let $a \in A$, $a \neq 0$. Let M be an A-module and suppose for each $b|a$ that the submodule $M[b] = \{m \in M \mid bm = 0\}$ has $q^{r \deg(b)}$ elements. Then*

$$M[a] \cong A/aA \oplus A/aA \oplus \cdots \oplus A/aA \quad r \text{ times} .$$

Proof. Consider the prime decomposition of a, $a = \alpha P_1^{e_1} P_2^{e_2} \cdots P_t^{e_t}$, where $\alpha \in \mathbb{F}^*$ and the P_i run through the monic, irreducible divisors of a. $M[a]$ is isomorphic to the direct sum of the submodules $M[P_i^{e_i}]$. Via the Chinese Remainder Theorem, it suffices to consider the case that $a = P^e$ is a prime power.

So, suppose $a = P^e$ is a prime power. Since $M[P]$ is a vector space over A/PA with $q^{r \deg(P)}$ elements, by hypothesis, it follows that the dimension of $M[P]$ over A/PA is r (recall that A/PA has $q^{\deg P}$ elements). It follows from the structure of modules over principal ideal domains that $M[P^e]$ is a sum of r cyclic submodules,

$$M[P^e] \cong A/P^{f_1} A \oplus A/P^{f_2} A \oplus \cdots \oplus A/P^{f_r} A .$$

One must have $f_i \le e$ for $1 \le i \le r$. The number of elements in the right-hand side of this isomorphism is q to the power $(f_1 + f_2 + \cdots + f_r) \deg(P)$. The number of elements in the left-hand side is, by hypothesis, q to the power $re \deg(P)$. These two numbers being equal implies that each $f_i = e$ and this concludes the proof.

Proposition 12.4. *Let ρ be a Drinfeld module of rank r, i.e., for each $a \in A$ the degree in τ of ρ_a is $r \deg(a)$. Then, for each $a \in A$, $a \ne 0$ we have*

$$\Lambda_\rho[a] \cong A/aA \oplus A/aA \oplus \cdots \oplus A/aA \quad r \text{ times .}$$

For the module Λ_ρ we have the isomorphism

$$\Lambda_\rho \cong k/A \oplus k/A \oplus \cdots \oplus k/A \quad r \text{ times .}$$

Proof. We apply Lemma 12.3 with $M = \bar{k}_\rho$. We have to check that for each $a \ne 0$ in A that $\Lambda_\rho[a]$ has $q^{r \deg(a)}$ elements. From our previous work we see that $\rho_a(x)$ has the form

$$\rho_a(x) = ax + b_1 x^q + b_2 x^{q^2} + \cdots + b_{r \deg(a)} x^{q^{r \deg(a)}} ,$$

where the $b_i \in k$ and $b_{r \deg(a)} \ne 0$. The derivative of $\rho_a(x)$ with respect to x is $a \ne 0$. Thus, $\rho_a(x)$ is a separable polynomial and in $\bar{k}$ has $q^{r \deg(a)}$ distinct roots. These roots are exactly the elements of $\Lambda_\rho[a]$ so the first part of the proof is complete.

The second assertion is a formal consequence of the first. Since we won't use it in what follows, we merely sketch the proof. Note first that Λ_ρ is the union of the submodules $\Lambda_\rho[a]$ as a runs through the non-zero elements of A. Secondly, since $A/aA \cong a^{-1}A/A$ we can rewrite the first isomorphism as

$$\Lambda_\rho[a] \cong a^{-1}A/A \oplus a^{-1}A/A \oplus \cdots \oplus a^{-1}A/A \quad r \text{ times .}$$

Order the non-zero elements of A by divisibility. The result would follow if we could pass to the direct limit and this process could be done in such a way that the direct limit could be interchanged with the direct sums. One can arrange the direct sum decompositions so that this is possible. However, we will omit the details.

Suppose we adjoin the elements of $\Lambda_\rho[a]$ to k to form the field $K_{\rho,a} := k(\Lambda_\rho[a])$. Since, as we have seen, $\rho_a(x)$ is a separable polynomial and $\Lambda_{\rho,a}$ is the set of roots of this polynomial, it follows that $K_{\rho,a}/k$ is a Galois extension. Since $\rho_a(x) \in k[x]$ we see $\rho_a(\lambda) = 0$ implies $\rho_a(\sigma\lambda) = 0$ for all $\sigma \in \mathrm{Gal}(K_{\rho,a}/k)$. For such σ it is easy to check that not only does σ map $\Lambda_\rho[a]$ into itself, it actually induces an automorphism of the A/aA-module structure. We thus get a map, in fact a homomorphism, from

$$\mathrm{Gal}(K_{\rho,a}/k) \to \mathrm{Aut}_{A/aA}(\Lambda_\rho[a]) \ .$$

Since $\Lambda_\rho[a]$ generates $K_{\rho,a}$, any automorphism inducing the identity map on $\Lambda_\rho[a]$ must be the identity automorphism. Thus, the kernel of the above map is trivial.

Finally, by the first assertion of Proposition 12.4, we see that

$$\mathrm{Aut}_{A/aA}(\Lambda_\rho[a]) \cong GL_r(A/aA) \ .$$

We have proved the following proposition.

Proposition 12.5. *Define* $K_{\rho,a}$ *to be the field* $k(\Lambda_\rho[a])$. *Then* $K_{\rho,a}/k$ *is a Galois extension and there is a monomorphism*

$$\mathrm{Gal}(K_{\rho,a}/k) \to GL_r(A/aA) \ .$$

Corollary. *If* ρ *has rank 1, then* $K_{\rho,a}/k$ *is an abelian extension.*

Proof. This is immediate from the Proposition since $GL_1(A/aA) = (A/aA)^*$ which is abelian.

One can ask about the size and nature of the image of the maps given in the Proposition. This is a very difficult question in general. Much remains to be discovered. Using a lot of sophisticated machinery some answers have recently been given by Richard Pink (see Pink [1]). Here, we will be concerned with a very special, but interesting case. Namely, the case of the Carlitz module.

Recall that the Carlitz module is characterized by $C_T = T + \tau$ or equivalently $C_T(x) = Tx + x^q$. Clearly the Carlitz module has rank 1 and so, by the corollary to Proposition 12.5, adjoining torsion points for the Carlitz module to k gives rise to abelian extensions. We will investigate these extensions in some detail and show that they have remarkably similar behavior to cyclotomic extensions of $\mathbb{Q}$.

Since the Carlitz module will be the focus of our work for the rest of this chapter, we will write Λ for Λ_C, Λ_a for $\Lambda_C[a]$, and K_a for $K_{C,a}$. Also, to emphasize the relation to our discussion of cyclotomic fields we will use the letter "m" from now on rather than "a" as our typical non-zero element of A. The fields $K_m = k(\Lambda_m)$ will be the analogues of cyclotomic number fields. We define them to be cyclotomic function fields.

By Proposition 12.5 and its corollary we see that $\mathrm{Gal}(K_m/k)$ is isomorphic to a subgroup of $(A/mA)^*$. Our first goal will be to show it is isomorphic to all of $(A/mA)^*$. Before doing this it will be necessary to make a number of preliminary observations.

Notice that $C_{\alpha m} = C_\alpha C_m = \alpha C_m$ for all $\alpha \in \mathbb{F}$. It follows thats that $\Lambda_{\alpha m} = \Lambda_m$ for all $\alpha \in \mathbb{F}^*$. Another way of saying this is that the torsion points Λ_m depend only on the ideal mA and not on any particular generator of this ideal.

Let m be a polynomial of degree d. Then

$$C_m(x) = [m,0]x + [m,1]x^q + [m,2]x^{q^2} + \cdots + [m,d]x^{q^d}, \tag{3}$$

where $[m,i] \in A$ for every i, $[m,0] = m$, and $[m,d]$ is the leading coefficient of m. Note that if m is monic, then $[m,d] = 1$. The degree of $C_m(x)$ is $q^d = |m|$ (see Chapter 1 for a discussion of this notation). Later we will show that as a polynomial in T, $[m,i]$ has degree $q^i(d-i)$. It is a good exercise to compute $C_m(x)$ explicitly for a few polynomials m of small degree to get a feel for the nature of the coefficients $[m,i]$.

It will turn out that if $m = P$, a irreducible polynomial, then $C_P(x)/x$ is an Eisenstein polynomial at P (i.e., the leading coefficient is not divisible by P, all the other coefficients are divisible by P, and the constant term is not divisible by P^2). Thus, $C_P(x)/x \in A[x]$ is analogous to $[(1+x)^p - 1]/x \in \mathbb{Z}[x]$ in the classical cyclotomic theory. It follows that $0 \neq \lambda_P \in \Lambda_P$ is analogous to $\zeta_p - 1$, not ζ_p.

From Proposition 12.5, we know that $\Lambda_m \cong A/mA$ as an A-module. Let λ_m be a generator of this module. Then, it is easy to see that $C_a(\lambda_m)$ is a generator if and only if $(a,m) = 1$. This shows that Λ_m has $\Phi(m)$ generators where $\Phi(m)$ is the analogue of the Euler ϕ-function for the ring A. By definition, $\Phi(m)$ is the number of non-zero polynomials in A of degree less than that of m and relatively prime to m. Alternatively, $\Phi(m)$ is the number of elements in $(A/mA)^*$ (see Chapter 1).

Since λ_m is a generator of Λ_m it follows that $K_m = k(\lambda_m)$. Let $\mathcal{O}_m$ denote the integral closure of A in K_m.

Proposition 12.6. *Let $\lambda_m \in \Lambda_m$ be a generator and suppose $a \in A$ is relatively prime to m. Then, $C_a(\lambda_m)/\lambda_m$ is a unit in $\mathcal{O}_m$. If m is divisible by two or more primes, then λ_m is itself a unit.*

Proof. From Equation 3 we see that λ_m is integral over A. From the same equation, replacing m by a, d by $\deg(a)$, and substituting $x = \lambda_m$, we see that $C_a(\lambda_m)/\lambda_m \in \mathcal{O}_m$. We must show the reciprocal of this element is also in $\mathcal{O}_m$.

Let $b \in A$ be such that $ba \equiv 1 \pmod{m}$. There is an element $f \in A$ such that $ba = 1 + fm$ and we have $C_b C_a = 1 + C_f C_m$. Applying this to λ_m yields $C_b(C_a(\lambda_m)) = \lambda_m$. Thus,

$$\frac{\lambda_m}{C_a(\lambda_m)} = \frac{C_b(C_a(\lambda_m))}{C_a(\lambda_m)} \in \mathcal{O}_m.$$

To prove the second assertion, it is no loss of generality to assume m is monic. Suppose $m = m_1 m_2$, where m_1 and m_2 are monic and relatively prime. Set $\lambda_{m_1} = C_{m_2}(\lambda_m)$ and $\lambda_{m_2} = C_{m_1}(\lambda_m)$. Then λ_{m_i} is a primitve m_i-th torsion point for $i = 1, 2$. For all $a \in A$, $C_a(x)$ is divisible by x. Consider the factorization

$$\lambda_{m_1} = \lambda_m \frac{C_{m_2}(\lambda_m)}{\lambda_m} .$$

This shows that λ_m divides λ_{m_1} and similarly λ_m divides λ_{m_2} in $\mathcal{O}_m$. Taking norms from K_m to k shows that the norm of λ_m divides a power of $N_{K_{m_i}/k}(\lambda_{m_i})$ for $i = 1, 2$.

To finish the proof one does induction on the number of distinct primes dividing m. We will need the corollary to Proposition 12.7, which will be proven shortly. Its proof is independent of what we are doing here, so it is legitimate to use it. The corollary implies that if $m = P^e$ is a prime power, then the norm of λ_{P^e} is P. Suppose m is a product of two prime powers $P_1^{e_1}$ and $P_2^{e_2}$. Then, from what we have proven, it follows that the norm of λ_m divides a power of P_1 and a power of P_2. This implies the norm of λ_m is a non-zero constant and so λ_m is a unit. If m is divisible by $t > 2$ distinct primes, set

$$m_1 = P_1^{e_1} \quad and \quad m_2 = \prod_{i=2}^{t} P_i^{e_i} .$$

Then, by induction, λ_{m_2} is a unit and its norm is a non-zero constant. By what we have proven above, it follows that the norm of λ_m is a non-zero constant. Thus, λ_m is a unit, and we are done.

With the aid of these units we will imitate some of the deductions of the classical theory. As there, we begin by considering the case when $m = P^e$ is a power of an irreducible polynomial P. Since $\Lambda_{P^e} \cong A/P^e A$ an element $\lambda \in \Lambda_{P^e}$ is a primitive generator iff $C_{P^e}(\lambda) = 0$ and $C_{P^{e-1}}(\lambda) \neq 0$. Thus, the primitive generators are precisely the roots of the polynomial

$$\frac{C_{P^e}(x)}{C_{P^{e-1}}(x)} = \frac{C_P(C_{P^{e-1}}(x))}{C_{P^{e-1}}(x)}$$

$$= P + [P, i] C_{P^{e-1}}(x)^{q-1} + \cdots + [P, d] C_{P^{e-1}}(x)^{q^d-1} , \quad (4)$$

where $d = \deg(P)$. The degree in x of the polynomial in Equation 4 is $|P|^{e-1}(q^d - 1) = |P|^{e-1}(|P| - 1) = \phi(P^e)$ as it should be.

Proposition 12.7. *Let $P \in A$ be a monic irreducible polynomial and $e \in \mathbb{Z}$, $e > 0$. Then, K_{P^e} is unramified at every prime ideal QA with $QA \neq PA$. The prime PA is totally ramified with ramification index $\Phi(P^e)$. Consequently,*

$$[K_{P^e} : k] = \Phi(P^e) \quad and \quad \mathrm{Gal}(K_{P^e}/k) \cong (A/P^e A)^* .$$

Finally, the prime ideal above PA is $(\lambda) = \lambda \mathcal{O}_{P^e}$ where λ is any generator of Λ_{P^e}.

Proof. Let λ be a primitive generator of Λ_{P^e} and let $g(x) \in k[x]$ be the monic irreducible polynomial it satisfies. Then $g(x)$ must divide $C_{P^e}(x)$. Write $C_{P^e}(x) = f(x)g(x)$. Differentiate both sides and substitute $x = \lambda$. The result is $P^e = f(\lambda)g'(\lambda)$. Since $K_{P^e} = k(\lambda)$, it follows that $g'(\lambda)$ is contained in the different of $\mathcal{O}_{P^e}/A$. Thus any prime ideal of $\mathcal{O}_{P^e}$ dividing the different, must contain a power of P and thus P itself. This shows that PA is the only possible prime ideal in A ramified in $\mathcal{O}_{P^e}$.

Let $d = \deg(P)$. As we have seen, the other primitive generators of Λ_{p^e} are

$$\{ C_a(\lambda) \mid 0 \le \deg(a) < \deg(P^e) = ed \text{ and } (a, P) = 1 \} .$$

These are the roots of the polynomial in Equation 4, which is monic (since P is assumed to be monic) and has constant term P. By the first part of Proposition 12.6, we deduce

$$P = \prod_{\substack{a, \deg(a) < \ ed \\ (a,P)=1}} C_a(\lambda) = \lambda^{\Phi(P^e)} \times \text{unit} .$$

It follows that $PA = (\lambda)^{\Phi(P^e)}$. Let $\mathfrak{P}$ be a prime ideal of $\mathcal{O}_{P^e}$ dividing (λ). Then, the ramification index of $\mathfrak{P}/P$ is divisible by $\Phi(P^e)$. Since λ is a root of the polynomial in Equation 4, we know that $[K_{P^e} : k] \le \Phi(P^e)$. It follows that the ramification index of $\mathfrak{P}/P$ is precisely $\Phi(P^e)$, that $P\mathcal{O}_{P^e} = \mathfrak{P}^{\Phi(P^e)}$, and that $\mathfrak{P} = (\lambda)$. The remaining assertions are now clear.

Corollary. *Let $P \in A$ be monic irreducible of positive degree, and $e \in \mathbb{Z}$, $e > 0$. Let λ be a generator of Λ_{P^e} and $g(x) \in k[x]$ its irreducible polynomial. Then $g(x)$ is an Eisenstein polynomial at P.*

Proof. By what has been proven in the proposition, $g(x)$ is the polynomial which appears in Equation 4. We have

$$g(x) = \prod_{(a,P)=1} (x - C_a(\lambda)) ,$$

where the product is over all primitive generators of Λ_{P^e}.

Except for the leading coefficient, which is 1, the coefficients of g are the elementary symmetric functions of the primitive elements in Λ_{P^e}. The proposition shows these are all in the ideal (λ). Thus, all the coefficients of $g(x)$, except the leading coefficient, are in $(\lambda) \cap A = PA$. Since the constant term is P, it follows that $g(x)$ is an Eisenstein polynomial.

Using the above corollary and an easy induction on e, we see that $C_{P^e}(x)$ is a product of Eisenstein polynomials at P. Consequently, all its non-leading coefficients must be divisible by P. In other words, we have shown $P \mid [P^e, i]$ for all $0 \le i < ed$ where $d = \deg(P)$.

Having dealt with the case $m = P^e$, we now pass on to the general case. Let $m \in A$ be a polynomial of positive degree and let $m = \alpha P_1^{e_1} P_2^{e_2} \cdots P_t^{e_t}$ be its prime decomposition.

Theorem 12.8. K_m *is the compositum of the fields* $K_{P_i^{e_i}}$. *The only ideals in* A *ramified in* $\mathcal{O}_m$ *are* $P_i A$ *with* $1 \le i \le t$. *We have* $[K_m : k] = \Phi(m)$. *More precisely,*

$$\mathrm{Gal}(K_m/k) \cong (A/mA)^* \ .$$

Proof. Define m_i to be m divided by $P_i^{e_i}$. Let λ_m be a generator of Λ_m as an A-module. It is clear that $C_{m_i}(\lambda_m)$ is a generator of $\Lambda_{P_i^{e_i}}$. Set $\lambda_{P_i^{e_i}} = C_{m_i}(\lambda_m)$.

Clearly, $K_{P_i^{e_i}} = k(\lambda_{P_i^{e_i}}) \subset k(\lambda_m) = K_m$. Thus, K_m contains the compositum of the fields $K_{P_i^{e_i}}$, for $1 \le i \le t$.

Since the greatest common divisor of the set $\{m_i \mid 1 \le i \le t\}$ is just 1, there exist polynomials $a_i \in A$ such that $1 = \sum_{i=1}^t a_i m_i$. It follows that $1 = \sum_{i=1}^t C_{a_i} C_{m_i}$. Applying this relation to the element λ_m yields

$$\lambda_m = \sum_{i=1}^t C_{a_i}(\lambda_{P_i^{e_i}}) \ . \tag{5}$$

This shows that λ_m is in the compositum of the fields $K_{P^{e_i}}$, which completes the proof that K_m is the compositum of these fields.

If P is a prime element such that $PA \ne P_i A$ for any i, then by Proposition 12.7, PA is unramified in every $K_{P_i^{e_i}}$ and so must be unramified in their compositum K_m. On the other hand, $P_i A$ is totally ramified in $K_{P_i^{e_i}}$ by the same proposition. Thus, all the ideals $P_i A$ are ramified in K_m.

We will prove that $[K_m : k] = \Phi(m)$ by induction on t. For $t = 1$ this assertion is part of Proposition 12.7. Assume the result is true for $t - 1$. Then, $[K_{m_t} : k] = \Phi(m_t)$. Now, $K_{m_t} \cap K_{P_t^{e_t}} = k$ because K_{m_t} is unramified at $P_t A$ and $K_{P_t^{e_t}}$ is totally ramified at $P_t A$. It follows that

$$[K_m : k] = [K_{m_t} : k][K_{P_t^{e_t}} : k] = \Phi(m_t)\Phi(P_t^{e_t}) = \Phi(m) \ .$$

Finally, we know from the corollary to Proposition 12.5 that there is a monomorphism from $\mathrm{Gal}(K_m/k)$ to $(A/mA)^*$. Since we now know that both of these groups have the same order, $\Phi(m)$, it follows that this monomorphism is an isomorphism.

Our next task is to investigate how the primes in A split in $\mathcal{O}_m$. To do this we have to look at the isomorphism $\mathrm{Gal}(K_m/k) \cong (A/mA)^*$ more closely.

We first recall how this isomorphism is defined. As usual, let λ_m denote a generator of Λ_m as an A-module. If $\sigma \in \mathrm{Gal}(K_m/k)$, then clearly $\sigma\lambda_m$ is another such generator. Thus, there is an $a \in A$ with $(a, m) = 1$ such that $\sigma(\lambda_m) = C_a(\lambda_m)$. The automorphism σ is completely determined by this relation since λ_m generates K_m over k. Note that a is determined only up to a multiple of m. We write $\sigma = \sigma_a$. The map $\sigma \to a$ is the isomorphism from $\mathrm{Gal}(K_m/k) \to (A/mA)^*$ which we have been discussing. The content of Theorem 12.8 is that for any $a \in A$, relatively prime to m, there is a unique automorphism $\sigma_a \in \mathrm{Gal}(K_m/k)$ such that $\sigma_a \lambda_m = C_a(\lambda_m)$. The important fact that we are after is that when P is a monic, irreducible polynomial not dividing m, then $\sigma_P = (PA, K_m/k)$, the Artin automorphism for the prime ideal PA. The next proposition, interesting in itself, will be a useful tool.

Proposition 12.9. *Let* $\mathcal{O}_m$ *be the integral closure of* A *in* K_m. *Then,* $\mathcal{O}_m = A[\lambda_m]$.

Proof. We first consider the case when $m = P^e$, i.e., when m is a prime power.

For the moment, let's drop the subscript and set $\lambda_{p^e} = \lambda$. We have that $A[\lambda] \subseteq \mathcal{O}_{P^e}$ and we want to show equality holds. Let $g(x) \in k[x]$ be the irreducible polynomial for λ. We showed at the beginning of the proof of Proposition 12.7 that $g'(\lambda)$ divided a power of P in $\mathcal{O}_{P^e}$. It is a standard fact about Dedekind domains that the discriminant of the A-order $A[\lambda] \subset K_{P^e}$ is the norm from K_{P^e} to k of the element $g'(\lambda)$ (see Serre [2]). It follows that the discriminant of $A[\lambda]$ is a constant times a power of P.

Let $\omega \in \mathcal{O}_{P^e}$. Then

$$\omega = \sum_{i=0}^{\Phi(P^e)-1} a_i \lambda^i ,$$

where $a_i \in k$ for $0 \le i < \Phi(P^e)$. Using the fact that the discriminant of $A[\lambda]$ is a constant times a power of P, we see in the usual way that each a_i is if the form b_i/P^n, where $b_i \in A$ and n can be chosen so that at least one of the b_i is not divisible by P. We want to show that under these circumstances, $n = 0$, so that $\omega \in A[\lambda]$. From the last equation, we find

$$P^n \omega = \sum_{i=0}^{\Phi(P^e)-1} b_i \lambda^i . \tag{6}$$

Extend the additive valuation ord_P from k to K_m by the equation $\mathrm{ord}_P = \Phi(P^e)^{-1} \mathrm{ord}_{(\lambda)}$. Recall that by Proposition 12.7, (λ) is a prime ideal and it is totally ramified over PA. Thus, this procedure makes good sense. Also, $\mathrm{ord}_P(\lambda) = 1/\Phi(P^e)$.

In Equation 6, ord_P of the left-hand side is $\ge n$. Let $i_0 < \Phi(P^e)$ be the smallest non-negative integer such that $\mathrm{ord}_P(b_{i_0}) = 0$. A moment's reflection will then establish that the i_0-th term on the right-hand side of Equation 6 has the smallest valuation of all the terms. Thus, the ord_P of the right-hand side is $i_0/\Phi(P^e) < 1$. It follows that $n = 0$ and so $\omega \in A[\lambda]$ as asserted.

To handle the general case suppose $m = \alpha P_1^{e_1} P_2^{e_2} \cdots P_t^{e_t}$ is the prime decomposition of m in A. One can show that $\mathcal{O}_m$ is the ring compositum of the subrings $\mathcal{O}_{P_i^{e_i}} = A[\lambda_{P_i^{e_i}}]$. We omit the details of this, but refer the reader to Proposition 17, Chapter 3 of Lang [5] where the analogous result is proven in the number field case. Since $A[\lambda_{P_i^{e_i}}] \subseteq A[\lambda_m]$ for all $1 \leq i \leq t$ we have $\mathcal{O}_m \subseteq A[\lambda_m] \subseteq \mathcal{O}_m$.

We are now in a position to give a fairly short proof of the following important result.

Theorem 12.10. *Let $m \in A$ have positive degree and let $P \in A$ be a monic, irreducible polynomial not dividing m. Then, the Artin automorphism of the prime ideal PA in the extension K_m/k is the automorphism σ_P which takes λ_m to $C_P(\lambda_m)$. Let f be the smallest positive integer such that $P^f \equiv 1 \pmod{m}$. Then, $P\mathcal{O}_m$ is the product of $\Phi(m)/f$ prime ideals each of degree f. In particular, PA splits completely iff $P \equiv 1 \pmod{m}$.*

Proof. Since P does not divide m, PA is unramified in K_m. Let $\mathfrak{P}$ be any prime ideal in $\mathcal{O}_m$ lying above PA. Then, the Artin automorphism is characterized by

$$(PA, K_m/k)\omega \equiv \omega^{|P|} \pmod{\mathfrak{P}} \quad \forall \omega \in \mathcal{O}_m .$$

This is because the norm of the ideal PA is the number of elements in A/PA which is $q^{\deg(P)} = |P|$.

As we have seen, the irreducible polynomial for λ_P is $C_P(x)/x$. By the Corollary to Proposition 12.7, $C_P(x)/x$ is an Eisenstein polynomial. Also, it is monic since we are assuming that P is monic. Thus,

$$C_P(x) \equiv x^{|P|} \pmod{P} ,$$

and this congruence continues to hold modulo $\mathfrak{P}$ since $P \in \mathfrak{P}$. Consequently,

$$\sigma_P \lambda_m = C_P(\lambda_m) \equiv \lambda_m^{|P|} \pmod{\mathfrak{P}} .$$

Let $\omega \in \mathcal{O}_m$. By Proposition 12.9, $\omega = \sum a_i \lambda_m^i$, where $a_i \in A$ and $0 \leq i < \Phi(m)$. Thus,

$$\sigma_P \omega = \sum_i a_i (\sigma_P \lambda_m)^i \equiv \sum_i a_i \lambda_m^{|P|i} \equiv \left(\sum_i a_i \lambda_m^i\right)^{|P|} \equiv \omega^{|P|} \pmod{\mathfrak{P}} .$$

$$(7)$$

We have used the fact that $|P|$ is a power of the characteristic p of k and Fermat's little theorem for polynomials; i.e., $a^{|P|} \equiv a \pmod{P}$ for all $a \in A$ (see the corollary to Proposition 1.8).

Equation 7 establishes the first part of the theorem, namely, $(PA, K_m/k) = \sigma_P$. For emphasis, we mention again that this equality is only true when P is a *monic* irreducible.

The last part of the Theorem follows from the standard property of the Artin automorphism of a prime. Namely, if f is the order of the Artin automorphism, then the prime splits into $\Phi(m)/f$ primes of degree f. From the isomorphism $\mathrm{Gal}(K_m/k) = (A/mA)^*$, we see that the order of σ_P is the smallest positive integer f such that $P^f \equiv 1 \pmod{m}$. This completes the proof of the theorem.

The last task in this chapter is to investigate the way in which the prime at infinity of k splits in the extension K_m. To do this we will need some preliminary work. First we need to know the degrees of the polynomial $[m, i] \in A$ which are the coefficients of $C_m(x)$. Secondly, we need a description of the completion of k at ∞. Finally, we will need an elementary, but powerful, technique of non-archimedean analysis, the Newton polygon.

Proposition 12.11. *Let* $C_m(x) = \sum_{i=0}^{d}[m, i]x^{q^i}$ *where each coefficient* $[m, i] \in A$ *and* $d = \deg(m)$. *Then the degree of* $[m, i]$ *as a polynomial in* T *is* $q^i(d - i)$.

Proof. If $i > d$, we set $[m, i] = 0$. Notice that $[m, 0] = m$ which has degree $d = q^0(d - 0)$. This shows the result is true for $i = 0$. For the rest of the proof we assume $i > 0$.

We first investigate the special case $m = T^n$ and proceed by induction on n. For $n = 1$ we have $[T, 0] = T$ of degree $1 = q^0(1 - 0)$ and $[T, 1] = 1$ of degree $0 = q^1(1 - 1)$. Thus, the result is true for $n = 1$.

To go further, we first derive a recursion formula for the coefficients $[T^n, i]$. Consider the equation

$$C_{T^n}(x) = C_T(C_{T^{n-1}}(x)) = TC_{T^{n-1}}(x) + C_{T^{n-1}}(x)^q .$$

By isolating the coefficient of x^{q^i} on both sides, we find

$$[T^n, i] = T[T^{n-1}, i] + [T^{n-1}, i - 1]^q .$$

By induction, the degree of the first term on the right-hand side is $1 + q^i(n - 1 - i)$ and the degree of the second term on the right-hand side is $q^i(n - i)$. Since we are assuming $i > 0$, the second term has larger degree and it follows that $[T^n, i]$ has degree $q^i(n - i)$.

Finally, if $m = \sum_{j=0}^{d} \alpha_j T^j$ with each $\alpha_j \in \mathbb{F}$ and $\alpha_d \neq 0$, then

$$[m, i] = \sum_{j=0}^{d} \alpha_j [T^j, i] ,$$

from which one sees that $\deg_T([m, i]) = q^i(d - i)$ since $\alpha_d[T^d, i]$ is the non-zero term of largest degree.

We now turn our attention to the completion of k at infinity. It is useful to first give a discussion of the completion of k at the prime corresponding to the monic irreducible polynomial T, i.e., the completion of k at zero. Every element of $h \in k$ can be written as a power of T, T^n, say, times a quotient $f(T)/g(T)$ of polynomials, both not divisible by T. Under these circumstances, $\mathrm{ord}_T(h) = n$. We can give k a metric space structure by setting $|h_1 - h_2|_T = q^{-\mathrm{ord}_T(h_1-h_2)}$, where $q = \#\mathbb{F}$. Two polynomials will be close in the resulting topology if their difference is divisible by a high power of T, i.e. if their initial coefficients coincide. Thus, a Cauchy sequence of polynomials will give rise to a uniquely defined power series in T. The completion of A in this topology, algebraically and topologically, is the ring of formal power series, $\mathbb{F}[[T]]$, where the topology is given by the powers of the maximal ideal (T). The completion of k is just the quotient field of $\mathbb{F}[[T]]$, which is called the field of formal Laurent series. We denote this field, $\mathbb{F}((T))$. A typical element has the form

$$\sum_{i=-N}^{\infty} \alpha_i T^i, \quad \text{where} \quad \alpha_i \in \mathbb{F} .$$

To get a good description of the completion of k at infinity the trick is to replace T by $1/T$ in the above analysis. To see this, recall that for a polynomial $f(T)$ in T we have, by definition, $\mathrm{ord}_\infty f(T) = -\deg f(T)$. Let $d = \deg f(T)$. We can write $f(T) = T^d h(1/T)$ where h is a polynomial with non-zero constant term. If we set $U = 1/T$, then the monic irreducible U of the ring $\mathbb{F}[U] = \mathbb{F}[1/T]$ defines a discrete, rank 1 valuation of k and clearly, $\mathrm{ord}_U f(T) = \mathrm{ord}_U U^{-d} h(U) = -d$. It follows that the two valuations ord_∞ and ord_U coincide on A and so they must coincide on k, i.e., they are equal. As a consequence of our previous discussion of the completion of k at zero we can now assert that the completion of k at infinity is the ring of formal Laurent series in U, i.e.,

$$k_\infty = \mathbb{F}((U)) = \mathbb{F}((1/T)) .$$

The elements which are regular at infinity are the power series in $1/T$, $\mathbb{F}[[1/T]]$, and the units at infinity are the power series in $1/T$ with non-zero constant term. If $0 \neq g \in \mathbb{F}((1/T))$, then we can write $g = (1/T)^N h$, where h is a unit in $\mathbb{F}[[1/T]]$. In this situation, $\mathrm{ord}_\infty g = N$.

We shall return to k_∞ shortly, but first we will describe the method of the Newton polygon. This method enables us to find information about the roots of a polynomial with coefficients in a field L, which is complete with respect to a discrete rank 1 valuation v. We denote by ord_v the corresponding ord function. Let

$$f(x) = \sum_{j=0}^{d} a_j x^j \in L[x] ,$$

be a polynomial and assume that $a_0 a_d \neq 0$. Consider the set of points

$$S_f = \{(j, \mathrm{ord}_v a_j) \in \mathbb{R}^2 \mid 0 \leq j \leq d, \ a_j \neq 0\} .$$

Above each point in S_f erect a vertical ray and then take the convex hull of the resulting set. This convex hull is bounded on the sides by two vertical rays and below by a polygonal path connecting $(0, \mathrm{ord}_v a_0)$ with $(d, \mathrm{ord}_v a_d)$. This polygonal path is defined to be the Newton polygon of f, N_f.

Let $\bar{L}$ be an algebraic closure of L. We continue to use the notation ord_v for the unique extension of ord_v to $\bar{L}$.

Theorem 12.12. *Let L be a field which is complete with respect to a discrete, rank 1 valuation v. Let ord_v be the corresponding ord function extended to $\bar{L}$, an algebraic closure of L. Let $f(x) = \sum_{j=0}^{d} a_j x^j \in L[x]$ be a polynomial with $a_0 a_d \neq 0$. Let l be a line segment of the Newton polygon of f joining $(j, \mathrm{ord}_v a_j)$ with $(h, \mathrm{ord}_v a_h)$ with $j < h$. Then $f(x)$ has exactly $h - j$ roots γ in $\bar{L}$ such that $\mathrm{ord}_v \gamma$ is the negative of the slope of l.*

We shall postpone the proof of this theorem until the end of the chapter.

To give some idea of its usefulness, let's apply it to the case where $f(x)$ is an Eisenstein polynomial, i.e., $\mathrm{ord}_v a_d = 0$, $\mathrm{ord}_v a_i > 0$ for $0 \leq i < d$, and $\mathrm{ord}_v a_0 = 1$. One sees, without effort, that the Newton polynomial of f is just the line segment joining $(0, 1)$ with $(d, 0)$. It follows that f has d roots γ such that $\mathrm{ord}_v \gamma = 1/d$. Since an Eisenstein polynomial is irreducible, it follows that adjoining any root of f to L results in a totally ramified extension of degree d. Other applications of this nature are easy to produce, but we leave these aside and proceed to apply the method to the case $f(x) = C_m(x) \in k_\infty[x]$.

Proposition 12.13. *Let $C_m(x) \in k[x] \subset k_\infty[x]$ be the Carlitz polynomial corresponding to $m \in \mathbb{F}[T]$ of degree d. Let $\bar{k}_\infty$ be an algebraic closure of k_∞ and continue to denote by ord_∞ the unique extension of ord_∞ to $\bar{k}_\infty$. Then, for each $1 < i \leq d$, there exist exactly $q^i - q^{i-1}$ roots $\tilde{\lambda}$ of $C_m(x)$ such that*

$$\mathrm{ord}_\infty \tilde{\lambda} = d - i - \frac{1}{q-1}.$$

Proof. Recall that $C_m(x) = \sum_{i=0}^{d} [m, i] x^{q^i}$. By Proposition 12.11, $\mathrm{ord}_\infty [m, i] = -\deg [m, i] = -q^i(d - i)$.

To apply Theorem 12.12, we first divide $C_m(x)$ by x to get a polynomial with non-zero constant term. The points to consider in the construction of the Newton polygon of $C_m(x)/x$ are

$$\{(q^j - 1, -q^j(d - j)) \mid 0 \leq j \leq d\} .$$

The lines connecting successive points all have different (and increasing) slopes, so the Newton polygon of $C_m(x)/x$ consists of just these line segments. Connecting the $i - 1$'st point with the i-th point gives the slope:

$$\frac{-q^i(d - i) - (-q^{i-1}(d - i + 1))}{q^i - q^{i-1}} = -(d - i) + \frac{1}{q-1} .$$

The proposition now follows from Theorem 12.12.

Corollary. *There are $q - 1$ roots $\tilde{\lambda}$ of $C_m(x)$ in $\bar{k}_\infty$ such that $\mathrm{ord}_\infty \tilde{\lambda} = d - 1 - 1/(q - 1)$. For each such root, we have $\tilde{\lambda}^{q-1} \in k_\infty$.*

Proof. The first assertion is a special case of the proposition corresponding to $i = 1$.

The monomials which occur in $C_m(x)/x$ with non-zero coefficients all have the form $x^{q^i - 1}$. Thus, $C_m(x)/x = f(w)$, where $w = x^{q-1}$ and $f(w) \in k[w]$. The roots of $f(w)$ in $\bar{k}_\infty$ are $\{\tilde{\lambda}^{q-1}\}$ where $\tilde{\lambda}$ runs through the roots of $C_m(x)$. The map $\tilde{\lambda} \to \tilde{\lambda}^{q-1}$ is $q - 1$ to 1, since whenever $\tilde{\lambda}$ is a root of $C_m(x)$ so is $\alpha\tilde{\lambda}$ for any $\alpha \in \mathbb{F}^*$. It follows that $f(w)$ has exactly one root γ with $\mathrm{ord}_\infty \gamma = (q - 1)(d - 1) - 1$.

Let σ be any automorphism of $\bar{k}_\infty$ leaving k_∞ fixed. Then, $\sigma\gamma$ is also a root of $f(w)$. Since $\mathrm{ord}_\infty \gamma = \mathrm{ord}_\infty \sigma\gamma$, we must have $\sigma\gamma = \gamma$. Since σ is arbitrary, it follows that $\gamma \in k_\infty$, as asserted.

Using the Carlitz action, $\bar{k}_\infty$ can be made into an A-module in exactly the same way that we made $\bar{k}$ into an A-module. Namely, if $a \in A$ and $u \in \bar{k}_\infty$ then we define $a \cdot u = C_a(u)$. If $m \in A$ is of positive degree, we denote the m-torsion points, $\bar{k}_\infty[m]$, by $\tilde{\Lambda}_m$.

Let ι denote a fixed field isomorphism over k from K_m to $\bar{k}_\infty$. Since K_m/k is a Galois extension, all field isomorphisms over k from K_m to $\bar{k}_\infty$ are of the form $\iota \circ \sigma$ with $\sigma \in \mathrm{Gal}(K_m/k)$.

The isomorphism ι corresponds to a prime $\mathfrak{P}_\infty$ of K_m lying over ∞. To see this, let $\mathcal{O}_\infty = \{\omega \in K_m \mid \mathrm{ord}_\infty \iota\omega \geq 0\}$. It is easy to see that $\mathcal{O}_\infty$ is a discrete valuation ring inside K_m which contains $\mathbb{F}$ and has quotient field K_m. By definition this is a prime of K_m which we denote by $\mathfrak{P}_\infty$, its maximal ideal. The proof of the fact that $\mathfrak{P}_\infty$ lies above ∞ is straightforward.

Suppose that λ is a root of $C_m(x)$ in $\bar{k}$. Since $C_m(\lambda) = 0$ implies $C_m(\iota\lambda) = 0$, we see that ι maps Λ_m to $\tilde{\Lambda}_m$. This map is an A-module isomorphism. By Proposition 12.13, there is an element $\tilde{\lambda}_m \in \tilde{\Lambda}_m$ such that $\mathrm{ord}_\infty \tilde{\lambda}_m = d - 1 - 1/(q - 1)$. Let $\lambda_m \in \Lambda_m$ be such that $\iota\lambda_m = \tilde{\lambda}_m$.

Theorem 12.14. *Let $J = \{\sigma_\alpha \in \mathrm{Gal}(K_m/k) \mid \alpha \in \mathbb{F}^*\}$ and set K_m^+ equal to the fixed field of J. Then ∞ splits completely in K_m^+ and every prime above ∞ in K_m^+ is totally and tamely ramified in K_m.*

Proof. The proof will proceed in steps.

Step 1. The first thing to do is to show that $\tilde{\lambda}_m$ is an A generator of $\tilde{\Lambda}_m$. Suppose $a \in A - \{0\}$ of degree less than d. Then,

$$\mathrm{ord}_\infty C_a(\tilde{\lambda}_m) = \mathrm{ord}_\infty \left(\sum_{i=0}^{\deg a} [a, i] \tilde{\lambda}_m^{q^i} \right) = \mathrm{ord}_\infty a\tilde{\lambda}_m = d - \deg a - 1 - \frac{1}{q - 1}.$$

$$(8)$$

To justify the second equality, one has to show that the term $a\tilde{\lambda}_m$ in the sum is the one with lowest ord. We leave this straightforward calculation to the reader.

If $\tilde{\lambda}_m$ were not an A-generator of $\tilde{\Lambda}_m$ there would be a proper divisor m_1 of m such that $C_{m_1}(\tilde{\lambda}_m) = 0$. Equation 8 shows this is impossible. Thus, $\tilde{\lambda}_m$ is an A-generator of $\tilde{\Lambda}_m$, as asserted. Moreover, since Λ_m and $\tilde{\Lambda}_m$ are A-isomorphic via ι, it follows that λ_m is an A-generator of Λ_m. It follows that $k_\infty(\tilde{\Lambda}_m) = k_\infty(\tilde{\lambda}_m)$ and $K_m = k(\Lambda_m) = k(\lambda_m)$.

Step 2. We show that $K_m^+ = k(\lambda_m^{q-1})$. Let $\sigma_\alpha \in J$. Then, $\sigma_\alpha(\lambda_m^{q-1}) = (\sigma_\alpha\lambda_m)^{q-1} = (\alpha\lambda_m)^{q-1} = \lambda_m^{q-1}$. Since $\alpha \in \mathbb{F}^*$ is arbitrary, it follows that $\lambda_m^{q-1} \in K_m^+$. Since λ_m is a root of $x^{q-1} - \lambda_m^{q-1} \in K_M^+[x]$ we have $[K_m : k(\lambda_m^{q-1})] \leq q - 1$. By Galois theory, $[K_m : K_m^+] = q - 1$. It follows that $K_m^+ = k(\lambda_m^{q-1})$, as claimed.

Step 3. From Step 2, we see that $\iota(K_m^+) = \iota(k(\lambda_m^{q-1}) \subset k_\infty(\tilde{\lambda}_m^{q-1})$. By the corollary to Proposition 12.13, we have $\tilde{\lambda}_m^{q-1} \in k_\infty$. Thus, $\iota(K_m^+) \subset k_\infty$. It follows that ∞ splits completely in K_m^+, which proves the first part of the theorem.

Step 4. We claim that the extension $k_\infty(\tilde{\lambda}_m)/k_\infty$ is totally and tamely ramified of degree $q-1$. Let $\gamma = \tilde{\lambda}_m^{q-1}$. Then, $\tilde{\lambda}_m$ satisfies $x^{q-1} - \gamma \in k_\infty[x]$. Thus, $[k_\infty(\tilde{\lambda}_m) : k_\infty] \leq q-1$. On the other hand, $\text{ord}_\infty\tilde{\lambda}_m = d-1-1/(q-1)$ so the ramification index of the extension is at least $q - 1$. It follows that the degree of the extension is $q - 1$ and the ramification index is $q - 1$, which is what was to be proven. The ramification is tame, since $q - 1$ is prime to the residue field characteristic which is p.

Step 5. Let $\mathfrak{P}_\infty$ be the prime of K_m discussed earlier and $\mathfrak{p}_\infty$ be the prime of K_m^+ lying below it. The completion of K_m^+ at $\mathfrak{p}_\infty$ is k_∞ by Step 3. The completion of K_m at $\mathfrak{P}_\infty$ is $k_\infty(\tilde{\lambda}_m)$. Thus, by Step 4, $\mathfrak{P}_\infty/\mathfrak{p}_\infty$ is totally and tamely ramified of ramification degree $q - 1$. The other primes over ∞ behave the same way since K_m/k is a Galois extension.

Corollary. *For all $m \in A$, $m \neq 0$, the constant field of K_m is $\mathbb{F}$, i.e., K_m/k is a geometric extension.*

Proof. Since $f(\mathfrak{p}_\infty/\infty) = 1$, the residue class field at $\mathfrak{p}_\infty$ is $\mathbb{F}$. We have $f(\mathfrak{P}_\infty/\mathfrak{p}_\infty) = 1$ since $\mathfrak{P}_\infty/\mathfrak{p}_\infty$ is totally ramified. Thus the residue class field of $\mathfrak{P}_\infty$ is also $\mathbb{F}$. Since the constant field of K_m injects into the residue class field of $\mathfrak{P}_\infty$, the result follows.

Since the properties of K_m^+ are so similar to those of $\mathbb{Q}_m^+$ we call K_m^+ the maximal real subfield of K_m. The point is that the prime at infinity of k splits completely in K_m^+ and every prime above it in K_m^+ ramifies totally in K_m. This is just the behavior of the prime at infinity of $\mathbb{Q}$, the only archimedean prime. It splits completely in $\mathbb{Q}_m^+$ and every prime above it ramifies totally in $\mathbb{Q}_m$. Also notice that the Galois group of K_m/K_m^+ is isomorphic to $\mathbb{F}^*$, the non-zero units of A, whereas the Galois group of $\mathbb{Q}_m/\mathbb{Q}_m^+$ is isomorphic to $\{\pm 1\}$ the non-zero units of $\mathbb{Z}$.

In general, we will call a finite extension K of k real if ∞ splits completely in K. For example, the theory of quadratic function fields (quadratic extensions of k) is divided up into the theory of real quadratic function fields, the case where ∞ splits, and complex quadratic function fields, the case where ∞ is either inert or ramifies. We will discuss this in greater detail later.

Having described in some detail the cyclotomic function fields, $K_m = k(\Lambda_m)$, we will give a sketch of the result of Hayes which is the function field analogue the Kronecker-Weber theorem. The latter theorem states that every finite abelian extension of the rational numbers $\mathbb{Q}$ is contained in a cyclotomic field $\mathbb{Q}(\zeta_n)$ for some positive integer n. For a discussion of the history of this theorem and a proof see Chapter 14 of Washington [1]. It cannot be true that every abelian extension of $k = \mathbb{F}(T)$ is contained in some field $K_m = k(\Lambda_m)$ because, among other reasons, the above Corollary shows that K_m/k is a geometric extension. Thus, it cannot contain a constant field extension of k (recall that all finite extensions of a finite field have a cyclic Galois group). Let's work within a fixed algebraic closure of k. Define $k(\Lambda)$ to be the union of all the fields K_m. Secondly, let $\bar{k} = \bar{\mathbb{F}}k$ be the maximal constant field extension of k. These fields are abelian and disjoint and one might think that every abelian extension of k is a subfield of the compositum of $k(\Lambda)$ and $\bar{k}$. However, this field is still not big enough since a moment's reflection shows that a subfield of this compositum must be tamely ramified at ∞. To construct abelian extensions of k that are wildly ramified at infinity, work with the parameter at infinity, i.e., $1/T$, rather than T. One considers the ideal $(1/T)$ in the ring $\mathbb{F}[1/T]$ and using the Carlitz construction for this situation produces the fields $k(\Lambda_{T^{-n-1}})$. These fields are abelian over k, totally ramified at ∞, and $[k(\Lambda_{T^{-n-1}}) : k] = q^n(q-1)$. Let L_n be the unique subfield such that $[L_n : k] = q^n$ and set L_∞ equal to the union of all the fields, L_n. These three fields, $k(\Lambda)$, $\bar{k}$, and L_∞, are disjoint and the main theorem states that every abelian extension of k is contained in their compositum. In this sense, the Carlitz module gives an explicit construction of the maximal abelian extension of k. Hayes' proof relies heavily on class field theory. In the case of the Kronecker-Weber theorem it is possible to produce more elementary proofs so it is certainly possible that a more elementary proof can also be given in the function field case. We leave this as a challenge to the interested reader.

To conclude this chapter, we sketch the proof of Theorem 12.12, the Newton polygon method. Our sketch will include enough details so that giving a complete proof will only involve setting up a formal induction step rather than the informal one given below.

Proof of Theorem 12.12. (Sketch)

We begin by noticing that we can assume that $\alpha_0 = 1$. This is because the Newton polygon of $\sum_{j=0}^{d} \alpha_j/\alpha_0 \, x^j$ is the same as that of $f(x)$ except that it is shifted vertically by $-\mathrm{ord}_v \alpha_0$. The roots remain the same and the length of the line segments and the slopes remain the same, so we may as well assume that $\alpha_0 = 1$.

Secondly, since zero is not a root, we may as well work with the inverse roots rather than the roots. We will thus connect the slopes of the line segments with the ord_vs of the inverse roots (no negative sign). Thus, write

$$f(x) = \prod_{i=1}^{d}(1 - \pi_i x) \, ,$$

and arrange the inverse roots such that

$$
\begin{aligned}
\mathrm{ord}_v \pi_1 \quad &= \cdots = \mathrm{ord}_v \pi_{r_1} = s_1 < \mathrm{ord}_v \pi_{r_1+1} \\
&= \cdots = \mathrm{ord}_v \pi_{r_1+r_2} = s_2 < \mathrm{ord}_v \pi_{r_1+r_2+1} \\
&= \cdots = \mathrm{ord}_v \pi_{r_1+r_2+r_3} = s_3 < \mathrm{ord}_v \pi_{r_1+r_2+r_3+1} = \text{etc} \, .
\end{aligned}
$$

Having ordered the inverse roots in this fashion, we claim that the vertices of the Newton polygon are

$$
\begin{aligned}
P_0 \;&=\; (0,0), P_1 = (r_1, r_1 s_1), P_2 = (r_1 + r_2, r_1 s_1 + r_2 s_2), \\
P_3 \;&=\; (r_1 + r_2 + r_3, r_1 s_1 + r_2 s_2 + r_3 s_3),
\end{aligned}
$$

etc. Assuming for the moment that this is the case, we see that the theorem is established since the difference of the x-coordinates of P_{i-1} and P_i is r_i and the slope of the line connecting them is s_i. By the way, we have grouped the inverse roots, we see there are precisely r_i of them such that $\mathrm{ord}_v \pi = s_i$.

To prove our assertion about the vertices, notice that the j-th coefficient of $f(x)$, namely a_j, is the j-th elementary symmetric function of the inverse roots π_i. If $0 \leq j < r_1$, then from the form of the j-th elementary symmetric function we see that $\mathrm{ord}_v a_j \geq j s_1$. It follows that the slopes of the lines connecting $(j, \mathrm{ord}_v a_j)$ to $(0,0)$ are all greater than or equal to s_1 for j in this range. However, we must have $\mathrm{ord}_v a_{r_1} = r_1 s_1$, since only one term in the r_1-st symmetric function has this order, namely, $\pi_1 \pi_2 \cdots \pi_{r_1}$, all the other terms having greater order. By exactly the same reasoning we see that for h in the range $0 \leq h < r_2$ we have $\mathrm{ord}_v a_{r_1+h} \geq r_1 s_1 + h s_2$, whereas $\mathrm{ord}_v a_{r_1+r_2} = r_1 s_1 + r_2 s_2$ (the term of the $r_1 + r_2$-th elementary symmetric function having the smallest order being $\pi_1 \cdots \pi_{r_1} \pi_{r_1+1} \cdots \pi_{r_1+r_2}$). Thus, the lines connecting these points to $(r_1, \mathrm{ord}_v a_{r_1}$ all have slopes greater than or equal to s_2, whereas the line connecting $(r_1, r_1 s_1)$ to $(r_1 + r_2, r_1 s_1 + r_2 s_2)$ has slope exactly s_2. In general, let h vary in the range $0 \leq h < r_i$ and consider the point with index $r = \sum_{m=0}^{i-1} r_m + h$. Looking at the r-th elementary symmetric function of the inverse roots we see

$$\mathrm{ord}_v a_r \geq \sum_{m=0}^{i-1} r_m s_m + h s_i \, .$$

Thus, the slopes connecting these points to P_{i-1} are greater than or equal to s_i, whereas the slope connecting P_i to P_{i-1} is exactly s_i. Since the slopes s_i are monotone increasing, this is sufficient information to conclude the proof.

Exercises

Throughout the exercises, $\mathbb{F}$ will denote a finite field with q elements, $A = \mathbb{F}[T]$, the polynomial ring over $\mathbb{F}$, and $k = \mathbb{F}(T)$ the quotient field of A. If $m \in A$ we set $|m| = q^{\deg m}$. If $m \in A$, Λ_m will denote the m-torsion points on the Carlitz module.

1. Let $P \in A$ be a monic irreducible polynomial and λ a generator of Λ_P. Show that $P = \prod_{\deg a < \deg P} C_a(\lambda)$, where the product is over all non-zero polynomials of degree less than $\deg P$.

2. Let $\mathcal{M}$ denote all monic polynomials of degree less than $\deg P$. Set $\pi = \prod_{a \in \mathcal{M}} C_a(\lambda)$. Use Exercise 1 to show that $P = (-1)^{\deg P} \pi^{q-1}$. Set $P^* = (-1)^{\deg P} P$. Then $P^* = \pi^{q-1}$.

3. (Continuation) Let $Q \neq P$ be another monic irreducible polynomial. Recall the symbol (a/Q), which is defined to be the unique element of $\mathbb{F}$ such that
$$a^{\frac{|Q|-1}{q-1}} \equiv \left(\frac{a}{Q}\right) \pmod{Q} .$$

   Use the fact that σ_Q is the Artin automorphism at Q in the field $K_P = k(\Lambda_P)$ to prove that $\sigma_Q(\pi) = (P^*/Q)\pi$.

4. (Continuation) Use Theorem 12.10 to show $\sigma_Q(\pi) = \prod_{a \in \mathcal{M}} C_{Qa}(\lambda)$. Now, use Gauss' criterion (see Exercise 10 of Chapter 3) to show that $\sigma_Q(\pi) = (Q/P)\pi$.

5. (Continuation) Combine Exercises 3 and 4 to prove the reciprocity law; i.e., if $P \neq Q$ are two monic irreducibles, then
$$\left(\frac{P}{Q}\right)\left(\frac{Q}{P}\right)^{-1} = (-1)^{\deg P \deg Q} .$$

   This nice proof is due to Carlitz.

6. Let P be a monic irreducible of degree d. Use the Riemann-Hurwitz formula to prove that the genus of $K_P = k(\Lambda_P)$ is $(d-1)q^d + 1 - 2d - (1 + q + \cdots + q^{d-1})$.

7. Let $e > 0$ be an integer. Compute the genus of K_{P^e}. (When $e \geq 2$ the calculation is more difficult because the extension is not tamely ramified. One needs to compute the exponent of the different at the prime above P by local considerations. See Hayes [1].)

8. We continue to assume that P is a monic irreducible of degree d. Let $e \geq 1$ and let λ be an A-generator of λ_{P^e}. Let $(\lambda)_\infty$ be the polar divisor of λ. Show $\deg (\lambda)_\infty = q^{d-1}$ if $e = 1$ and $\deg (\lambda)_\infty = q^{(e-1)d-1}(q^d - 1)$ if $e \geq 2$.

9. Let $m \in A$ be a monic polynomial, λ a non-zero element of Λ_m, and σ an element of the Galois group of K_m/k. Prove that $\sigma\lambda/\lambda$ is a unit in $\mathcal{O}_m$. (These units are called cyclotomic units since they are analogous to cyclotomic units in $\mathbb{Q}(\zeta_m)$. We will encounter them again in Chapter 16.)

10. Let $m \in A$ be a monic polynomial. Define Q_0 to be the index $[\mathcal{O}_m^* : \mathcal{O}_m^{+*}]$. Show that $Q_0 = 1$ if m is a prime power and that $Q_0 = q - 1$ if m is not a prime power. Hint: Try imitating the proof of the corresponding fact in cyclotomic number fields.

13
Drinfeld Modules: An Introduction

In the last chapter we introduced a special class of Drinfeld modules for the ring $A = \mathbb{F}[T]$ defined over the field $k = \mathbb{F}(T)$ and discussed some of their properties. By considering the Carlitz module, in particular, we were able to construct a family of field extensions of k with properties remarkably similar to those of cyclotomic fields. In this chapter we will give a more general definition of a Drinfeld module. The definition and theory of these modules was given by V. Drinfeld in the mid-seventies, see Drinfeld [1, 2]. The application of the rank 1 theory to the class field theory of global function fields is due to Drinfeld and independently to D. Hayes [2]. The article by Hayes [6] provides a compact introduction to this material. A comprehensive treatment of Drinfeld modules (and, even more generally, T-modules) can be found in the treatise of Goss [4].

In this chapter we will develop the beginnings of the general theory, but will not pursue it further. Our aim is to supply the reader with some of the basic ideas and, hopefully, the stimulus to pursue the study of these modules further. Many beautiful and deep applications have already been discovered. However, the subject remains young and is under active development.

Let $k/\mathbb{F}$ be a function field with exact field of constants $\mathbb{F}$, a finite field with q elements. Let ∞ be a fixed prime of k and let $A \subset k$ be the ring of all elements of k whose only poles are at ∞. It is well known that A is a Dedekind domain whose non-zero prime ideals are in one-to-one correspondence with the primes of k different from ∞. If $(\mathcal{O}, \mathcal{M})$ is a prime of k such that $A \subset \mathcal{O}$, the corresponding prime ideal of A is $\mathcal{M} \cap A$. On the other hand, if $P \subset A$ is a non-zero prime ideal of A, then (A_P, PA_P) is the corresponding prime of k. It is clear that the polynomial ring $A = \mathbb{F}[T]$ considered in the last chapter is a special case of this construction. In this special case, the degree of the prime at infinity is 1, but that might not be so in the general case. We set d_∞ to be the degree of the prime at infinity.

Let L be a field containing $\mathbb{F}$ and $G_{a/L}$ the additive group scheme over L. If Ω is an L-algebra, $G_{a/L}$ assigns to Ω its additive group Ω_+, i.e., Ω considered solely as an additive group under addition. In homological algebra this is sometimes referred to as a "forgetful functor." The additive group scheme over L assigns to every L-algebra its structure as an additive group, forgetting about the multiplicative structure. The endomorphism ring of $G_{a/L}$ over $\mathbb{F}$, $\operatorname{End}_\mathbb{F}(G_{a/L})$, assigns to every L-algebra Ω the algebraic endomorphisms of Ω_+ which commute with the action of $\mathbb{F}$. Using Proposition 12.2, one can show that $\operatorname{End}_\mathbb{F}(G_{a/L}) \cong L < \tau >$ where τ is the map which raises an element to the q-th power. So, if $u \in \Omega$ and $\sum c_i \tau^i \in L < \tau >$, then

$$\left(\sum c_i \tau^i\right)(u) = \sum c_i u^{q^i} .$$

The right-hand side of this equation is an additive polynomial with coefficients in L. The endomorphism ring of $G_{a/L}$ over $\mathbb{F}$ can thus be considered either as the non-commutative polynomial ring $L < \tau >$ with the key relation $\tau a = a^q \tau$, or as the ring of additive polynomials over L with multiplication being given by composition. Both descriptions are useful and, after a little experience, no confusion is likely to result from employing them both.

The map $D : L < \tau > \to L$ given by $D(\sum c_i \tau^i) = c_0$ is a homomorphism. It will play a role in the definition of a Drinfeld module, which we are about to give. In the alternate world of additive polynomials, D applied to $\sum c_i x^{q^i}$ is just differentiation with respect to x.

Definition. A Drinfeld A-module over L consists of an $\mathbb{F}$-algebra homomorphism δ from A to L, together with an $\mathbb{F}$-algebra homomorphism $\rho : A \to L < \tau >$ such that for all $a \in A$, $D(\rho_a) = \delta(a)$. Moreover, we require that the image of ρ not be contained in L. The notation $\operatorname{Drin}_A(L)$ will denote the set of all Drinfeld A-modules over L, the structural map δ being assumed fixed.

In practice the map δ is often just containment in a field L, but it is also occurs as reduction modulo a prime ideal. In the last chapter, it was just the containment of A in k.

A simple but useful, remark is that $\rho_\alpha = \alpha \tau^0$, the identity of $L < \tau >$, for all $\alpha \in \mathbb{F}$. This is because ρ is an $\mathbb{F}$-algebra homomorphism taking 1 to τ^0. Thus, $\rho_\alpha = \alpha \rho_1 = \alpha \tau^0$ for all $\alpha \in \mathbb{F}$.

If Ω is an L-algebra, then δ makes Ω into an A-module in the obvious way, namely, $a \cdot u = \delta(a)u$. The idea of a Drinfeld module is that it makes

Ω into an A-module in a new way which is a "deformation" of the standard one. Namely, if we define $a * u = \rho_a(u)$, it is straightforward to check that this makes Ω into an A-module. This is a deformation of the standard action since $a \cdot u$ and $a * u$ both have the same linear term, $\delta(a)u$. It is a new action since from the definition of a Drinfeld module, $a \cdot u \neq a * u$ for at least one $a \in A$. When considering Ω as an A-module under the action of ρ we shall use the notation Ω_ρ. In this chapter, the only algebras we will consider are field extensions of L.

Let $\rho \in \mathrm{Drin}_A(L)$ and M/L be an algebraically closed field extension. Consider the A-module, M_ρ. If $0 \neq a \in A$, we want to investigate the structure of the torsion submodule $M_\rho[a] = \{u \in M_\rho \mid \rho_a(u) = 0\}$. More generally, if $(0) \neq I \subset A$ is an ideal in A, we want to investigate the structure of $M_\rho[I] = \{u \in M \mid \rho_a(u) = 0, \forall a \in I\}$. In the course of doing this we will have to define and explain the notions of the rank and height of a Drinfeld module.

We begin by defining another notion, the A-characteristic of L, considered as an A-module via δ. If δ is one to one, we say that L has A-characteristic 0. If δ is not one to one, its kernel is a non-zero prime ideal Q of A. In this case we call Q the A-characteristic of L. This notion is not to be confused with the characteristic of L as a $\mathbb{Z}$ module. For all fields under consideration in this chapter, this characteristic is p, a non-zero prime number in $\mathbb{Z}$. The A-characteristic of L is a completely different notion.

A reader familiar with the arithmetic of elliptic curves or, more generally, abelian varieties, will find the form of the following result and its corollary quite familiar.

Theorem 13.1. *Let $\rho \in \mathrm{Drin}_A(L)$ as above. Let Q be the A-characteristic of L. Finally, let M be an algebraically closed field containing L. If $P \neq Q$ is a non-zero prime ideal of A, and $e \geq 1$ an integer, then there is a positive integer r independent of P and e such that*

$$M_\rho[P^e] \cong (A/P^e)^{(r)} .$$

If $Q \neq (0)$ and $e \geq 1$ is an integer, there is another integer h, independent of e, such that

$$M_\rho[Q^e] \cong (A/Q^e)^{(r-h)} .$$

Corollary. *Let $I \subset A$ be an ideal relatively prime to the A-characteristic of L. Let $\rho \in \mathrm{Drin}_A(L)$ and let M be an algebraically closed field containing L. Then there is an integer r, independent of I, such that*

$$M_\rho[I] \cong (A/I)^{(r)} .$$

We will not prove these results now, but we will do so after developing some preliminary machinery. The integer r will turn out to be the rank of ρ and the integer h will turn out to be the height of ρ, concepts whose definition will be given shortly.

The next few lemmas will be very general results about torsion modules over a Dedekind domain. Since these facts are fairly standard, we will only sketch the proofs, assuming that the reader either already knows these results or can fill in the details without difficulty. In these lemmas, A will be a Dedekind domain and M a module over A. If $I \subset A$ is a non-zero ideal, then $M[I] = \{m \in M \mid ax = 0, \forall a \in I\}$.

Lemma 13.2. *Suppose J and H are non-zero ideals of A which are relatively prime; i.e., $J + H = A$. Then $M[JH] = M[J] \oplus M[H]$.*

Proof. By hypothesis, there exist $a \in J$ and $b \in H$ such that $a + b = 1$. For $m \in M[JH]$ we have $m = am + bm$. Since $am \in M[H]$ and $bM \in M[J]$, we have shown $M[JH] = M[J] + M[H]$. To prove that the sum is a direct sum, let $m \in M[I] \cap M[J]$. Then $m = am + bm = 0 + 0 = 0$, and we're done.

Corollary. *Let $I = P_1^{e_1} P_2^{e_2} \cdots P_t^{e_t}$ be the prime decomposition of the ideal $I \neq (0)$. Then,*

$$M[I] = M[P_1^{e_1}] \oplus M[P_2^{e_2}] \oplus \cdots M[P_t^{e_t}] .$$

Proof. This follows from the lemma by a simple induction on t.

If P is a maximal ideal of A, let's define $M(P) = \cup_{e=1}^{\infty} M[P^e]$. This is called the P-primary component of M.

Lemma 13.3. *Let M be a torsion A-module. Then*

$$M = \bigoplus_P M(P) ,$$

where the sum is over all maximal ideals of A.

Proof. Let $m \in M$. Since M is a torsion module, there is a non-zero $a \in A$ such that $am = 0$. Consider the prime decomposition of the principal ideal (a) and apply Corollary 1 to Lemma 13.2. This shows that $M = \sum_P M(P)$.

To show the sum is direct, suppose $0 = \sum_{i=1}^{t} m_i$, where $m_i \in M(P_i)$. For each i with $1 \leq i \leq t$, there is an $e_i > 0$ such that $m_i \in M[P_i^{e_i}]$. Now apply the corollary to Lemma 13.2.

Lemma 13.4. *If $(0) \to M_1 \to M_2 \to M_3 \to (0)$ is an exact sequence of torsion A-modules and $P \subset A$ is a maximal ideal, then $(0) \to M_1(P) \to M_2(P) \to M_3(P) \to (0)$ is also exact.*

Proof. This follows easily from Lemma 13.3.

Lemma 13.5. *Let P be a maximal ideal of A and select $\pi \in P - P^2$. Then*

$$M[P^e] = M[\pi^e](P) .$$

Proof. We have $(\pi^e) = P^e J$ where P and J are relatively prime. By Lemma 13.2,

$$M[\pi^e] = M[P^e] \oplus M[J] .$$

Taking the P-primary component of both sides gives the result, since $M[J](P) = (0)$.

Lemma 13.6. *Suppose M is a divisible A-module, $P \subset A$ a maximal ideal, and $e > 1$ and integer. The following sequence is exact:*

$$(0) \to M[P] \to M[P^e] \to M[P^{e-1}] \to (0) ,$$

where the third arrow is given by multiplication by $\pi \in P - P^2$.

Proof. Choose $\pi \in P - P^2$. Using the divisibility of M, we see that the following sequence is exact:

$$(0) \to M[\pi] \to M[\pi^e] \to M[\pi^{e-1}] \to (0) .$$

The result follows by taking P-primary components and using Lemmas 13.4 and 13.5.

Corollary. *Suppose that $M[P]$ is finite. Then $M[P^e]$ is finite for all $e > 0$ and $\#M[P^e] = \#M[P]^e$.*

Proof. This is immediate from the lemma and a simple induction.

We will now give a result which provides the basis for the definition of the rank of a Drinfeld module.

Proposition 13.7. *Let ρ be an A-Drinfeld module defined over a field L. Define $\mu(a) = -\deg_\tau \rho_a$ for all $a \in A$. Then there is a positive rational number r such that $\mu(a) = r \operatorname{ord}_\infty(a) d_\infty$ for all $a \in A$.*

Proof. If we define $\mu(0) = \infty$ we easily check that μ gives a map from A to $\mathbb{Z} \cup \infty$ such that $\mu(a) = \infty$ if and only if $a = 0$, $\mu(ab) = \mu(a) + \mu(b)$, and $\mu(a + b) \geq \min(\mu(a), \mu(b))$. Moreover, $\mu(a) \leq 0$ for all $a \in A$ and $\mu(a) < 0$ for some $a \in A$. These properties show that μ can be extended to an additive valuation on the quotient field k of A. It cannot be one of the valuations associated to maximal ideals of A since all these have non-negative ord on A. Thus, μ must be equivalent to the valuation at infinity. This shows there is a real number r such that $\mu(a) = r \operatorname{ord}_\infty(a) d_\infty$. Let $a \in A$ be such that $\deg_\tau(\rho_a) > 0$ (the existence of such an $a \in A$ is specified in the definition of Drinfeld module). Then, $\mu(a)$ is a negative integer. We claim $\operatorname{ord}_\infty(a)$ is also a negative integer. To see this, note that the remark immediately following the definition of a Drinfeld module shows that a is not a constant. Since a has no pole at primes corresponding to maximal ideals in A, it must have a pole at infinity (see Proposition 5.1). It follows that r is a positive rational number, as asserted.

Definition. The rank of a Drinfeld A-module ρ is defined to be the unique positive, rational number r , such that $\deg_\tau \rho_a = -r \operatorname{ord}_\infty(a) d_\infty$ for all $a \in A$.

We will soon see that the rank of a Drinfeld module is actually a positive integer.

It is illuminating to reformulate the definition somewhat. For $a \in A$ define $\deg a$ to be the dimension over $\mathbb{F}$ of A/aA. This is clearly a generalization of the degree of a polynomial in a polynomial ring over a field. (Caution! This is not the degree of the principal divisor (a) which we know is zero.) As we will see from the following proof, $\deg a$ coincides with the degree of the zero divisor of a.

Proposition 13.8. *For all $a \in A$ we have $\deg a = -\operatorname{ord}_\infty(a) d_\infty$. Thus, the rank of a Drinfeld A-module ρ, can also be defined to be the unique positive, rational number r such that $\deg_\tau(\rho_a) = r \deg a$ for all $a \in A$.*

Proof. Let $m_P = \operatorname{ord}_P(a)$ and $aA = \prod_P P^{m_P}$ be the prime decomposition of the principal ideal aA. By the Chinese Remainder Theorem, we have

$$A/aA \cong \bigoplus_P A/P^{m_P} .$$

For any maximal ideal $P \subset A$, we have $A/P^m \cong A_P/(PA_P)^m$. Since PA_P is a principal ideal, $(PA_p)^{i-1}/(PA_P)^i \cong A_P/PA_P$ for all $i \geq 1$. Thus,

$$\dim_{\mathbb{F}}(A/P^m) = m \dim_{\mathbb{F}}(A_P/PA_P) = m \deg P .$$

We conclude that $\deg a = \sum_{P \neq \infty} \operatorname{ord}_P(a) \deg P$. Since the degree of a principal divisor is zero (Proposition 5.1), it follows that

$$\deg a = \sum_{P \neq \infty} \operatorname{ord}_P(a) \deg P = -\operatorname{ord}_\infty(a) d_\infty ,$$

as asserted.

The next topic to consider is the height of a Drinfeld module. This is of interest only for Drinfeld A-modules ρ of non-zero characteristic. Recall that the A-characteristic of ρ is the kernel of the structural map $\delta : A \to L$, where L is the field of definition of ρ. Call this ideal Q and suppose that $Q \neq (0)$. For $a \in A$, let $\omega(a)$ be the index of the smallest power of τ in ρ_a with non-zero coefficient. In other words, if $\rho_a = \sum_{i=0}^m c_m \tau^m$ and $c_i = 0$ for $i < n$ but $c_n \neq 0$, then $\omega(a) = n$. Define $\omega(0) = \infty$.

Proposition 13.9. *There is a positive rational number h with the property that for all $a \in A$ we have $\omega(a) = h \operatorname{ord}_Q(a) \deg Q$.*

Proof. The map ω takes A to $\mathbb{Z} \cup \infty$ and has the following properties: $\omega(a) = \infty$ if and only if $a = 0$, $\omega(ab) = \omega(a) + \omega(b)$, and $\omega(a + b) \geq \min(\omega(a), \omega(b))$. Moreover, $\omega(a) \geq 0$ for all $a \in A$, and $\omega(a) > 0$ if and only if $a \in Q$. It follows easily from these facts that ω extends to an additive valuation of k which is equivalent to $\mathrm{ord}_Q(*)$. Thus, there is a real number h having the required property. To show that h is positive and rational, just choose $a \in Q$ with $a \neq 0$. Then $\omega(a)$ is a positive integer and so is $\mathrm{ord}_Q(a)$. Thus, h is a positive, rational number.

Definition. If ρ is an A-Drinfeld module with A-characteristic zero, define the height of ρ to be zero. If ρ has A-characteristic $Q \neq (0)$, define the height to be the unique positive, rational number h such that $\omega(a) = h \, \mathrm{ord}_Q(a) \deg Q$ for all $a \in A$.

We are now in a position to prove Theorem 13.1 and its corollary. In the course of the proof we will show that both the rank and the height of ρ are integers.

Proof of Theorem 13.1. We recall that M is an algebraically closed field containing L and that $\rho \in \mathrm{Drin}_A(L)$. Let $P \subset A$ be a prime ideal different from the A-characteristic Q of L. Choose $b \in P$ with $b \neq 0$. The A-module $M_\rho[P]$ is finite since $M_\rho[P] \subset M_\rho[b]$ and the latter set is finite being the set of zeros of $\rho_b(u)$. If follows that $M_\rho[P]$ is a finite dimensional vector space over A/P of dimension d, say. Thus,

$$\#M_\rho[P] = q^{d \deg P} .$$

The class group of A is finite. We borrow this result from Chapter 14, (see Corollary 2 to Proposition 14.1 and take the set S in that proposition to be $S = \{\infty\}$). It follows that there is a positive integer m such that $P^m = aA$, a principal ideal. Thus, $\#M_\rho[P^m] = \#M[a]$. We compute the size of both sides. By the corollary to Lemma 13.6, we have

$$\#M_\rho[P^m] = \#M[P]^m = q^{md \deg P} .$$

If $a \in Q$, the A-characteristic of ρ, then $P^m \subseteq Q$, and it would follow that $P = Q$, contrary to assumption. Thus, $a \notin Q$ and $\rho_a(u)$ is a separable polynomial (its derivative with respect to u is $\delta(a)$). Thus,

$$\#M_\rho[a] = q^{\deg_\tau \rho_a} = q^{r \deg a} ,$$

where the last equality follows from Proposition 13.8. Here, r is the rank of ρ.

Since $P^m = aA$, we have $\deg a = m \deg P$. Thus,

$$md \deg P = r \deg a = rm \deg P .$$

Cancelling $m \deg P$ from both sides, we conclude $r = d$. This shows that the rank is an integer and proves the first assertion of Theorem 13.1 in the case where $e = 1$. The general case follows from what we have already proven, Lemma 13.6, and the structure theorems of finitely generated, torsion modules over Dedekind domains. We leave the details to the reader. (See Exercise 1 at the end of the chapter.)

Suppose now that the A-characteristic Q is not the zero ideal. Again, using the finiteness of the ideal class group we see there is a positive integer n such that $Q^n = bA$, a principal ideal. It follows that $\#M_\rho[Q^n] = \#M_\rho[b]$.

We first count the number of elements in $M_\rho[b]$. Since the polynomial $\rho_b(u)$ is inseparable in this case, we cannot simply use its degree. However, it is clear that by factoring out $\tau^{\omega(b)}$ on the right, we can write $\rho_b = \rho'_b \tau^{\omega(b)}$, where $\rho'_b(u)$ is separable and $\deg_\tau \rho'_b = \deg_\tau \rho_b - \omega(b)$. Since τ is a one to one and onto map from $M \to M$, we conclude

$$\#M[b] = q^{\deg_\tau \rho_b - \omega(b)} = q^{r \deg b - h \, \mathrm{ord}_Q(b) \deg Q} .$$

The last equality follows from Propositions 13.8 and 13.9.

On the other hand,

$$\#M_\rho[Q^n] = \#M_\rho[Q]^n = q^{nd' \deg Q} ,$$

where d' is the dimension of $M_\rho[Q]$ considered as a vector space over A/Q.

Using the last two equations together with the facts, $\deg b = n \deg Q$ and $\mathrm{ord}_Q(b) = n$, we find

$$nd' \deg Q = rn \deg Q - hn \deg Q .$$

Cancelling $n \deg Q$ from both sides, we conclude $d' = r - h$. This proves h is an integer and simultaneously proves the second assertion of Theorem 13.1 in the case when $e = 1$. As before, the general case follows without much difficulty .

Proof of the Corollary to Theorem 13.1. Since I is prime to Q, Q does not occur in the prime decomposition of $I = P_1^{e_1} P_2^{e_2} \dots P_t^{e_t}$. By the Theorem, we have for each i with $1 \le i \le t$,

$$M[P_i^{e_i}] \cong (A/P^{e_i})^{(r)} .$$

Sum both sides from 1 to t. The result follows from the Corollary to Lemma 13.2 together with the Chinese Remainder Theorem.

Having defined Drinfeld modules and discussed their torsion points and the notions of rank and height, we now proceed to define maps between Drinfeld modules, i.e., we want to study the category of such objects.

Definition. Let $\rho, \rho' \in \mathrm{Drin}_A(L)$. A morphism from ρ to ρ' is an element f of $L < \tau >$ with the property that $f\rho_a = \rho'_a f$ for all $a \in A$. The set of all such morphisms is denoted by $\mathrm{Hom}_L(\rho, \rho')$.

Under the addition in $L < \tau >$ it is easy to see that $\mathrm{Hom}_L(\rho, \rho')$ is an abelian group. Also, the product in $L < \tau >$ gives a bi-additive map

$(f, g) \rightarrow gf$ from $\text{Hom}_L(\rho, \rho') \times \text{Hom}_L(\rho', \rho'')$ to $\text{Hom}_L(\rho, \rho'')$. In particular, these two operations make $\text{Hom}_L(\rho, \rho) = \text{End}_L(\rho)$ into a ring.

Let Ω be an L-algebra and $\rho, \rho' \in \text{Drin}_A(L)$. Let $f \in \text{Hom}_L(\rho, \rho')$. Then, $u \rightarrow f(u)$ is a homomorphism of $\Omega \rightarrow \Omega$ as abelian groups and, as is easily checked from the definition, an A-module homomorphism from $\Omega_\rho \rightarrow \Omega_{\rho'}$.

If, as is often the case, we take $\Omega = M$, an algebraically closed field extension of L, we see that $u \rightarrow f(u)$ is an onto map from $M_\rho \rightarrow M_{\rho'}$ with finite kernel (the zeros of $f(u)$). For this reason we may refer to non-zero elements of $\text{Hom}_L(\rho, \rho')$ as isogenies. Also, we say that ρ and ρ' are isogenous over L if $\text{Hom}_L(\rho, \rho') \neq (0)$.

Proposition 13.10. *If ρ and ρ' are isogenous Drinfeld modules, then they have the same rank and height.*

Proof. Let $0 \neq f \in \text{Hom}_L(\rho, \rho')$ and choose a non-constant element $a \in A$. Then, $f\rho_a = \rho'_a f$. Taking the degree with respect to τ of both sides shows that $\deg_\tau \rho_a = \deg_\tau \rho'_a$. Thus, $r \deg(a) = r' \deg(a)$, where r and r' are the ranks of ρ and ρ', respectively. This shows $r = r'$.

That the heights of ρ and ρ' are equal follows from similar reasoning.

The identity of $\text{Hom}_L(\rho, \rho)$ is clearly τ^0 for all Drinfeld modules. What is an isomorphism? By definition, $f \in \text{Hom}_L(\rho, \rho')$ is an isomorphism if and only if there is a $g \in \text{Hom}_L(\rho', \rho)$ such that $fg = \tau^0 = gf$. In the twisted polynomial ring $L < \tau >$ this can only happen if $f = c\tau^0$ and $g = c^{-1}\tau^0$ for some non-zero element $c \in L$. Thus, ρ and ρ' are isomorphic if and only if there is a $c \in L^*$ such that $c\rho_a = \rho'_a c$ for all $a \in A$.

Suppose $I \subset A$ is a non-zero ideal and that $\rho \in \text{Drin}_A(L)$. We are going to construct a new Drinfeld module ρ' and a non-zero isogeny $\rho_I \in \text{Hom}_L(\rho, \rho')$. This construction plays an important role, especially in the arithmetic applications of the theory of rank 1 Drinfeld modules.

Lemma 13.11. *The ring $L < \tau >$ has a division algorithm on the right. More precisely, if $f, g \in L < \tau >$ and $g \neq 0$, there exist $s, r \in L < \tau >$ such that $f = sg + r$ with $r = 0$ or $\deg_\tau r < \deg_\tau g$.*

Proof. The proof is just about the same as in the case of a commutative polynomial ring. For details see Goss [4], Chapter 1.

Corollary. *Every left ideal in $L < \tau >$ is principal.*

Proof. If $J \subset L < \tau >$ is a non-zero left ideal, let $g \in J$ be an element of J with smallest degree in τ. If $f \in J$, then by the theorem we have $f = sg + r$ with either $r = 0$ or $\deg_\tau r < \deg_\tau g$. The latter alternative is impossible by the definition of g and the fact that $r \in J$. Thus, $r = 0$, so every element of J is a left multiple of g.

Definition. Let $I \subset A$ be an ideal and J the left ideal in $L < \tau >$ generated by the set $\{\rho_b \mid b \in I\}$. Define ρ_I to be the unique monic generator of the left ideal J.

Proposition 13.12. *Let M be an algebraically closed field containing L. Then,*
$$M_\rho[I] = \{\lambda \in M \mid \rho_I(\lambda) = 0\} .$$

Proof. Suppose λ is a root of $\rho_I(x)$. If $b \in I$, there is an $f_b \in L < \tau >$ such that $f_b \rho_I = \rho_b$. Thus, $0 = f_b(0) = f_b(\rho_I(\lambda)) = \rho_b(\lambda)$. It follows that the roots of ρ_I are contained in $M_\rho[I]$.

On the other hand, there exist $b_j \in I$ and $f_j \in L < \tau >$, with $1 \leq j \leq t$, such that
$$\rho_I = \sum_{j=1}^{t} f_j \rho_{b_j} .$$

From this it follows easily that every element of $M_\rho[I]$ is a root of $\rho_I(x)$.

Proposition 13.13. *Let $\rho \in \mathrm{Drin}_L(A)$ and $(0) \neq I \subset A$ be an ideal. Then there is a uniquely determined Drinfeld module $I * \rho \in \mathrm{Drin}_L(A)$ such that ρ_I is an isogeny from ρ to $I * \rho$.*

Proof. The left ideal $J \subset L < \tau >$ generated by the set, $\{\rho_b \mid b \in I\}$, is mapped into itself under right multiplication by ρ_a for any $a \in A$. Thus, for all $a \in A$, there is a $\rho'_a \in L < \tau >$ such that $\rho_I \rho_a = \rho'_a \rho_I$. A straightforward calculation shows that the map $a \to \rho'_a$ is an $\mathbb{F}$-algebra homomorphism from $A \to L < \tau >$.

Define $\delta'(a) = D(\rho'_a)$. This is easily seen to be an $\mathbb{F}$-algebra homomorphism from $A \to L$. If we knew that $\delta' = \delta$ it would follow that $\rho' \in \mathrm{Drin}_L(A)$ and by setting $I * \rho = \rho'$, the proof would be concluded. As a matter of fact, δ' is equal to δ, but to show this requires a little more work. We will postpone the proof for a while (see the corollary to Proposition 13.18). For the moment we simply note that $I * \rho$ is a Drinfeld module with, perhaps, a different structural map from that of ρ.

It is instructive, and useful, to understand ρ_I when I is a principal ideal. Let $I = (b)$. Then, by definition, ρ_I is the unique monic generater of the left ideal in $L < \tau >$ generated by ρ_b. Let $c \in L$ be the leading coefficient of ρ_b. Then, clearly, $\rho_{(b)} = c^{-1} \rho_b$.

Proposition 13.14. *If $\rho \in \mathrm{Drin}_L(A)$ and $0 \neq b \in A$, then $\rho_{(b)} = c^{-1}\rho_b$, where c is the leading coefficient of ρ_b. Moreover, $c[(b) * \rho]_a = \rho_a c$ for all $a \in A$; i.e., $(b) * \rho$ is isomorphic to ρ over L.*

Proof. We have already proven the first assertion. To prove the second, note that for all $a \in A$ we have $\rho_{(b)}\rho_a = [(b) * \rho]_a \rho_{(b)}$. Thus, $c^{-1}\rho_b\rho_a =$

$[(b) * \rho]_a c^{-1} \rho_b$. Since $\rho_b \rho_a = \rho_a \rho_b$ we can cancel ρ_b (because $L < \tau >$ has no zero divisors) and conclude that $c^{-1} \rho_a = [(b) * \rho]_a c^{-1}$.

The *-operation has a number of important properties. The following proposition provides two of the most useful of them. The proof is fairly straightforward, but somewhat tedious. We will leave the proofs as an exercise. A good reference is Goss [4], Lemma 4.9.2.

Proposition 13.15. *Let $\rho \in \mathrm{Drin}_L(A)$ and $I, J \subset A$ be non-zero ideals. Then*

$$\rho_{IJ} = (I * \rho)_J \rho_I \quad \text{and} \quad I * (J * \rho) = IJ * \rho .$$

The first of these relations will be especially useful in our next task, which is to generalize the relations $\deg_\tau \rho_a = r \deg a$ and $w(a) = h \deg Q \, \mathrm{ord}_Q(a)$ to the isogenies ρ_I. To do this we will need a new definition.

Let $\omega_\tau : L < \tau > \rightarrow \mathbb{Z}$ be the map which assigns to a non-zero element f of $L < \tau >$ the smallest index of the non-vanishing coefficients of f. In other words, if $f = \sum_{i=0}^d c_i \tau^i$, the $\omega_\tau(f) = i_0$ if $c_i = 0$ for $i < i_0$ and $c_{i_0} \neq 0$.

It is clear that $\omega_\tau(fg) = \omega_\tau(f) + \omega_\tau(g)$ and that $\omega_\tau(f) = 0$ if and only if $f(x)$ is a separable polynomial in x. If ρ is a Drinfeld module then, in our previous notation, $\omega(a) = \omega_\tau(\rho_a)$. We introduce this new mapping because, among other things, when dealing with more than one Drinfeld module, the notation "$\omega(a)$" is ambiguous.

Lemma 13.16. *Let $\rho \in \mathrm{Drin}_L(A)$ and let $I \subset A$ be an ideal prime to the A-characteristic Q of L. Then $\rho_I(x)$ is a separable polynomial.*

Proof. Let $a \in I$ with a not in the A-characteristic of L. We have $(a) = IJ$ for some ideal J. By Proposition 13.15, $\rho_{(a)} = (I * \rho)_J \rho_I$. Since a is not in Q, we know $\omega_\tau(\rho_a) = 0$ (see Proposition 13.9). Thus, $\omega_\tau(\rho_{(a)}) = \omega_\tau(\rho_a) = 0$. The result now follows, because

$$0 = \omega_\tau(\rho_{(a)}) = \omega_\tau((I * \rho)_J) + \omega_\tau(\rho_I) ,$$

which implies that $\omega_\tau(\rho_I) = 0$.

Proposition 13.17. *Let $\rho \in \mathrm{Drin}_L(A)$ be a Drinfeld module of rank r. Let $I \subset A$ be a non-zero ideal. Then, $\deg_\tau \rho_I = r \deg I$.*

Proof. To begin with, let us assume that I is prime to the A-characteristic of L. Let M be an algebraically closed field containing L and consider $M_\rho[I]$. By Lemma 13.16 and Proposition 13.12, we see that

$$\#M_\rho[I] = q^{\deg_\tau \rho_I} .$$

On the other hand, by the Corollary to Theorem 13.1, we see that

$$\#M_\rho[I] = q^{r \deg I} .$$

Combining these two facts gives the result in the present case.

Assume now that the A-characteristic, Q, is not zero, and that J is a non-zero ideal divisible by Q. By a standard result in the theory of Dedekind domains, there exist elements $a, b \in A$ and an ideal I, prime to Q, such that $aJ = bI$. Applying the first part of Proposition 13.15, we find

$$(J * \rho)_{(a)} \rho_J = (I * \rho)_{(b)} \rho_I \ . \tag{1}$$

Since isogenous Drinfeld modules have the same rank and using the fact that $\rho_{(a)}$ differs from ρ_a by multiplication by a non-zero element of L, we deduce from Equation 1 that

$$r \deg a + \deg_\tau \rho_J = r \deg b + r \deg I \ .$$

We have used Proposition 13.8 and the first part of the proof applied to I.

From the relation $aJ = bI$ we find that $\deg a + \deg J = \deg b + \deg I$. Putting all this together, we see that $\deg_\tau \rho_J = r \deg J$, as asserted.

Proposition 13.18. *Let $\rho \in \mathrm{Drin}_L(A)$. Assume that the A-characteristic of L, Q, is not zero. Let $J \subset A$ be a non-zero ideal. Then $\omega_\tau(\rho_J) = h \deg Q \, \mathrm{ord}_Q J$, where h is the height of ρ.*

Proof. If J is prime to Q, the $\omega_\tau(\rho_J) = 0$ by Proposition 13.16. We also have $\mathrm{ord}_Q J = 0$, so the proposition is proven in this case.

Now, assume Q divides J. Then as above we can write $aJ = bI$, where I is an ideal prime to Q. Applying ω_τ to both sides of Equation 1, we find

$$h \deg Q \, \mathrm{ord}_Q(a) + \omega_\tau(\rho_J) = h \deg Q \, \mathrm{ord}_Q(b) + 0 \ .$$

We have used the fact that isogenous Drinfeld modules have the same height, Proposition 13.9, and the first part of the proof applied to the ideal I.

Since $aJ = bI$, we have $\mathrm{ord}_Q(a) + \mathrm{ord}_Q J = \mathrm{ord}_Q(b) + 0$. Thus,

$$\omega_\tau(\rho_J) = h \deg Q \, \mathrm{ord}_Q(b/a) = h \deg Q \, \mathrm{ord}_Q J \ ,$$

and the proof is complete.

Corollary. *Let ρ be a Drinfeld module over a field L with structural map $\delta : A \to L$. Let $I \subset A$ be an ideal, and $\rho' = I * \rho$ with structural map $\delta' : A \to L$. Then $\delta = \delta'$.*

Proof. If ρ has A-characteristic zero, then by Lemma 13.16, $\rho_I(x)$ is a separable polynomial for all ideals I. This implies that the constant term of $\rho_I \in L < \tau >$, $c(I)$, is not zero. Setting $I * \rho = \rho'$, consider the equation $\rho_I \rho_a = \rho'_a \rho_I$. Comparing the constant terms on both sides yields $c(I)\delta(a) = \delta'(a)c(I)$. Since $c(I) \neq 0$, we conclude $\delta(a) = \delta'(a)$ for all $a \in A$.

Now assume that the A-characteristic of ρ, Q, is not zero. By the proposition, the first non-vanishing term of ρ_I is of the form $c\tau^m$ where $c \neq 0$ and

m is an integer divisible by $\deg Q$. Again consider the equation $\rho_I \rho_a = \rho_a' \rho_I$ and compare the coefficients of τ^m on both sides. We find

$$c\delta(a)^{q^m} = \delta'(a)c \quad \text{and so} \quad \delta(a)^{q^m} = \delta'(a) .$$

Now, δ induces an injection of A/Q into L and so $\delta(a)$ is an element of a finite field with $q^{\deg Q}$ elements. Since $\deg Q \mid m$ it follows that $\delta(a)^{q^m} = \delta(a)$. Thus, $\delta(a) = \delta'(a)$ for all $a \in A$ and we are done.

Having introduced the general notion of Drinfeld module, division points, rank and height, morphisms (isogenies), and some of their properties we now break off the general development to ask the possibly embarrassing question about whether Drinfeld modules exist. When A is a polynomial ring, there is no problem. As we observed in the last chapter, when $A = \mathbb{F}[T]$, we simply assign to T any element of $L < \tau >$ with constant term $\delta(T)$ and this automatically extends to a homomorphism $\rho : A \to L < \tau >$ with the property that $D(\rho_a) = \delta(a)$, i.e., a Drinfeld A-module over L. When A is more general, it is not clear that there are any elements of $\mathrm{Drin}_A(L)$. Indeed, why should the non-commutative ring $L < \tau >$ have a commutative subring isomorphic to A? To construct Drinfeld modules in the more general situation we follow Drinfeld and introduce analytic methods. The construction will be similar to the construction of elliptic curves over the complex numbers $\mathbb{C}$ by means of two dimensional $\mathbb{Z}$-lattices and the associated Weierstrass $\mathcal{P}$-functions.

Recall that k_∞ is the completion of k at the prime ∞. Let $\bar{k}_\infty$ be the algebraic closure of k_∞. The (normalized) valuation on k_∞

$$|b|_\infty = q^{-\mathrm{ord}_\infty(b)d_\infty} ,$$

extends to $\bar{k}_\infty$ uniquely by means of the formula

$$|\gamma|_\infty = |N_{E/k_\infty}(\gamma)|_\infty^{1/[E:k_\infty]} ,$$

where E is any intermediate field containing γ and of finite degree over k_∞.

We now define $\mathbf{C}$ to be the completion of $\bar{k}_\infty$ with respect to $\mid * \mid_\infty$. This valuation extends uniquely to $\mathbf{C}$ and $\mathbf{C}$ is complete. It is also well known that $\mathbf{C}$ is algebraically closed. The field $\mathbf{C}$ plays the role of the complex numbers in our context. The theory of infinite series and infinite products can be developed for functions defined on open subsets of $\mathbf{C}$ and the usual theorems continue to hold in even stronger form. In particular $\sum a_n$ converges if the terms a_n tend to zero and $\prod(1 + a_n)$ converges if $\sum a_n$ converges. A function from $\mathbf{C} \to \mathbf{C}$ is said to be entire if it can be represented by a power series $\sum a_n z^n$ which converges everywhere. The set of zeros of an entire function form a discrete subset of $\mathbf{C}$. Moreover, an entire function is determined by its zeros in a much stricter way than in the theory over the complex variables. For example, the exponential function in the complex theory is a highly non-trivial function, but it has no zeros at all. Over $\mathbf{C}$ we have the following, very different, type of result.

Proposition 13.19. *Let $f(x)$ be a non-constant entire function on* **C**. *Then $f(x)$ has at least one zero.*

Corollary. *Let $f(x)$ be a non-constant entire function on* **C**. *Then $f(x)$ is onto as a map from* **C** $\to$ **C**.

Proof. Let $c \in$ **C** and consider then entire function $-c + f(x)$. By the proposition this function has a zero, say, γ. Thus, $f(\gamma) = c$.

The proof of the proposition uses the Newton polygon in the context of power series. For a treatment of this and other results which we employ about analysis on **C**, see Goss [4], Chapter 2.

If γ is a zero of an entire function, there is a uniquely determined positive integer m such that $f(x) = (x - \gamma)^m g(x)$, where $g(x)$ is entire and $g(\gamma) \neq 0$. The integer m is called the multiplicity of the zero γ and is denoted by $\mathrm{ord}_{x=\gamma} f(x)$.

Theorem 13.19. *Let $f(x)$ be an entire function on* **C** *and let $\{\gamma_i \mid i = 1, 2, 3, \cdots\}$ be its zero set with 0 excluded if $f(0) = 0$. Let m_i be the multiplicity of γ_i. Then, $\lim_{i \to \infty} \gamma_i = \infty$ and there is a constant $c \neq 0$ such that*

$$f(x) = cx^n \prod_{i=1}^{\infty} \left(1 - \frac{x}{\gamma_i} \right)^{m_i} .$$

The integer n is equal to $\mathrm{ord}_{x=0} f(x)$. Conversely, if $\lim_{i \to \infty} \gamma_i = \infty$, then the above infinite product defines an entire function on **C**.

Definition. A lattice is a discrete, finitely generated, A-submodule of **C**. If $\Gamma \subset$ **C** is a lattice, the dimension of the vector space $k_\infty \Gamma$ over k_∞ is called the rank of the lattice.

One can show that lattices are formed in the following manner. Let $\{\omega_1, \omega_2, \cdots, \omega_r\} \subset$ **C** be a set of elements linearly independent over k_∞. Let $\{I_1, I_2, \cdots, I_r\}$ be a set of fractional ideals of A. Then,

$$\Gamma = I_1 \omega_1 + I_2 \omega_2 + \cdots + I_r \omega_r ,$$

is a lattice in **C** of rank r. In fact every lattice of rank r has this form. This shows that lattices exist in abundance and in every rank.

Let Γ_1 and Γ_2 be two lattices, and let $c \in$ **C** be such that $c\Gamma_1 \subseteq \Gamma_2$. Then $\phi : \Gamma_1 \to \Gamma_2$ given by $\phi(x) = cx$, is an A-module mapping. We define

$$\mathrm{Hom}(\Gamma_1, \Gamma_2) = \{c \in \mathbf{C} \mid c\Gamma_1 \subseteq \Gamma_2\} .$$

One can show that since a lattice Γ is discrete, we must have $|\gamma|_\infty \to \infty$ as γ varies over the elements of Γ. Thus, if we define

$$e_\Gamma(x) = x \prod_{\gamma \in \Gamma}' \left(1 - \frac{x}{\gamma} \right) ,$$

the result is an entire function on $\mathbf{C}$ (the product is over all non-zero elements of Γ). We call $e_\Gamma(x)$ the exponential function associated to Γ. It is characterized as being the unique entire function with simple zeros on the elements of Γ and with leading term x. It also has the remarkable property of being additive, as we now show.

Proposition 13.20. *Let Γ be a lattice in $\mathbf{C}$. Then for all $u, v \in \mathbf{C}$ and $\alpha \in \mathbb{F}$ we have*

$$e_\Gamma(u+v) = e_\Gamma(v) + e_\Gamma(v) \quad \text{and} \quad e_\Gamma(\alpha u) = \alpha e_\Gamma(u) .$$

Proof. For each positive integer M, define $\Gamma_M = \{\gamma \in \Gamma \mid |\gamma|_\infty \leq M\}$. This is readily checked to be a finite $\mathbb{F}$ vector space (as we've seen, $|\gamma|_\infty \to \infty$ for $\gamma \in \Gamma$). If we set

$$P_M(x) = x \prod_{\gamma \in \Gamma_M}' \left(1 - \frac{x}{\gamma}\right) ,$$

then $e_\Gamma(x) = \lim_{M \to \infty} P_M(x)$. The result then follows from the following lemma.

Lemma 13.21. *Let $V \subset \mathbf{C}$ be a finite, $\mathbb{F}$ vector space, and set*

$$f_V(x) = \prod_{\nu \in V} (x - \nu) .$$

Then, $P_V(x)$ is an $\mathbb{F}$-linear, additive polynomial in x.

Proof. We prove this by induction on the dimension of V. If $\dim V = 0$, then $V = (0)$ and $f_V(x) = x$, so the result is true in this case.

Now assume that the result is true for vector spaces of dimension less than n and that $\dim V = n$. Write $V = W + \mathbb{F}\mu$, where W is a $n - 1$ dimensional subspace of V. From the definition it is easy to see

$$f_V(x) = f_W(x) \prod_{0 \neq \alpha \in \mathbb{F}} f_W(x - \alpha\mu) .$$

By induction, $f_W(x - \alpha\mu) = f_W(x) - \alpha f_W(\mu)$. It follows that

$$f_V(x) = f_W(x)^q - f_W(\mu)^{q-1} f_W(x) .$$

It follows immediately that $f_V(x)$ has the required properties.

We see from these considerations that $e_\Gamma(x)$ can be written as an infinite series as follows:

$$e_\Gamma(x) = x + \sum_{i=1}^{\infty} c_i(\Gamma) x^{q^i} \quad \text{with} \quad c_i(\Gamma) \in \mathbf{C} .$$

The main reason for introducing these exponential functions is that, as we shall now show, to every lattice we can construct an element of $\mathrm{Drin}_A(\mathbf{C})$. Moreover, this assignment is an equivalence of categories if we use our definitions of $\mathrm{Hom}(\Gamma_1, \Gamma_2)$ for lattices and $\mathrm{Hom}(\rho_1, \rho_2)$ for Drinfeld modules. In particular, this will show that $\mathrm{Drin}_A(\mathbf{C})$ has lots of elements.

Let $\Gamma \subseteq \Gamma'$ be two lattices of the same rank. Then, Γ'/Γ is a finite A-module which maps isomorphically into a finite $\mathbb{F}$ vector subspace of $\mathbf{C}$ by means of the exponential function $e_\Gamma(x)$ (note that the exponential function is $\mathbb{F}$-linear by Proposition 13.20). Define

$$ P(x; \Gamma'/\Gamma) = x \prod_{\mu \in \Gamma'/\Gamma}{}' \left(1 - \frac{x}{e_\Gamma(\mu)} \right) . $$

Proposition 13.22. *The polynomial $P(x; \Gamma'/\Gamma)$ is $\mathbb{F}$-linear of degree $\#(\Gamma'/\Gamma)$. Its initial term is x. Moreover,*

$$ e_{\Gamma'}(u) = P(e_\Gamma(u); \Gamma'/\Gamma) . $$

Proof. The first assertion follows from Lemma 13.21, and the second assertion is clear from the definition.

To prove the identity, notice that $P(e_\Gamma(u); \Gamma'/\Gamma)$ is zero if and only if $e_\Gamma(u) = e_\Gamma(\mu)$ for some $\mu \in \Gamma'$, i.e., if and only if $e_\Gamma(u - \mu) = 0$. This is true if and only if $u - \mu \in \Gamma$, which in turn is true if and only if $u \in \Gamma'$. Thus, the right-hand side of the proposed identity is an entire function with simple zeros (the simplicity of the zeros is easily checked) at the elements of Γ' and the initial term is u. These conditions characterize $e_{\Gamma'}(u)$.

Theorem 13.23. *Let $\Gamma \subset \mathbf{C}$ be a lattice of rank r. For each $a \in A$, $a \neq 0$, define $\rho_a^\Gamma \in \mathbf{C} < \tau >$ by the formula*

$$ \rho_a^\Gamma(x) = aP(x, a^{-1}\Gamma/\Gamma) . $$

Then, if we send zero to zero and map $a \to \rho_a^\Gamma$ for $0 \neq a \in A$, the result is a Drinfeld A module over $\mathbf{C}$ of rank r.

Proof. In what follows we regard A is a subset of $\mathbf{C}$ via the inclusions $A \to k \to k_\infty \to \mathbf{C}$. Thus the structure map $\delta : A \to \mathbf{C}$ is just the inclusion, so the first thing we must show is that $D(\rho_a^\Gamma) = a$. This, however, is clear from the definition.

Next, we have to show that $\rho_{ab}^\Gamma = \rho_a^\Gamma \rho_b^\Gamma$ and $\rho_{a+b}^\Gamma = \rho_a^\Gamma + \rho_b^\Gamma$. We prove the first and leave the second as an exercise.

The idea is to work on the level of $\mathbb{F}$-linear polynomials and use the exponential functions. By Proposition 13.22,

$$ e_{a^{-1}\Gamma}(u) = P(e_\Gamma(u); a^{-1}\Gamma/\Gamma) . $$

By looking at the zero set and the initial term, it is easy to see $e_{a^{-1}\Gamma}(u) = a^{-1}e_\Gamma(au)$. Substituting this in the above equation, and using the definition of ρ_a^Γ, yields the following fundamental identity:

$$e_\Gamma(au) = \rho_a^\Gamma(e_\Gamma(u)) . \tag{2}$$

One now computes

$$\rho_{ab}^\Gamma(e_\Gamma(u)) = e_\Gamma(abu) = \rho_a^\Gamma(e_\Gamma(bu)) = \rho_a^\Gamma(\rho_b^\Gamma(e_\Gamma(u))) .$$

Since $e_\Gamma(u)$ maps $\mathbf{C}$ onto $\mathbf{C}$, we can conclude $\rho_{ab}^\Gamma = \rho_a^\Gamma \rho_b^\Gamma$. The proof of the additive identity is similar and even easier.

It remains to show that the rank of ρ^Γ is r. To do this we must show that $\deg_\tau \rho_a^\Gamma = r \deg a$. The degree of $\rho_a^\Gamma(x)$ as a polynomial in x is $\#(a^{-1}\Gamma/\Gamma)$. Recall that as an A module Γ is isomorphic to the direct sum of r fractional ideals. Since $a^{-1}I/I \cong a^{-1}A/A \cong A/aA$, for any non-zero fractional ideal I, it follows that $\#(a^{-1}\Gamma/\Gamma) = q^{r \deg a}$. Thus, $\deg_\tau \rho_a^\Gamma = r \deg a$. The proof is complete.

Theorem 13.24. *Let* $\mathrm{Lat}_A(\mathbf{C})$ *be the set of A-lattices inside* $\mathbf{C}$*. The map* $\Gamma \to \rho^\Gamma$ *from* $\mathrm{Lat}_A(\mathbf{C}) \to \mathrm{Drin}_A(\mathbf{C})$ *is one to one and onto.*

Proof. We will prove the map is one to one and only give a brief sketch of the proof that it is onto.

Suppose Γ and Γ' are two lattices such that $\rho^\Gamma = \rho^{\Gamma'}$. It is convenient to work inside the ring of twisted power series $\mathbf{C} << \tau >>$. This consists in formal power series $\sum_{i=0}^\infty c_i \tau^i$, with the usual addition and multiplication except for the non-commutativity relation $\tau c = c^q \tau$. Clearly, $\mathbf{C} < \tau >$ is a subring of $\mathbf{C} << \tau >>$. Every additive power series, such as $e_\Gamma(u)$, can be considered as an element of $\mathbf{C} << \tau >>$ applied to u where $\tau(u) = u^q$. The fundamental relation (Equation 2) given in the proof of Theorem 13.23 can be rewritten as

$$e_\Gamma a = \rho_a^\Gamma e_\Gamma .$$

Since $\rho_a^{\Gamma'} = \rho_a^\Gamma$, we also have

$$e_{\Gamma'} a = \rho_a^\Gamma e_{\Gamma'} .$$

Subtracting, we find that

$$(e_\Gamma - e_{\Gamma'})a = \rho_a^\Gamma(e_\Gamma - e_{\Gamma'}) .$$

We want to deduce from this that $e_\Gamma = e_{\Gamma'}$. Suppose that $e_\Gamma - e_{\Gamma'}$ is not zero. Since both exponential series have initial term τ^0, the first non-vanishing term of their difference has the form $c\tau^k$ where $0 \neq c \in \mathbf{C}$ and $k > 0$. Then, comparing the coefficients of τ^k on both sides of the above identity yields

$$ca^{q^k} = ac .$$

This shows $a^{q^k} = a$ for all $a \in A$. This is false if a is not a constant. We have arrived at a contradiction which implies that $e_\Gamma = e_{\Gamma'}$. Since Γ is the zero set of $e_\Gamma(x)$ and Γ' is the zero set of $e_{\Gamma'}(x)$, it follows that $\Gamma = \Gamma'$ as asserted.

To prove the onto-ness of our map, let $\rho \in \mathrm{Drin}_A(\mathbf{C})$, and choose an $a \in A$ with $a \notin \mathbb{F}$. Using the method of undetermined coefficients, one finds a power series $f \in \mathbf{C} << \tau >>$ with initial term τ^0, such that $fa = \rho_a f$. One then proves this relation must hold for all $a \in A$ and that $f(x)$ is an additive power series, which is, in fact, an entire function on $\mathbf{C}$. The zero set of this function turns out to be an A-lattice Γ. One then proves that $\rho^\Gamma = \rho$. For the details of this argument see Goss [4], Theorem 4.6.9.

It is of interest to pause at this point and ask what is the lattice corresponding to the Carlitz module, the first Drinfeld module to appear in the literature. Since the Carlitz module has rank 1, the corresponding lattice must be of rank 1 over $A = \mathbb{F}[T]$. Thus, it must be of the form $A\tilde{\pi}$ for some $\tilde{\pi} \in \mathbf{C}$. Carlitz found an explicit expression for $\tilde{\pi}$ as an infinite product. Set $[i] = T^{q^i} - T$ and $F_i = [i][i-1]^q \dots [1]^{q^{i-1}}$ (see the exercises to Chapter 1 where some of the properties of these polynomials are set forth). Define $\mathbf{i}$ to be any $q - 1$-st root of $T - T^q$. Then, we have

$$\tilde{\pi} = \mathbf{i} \prod_{i=0}^{\infty} \left(1 - \frac{[i]}{[i+1]} \right) .$$

Carlitz also computed the exponential function corresponding to the lattice $A\tilde{\pi}$. It is given by

$$e_{A\tilde{\pi}}(u) = u \prod_{a \in A}' \left(1 - \frac{u}{a\tilde{\pi}} \right) = \sum_{i=0}^{\infty} \frac{u^{q^i}}{F_i} .$$

Actually, Carlitz did not first define the Carlitz module and then work out this exponential function. He was first led to construct this exponential function and then proved the "complex multiplication" property $e_{A\tilde{\pi}}(Tu) = Te_{A\tilde{\pi}}(u) + e_{A\tilde{\pi}}(u)^q$. It was this remarkable property of the exponential function which led to the invention of what we now call the Carlitz module; see Carlitz [2, 3], and, also, Goss [4], Chapter 3 (in Carlitz's papers the notation is somewhat different and the module he works with is defined by $u \to Tu - u^q$ rather than $u \to Tu + u^q$).

Having made a short detour to discuss the special case of the Carlitz module, we now return to the general theory. We have set up a one-to-one correspondence between $\mathrm{Lat}_A(\mathbf{C})$ and $\mathrm{Drin}_A(\mathbf{C})$ (which is rank preserving). We now want to deepen this relationship by showing a correspondence between elements of $\mathrm{Hom}(\Gamma, \Gamma')$ and $\mathrm{Hom}(\rho^\Gamma, \rho^{\Gamma'})$.

Theorem 13.25. *Let* $\Gamma, \Gamma' \in \mathrm{Lat}_A(\mathbf{C})$ *be lattices of the same rank and suppose* $0 \neq c \in \mathrm{Hom}(\Gamma, \Gamma')$. *Define*

$$f_c(x) = cP(x; c^{-1}\Gamma'/\Gamma) .$$

Then, $f_c \in \mathrm{Hom}(\rho^\Gamma, \rho^{\Gamma'})$. Moreover, $c \to f_c$ is an isomorphism of $\mathrm{Hom}(\Gamma, \Gamma')$ with $\mathrm{Hom}(\rho^\Gamma, \rho^{\Gamma'})$ as abelian groups (and even as $\mathbb{F}$ vector spaces).

Proof. By Proposition 13.22, we know that $e_{c^{-1}\Gamma'}(u) = P(e_\Gamma(u); c^{-1}\Gamma'/\Gamma)$. As we have seen previously, $e_{c^{-1}\Gamma'}(u) = c^{-1}e_{\Gamma'}(cu)$. Thus, $e_{\Gamma'}(cu) = f_c(e_\Gamma(u))$. It is convenient to work inside $\mathbf{C} << \tau >>$ where this relation becomes $e_{\Gamma'}c = f_c e_\Gamma$. Multiply both sides on the right by any $a \in A$ and calculate. We find

$$e_{\Gamma'}ca = e_{\Gamma'}ac = \rho_a^{\Gamma'} e_{\Gamma'}c = \rho_a^{\Gamma'} f_c e_\Gamma ,$$

and

$$f_c e_\Gamma a = f_c \rho_a^\Gamma e_\Gamma .$$

Setting both expressions equal to one another and cancelling e_Γ on the right yields the identity $\rho_a^{\Gamma'} f_c = f_c \rho_a^\Gamma$. Since this is true for all $a \in A$ we have shown that $f_c \in \mathrm{Hom}(\rho^\Gamma, \rho^{\Gamma'})$.

It is easy to check that $c \to f_c$ is $\mathbb{F}$ linear. Since $D(f_c) = c$, it is also clear that this homomorphism is one to one. It remains to show that it is onto.

We will sketch the proof of the ontoness. Suppose $f \in \mathrm{Hom}(\rho^\Gamma, \rho^{\Gamma'})$. If $f = 0$ we may choose $c = 0$, so suppose $f \neq 0$. For each $a \in A$ we have $f\rho_a^\Gamma = \rho_a^{\Gamma'} f$. Multiply both sides of this identity on the right with e_Γ. We find $(f e_\Gamma)a = \rho_a^{\Gamma'}(f e_\Gamma)$. Let $c = D(f)$. One has to show that $c \neq 0$. This is not too hard using the fact that f intertwines ρ^Γ and $\rho^{\Gamma'}$ and that $\mathbf{C}$ has A-characteristic zero. Then, $e_{\Gamma'}c$ has the property that $e_{\Gamma'}ca = e_{\Gamma'}ac = \rho_a^{\Gamma'}e_{\Gamma'}c$. Thus,

$$(f e_\Gamma - e_{\Gamma'}c)a = \rho_a^{\Gamma'}(f e_\Gamma - e_{\Gamma'}c) .$$

By our choice of c, the coefficient of τ^0 in $f e_\Gamma - e_{\Gamma'}c$ is zero. By the same argument used at the end of the proof of Theorem 13.24, we can conclude that $f e_\Gamma = e_{\Gamma'}c$. As power series, this says that $f(e_\Gamma(u)) = e_{\Gamma'}(cu)$. If $\gamma \in \Gamma$ we se that γ is a root of the left hand side which implies $0 = e_{\Gamma'}(c\gamma)$ and so $c\gamma \in \Gamma'$. We conclude that $c\Gamma \subseteq \Gamma'$, i.e., $c \in \mathrm{Hom}(\Gamma, \Gamma')$. The proof is concluded by showing that $f = f_c$. The argument uses the fact that $D(f) = D(f_c)$ and that for all $a \in A$, $(f - f_c)\rho_a^\Gamma = \rho_a^{\Gamma'}(f - f_c)$.

The last two theorems make it possible to answer questions about the category of Drinfeld modules over $\mathbf{C}$ by considering the same question in the category of lattices which is much easier to analyze. As an example, we prove the following theorem about $\mathrm{Drin}_A^o(\mathbf{C}, 1)$, the set of rank 1 Drinfeld A-modules over $\mathbf{C}$ up to isomorphism. To be more precise, $\mathrm{Drin}_A^o(\mathbf{C}, 1)$ is the quotient of the set $\mathrm{Drin}(\mathbf{C}, 1)$ of rank 1 Drinfeld A-modules over $\mathbf{C}$ modulo the equivalence $\rho \sim \rho'$ if and only if ρ and ρ' are isomorphic over $\mathbf{C}$.

Theorem 13.26. *The set $\mathrm{Drin}_A^o(\mathbf{C}, 1)$ is finite with cardinality equal to the order of the class group of A.*

Proof. By Theorems 13.24 and 13.25, it is equivalent to consider the set of rank 1 A-lattices in $\mathbf{C}$ up to isomorphism. Note that two lattice Γ and Γ' are isomorphic if and only if there is a $c \in \mathbf{C}^*$ such that $\Gamma = c\Gamma'$.

Every rank 1 lattice has the form $I\omega$, where I is a fractional ideal of A and $\omega \in \mathbf{C}^*$. For every fractional ideal I of A, let $\bar{I}$ be the set of lattices equivalent to I. Clearly if two ideals are in the same ideal class, they go to the same class of lattices. By the first remark, this map from ideal classes to lattice classes is onto. Suppose I_1 and I_2 are two ideals such that $\bar{I}_1 = \bar{I}_2$. By definition, there is an $\omega \in \mathbf{C}^*$ such that $I_1 = I_2\omega$. From this equation we can deduce that $\omega \in k$ and this shows that I_1 and I_2 are in the same ideal class. Altogether then, we have produced a one-to-one, onto map from the class group of A to the isomorphism classes of rank one A-lattices. This proves the theorem.

We conclude this chapter with a few remarks on how to make Theorem 13.26 into a more structural theorem. The operation star operation $(I, \rho) \to I * \rho$ gives an operation of the group of fractional ideals of A on $\mathrm{Drin}_A(\mathbf{C})$. Since for a principal ideal (a) we have $(a) * \rho$ is isomorphic to ρ, this descends to an action of $Cl(A)$, the class group of A, on $\mathrm{Drin}_A^o(\mathbf{C})$. If we restrict this action to rank 1 Drinfeld modules, we claim this action is one to one and transitive, i.e., $\mathrm{Drin}_A^o(\mathbf{C}, 1)$ is a principal homogeneous space for $Cl(A)$. This is the more structural form of Theorem 13.26.

To prove this result we pass to the equivalent category of lattices, $\mathrm{Lat}_A(\mathbf{C})$. We have an obvious action $(I, \Gamma) \to I\Gamma$ of fractional ideals on lattices. By the way isomorphism between lattices is defined, it is clear that this action descends to an action of $Cl(A)$ on $\mathrm{Lat}_A^o(\mathbf{C})$, the isomorphism classes of A-lattices in $\mathbf{C}$. If we restrict attention to rank 1 lattices we easily see that $\mathrm{Lat}_A^o(\mathbf{C}, 1)$ is a principal homogeneous space for $Cl(A)$.

There is one subtlety, however, which must be addressed before applying this calculation with lattices to Drinfeld modules. Namely, if Γ is the lattice associated with ρ, what is the lattice associated with $I * \rho$? The answer is not $I\Gamma$, which is a good first guess. Let $c(I) = D(\rho_I)$. Then the lattice associated with $I * \rho$ is $c(I)I^{-1}\Gamma$. The proof of this is not too hard to give using the techniques introduced in this chapter. Although the right answer is a little more complicated than expected, it nevertheless leads to the final result.

Theorem 13.27. *The set* $\mathrm{Drin}_A^o(\mathbf{C}, 1)$ *is a principal homogeneous space for* $Cl(A)$ *under the action induced by* $(I, \rho) \to I * \rho$, *where I is a fractional ideal of A and $\rho \in \mathrm{Drin}_A(\mathbf{C})$.*

We could go on to consider isomorphism classes of rank 2 Drinfeld modules, a question which leads to the theory of Drinfeld modular curves. Instead, we will stop here and go to other topics. A good introduction to Drinfeld modular curves and their properties is found in Gekeler [2].

Exercises

1. Fill in the details of the proof of Theorem 13.1 in the case $M[P^e]$ where $e \geq 2$ and $P \neq Q$. You may use the following fact. If M is a finitely generated module over a Dedekind domain A which is annihilated by a power of a maximal ideal, say, P^e, then M is a direct sum of cyclic modules of the form A/P^f where $f \leq e$.

2. Similarly, fill in the details of the proof of Theorem 13.1 in the case $M_\rho[Q^e]$ where $e \geq 2$.

3. Let $\rho, \rho' \in \mathrm{Drin}_A(L)$. If ρ and ρ' are isogenous show that they have the same height.

4. Prove Lemma 13.11.

5. Prove Proposition 13.15.

6. In Proposition 13.22 we showed $P(e_\Gamma(u), \Gamma'/\Gamma)$ has a zero at each element of Γ'. Show that each such zero is simple.

7. Prove that $\rho^\Gamma_{a+b} = \rho^\Gamma_a + \rho^\Gamma_b$ (part of Theorem 13.23).

8. Show that in a neighborhood of zero in $\mathbf{C}$ we have

$$\frac{u}{e_\Gamma(u)} = 1 - \sum_{\substack{n>0 \\ (q-1)|n}} G_n(\Gamma) u^n \; ,$$

   where $G_n(\Gamma) = \sum'_{\gamma \in \Gamma} \gamma^{-n}$, where the prime indicates that 0 is to be omitted. Show that the sums $G_n(\Gamma)$ converge and discuss the convergence of the expression given above for $u/e_\Gamma(u)$.

9. Show that the lattice associated to the Drinfeld module $I * \rho^\Gamma$ is $D(I)I^{-1}\Gamma$ where $D(I)$ is the constant term of $\rho^\Gamma_I(\tau)$.

10. In the text we discussed the lattice corresponding to the Carlitz module and the corresponding exponential function. The lattice is $A\tilde{\pi}$ and we gave an Carlitz's explicit formula for $\tilde{\pi}$. For convenience, set $e_{A\tilde{\pi}}(u) = e_C(u)$. If $m \in A$ is a monic polynomial show that the set of m-division points in $\mathbf{C}$ for the Carlitz module is given by

$$\tilde{\Lambda} = \{ e_C(a\tilde{\pi}/m) \mid a \in A, \ \deg a < \deg m \} \cup \{0\} \; .$$

11. (Continuation) Set $\tilde{\lambda}_m = e_C(\tilde{\pi}/m)$. Let $a \in A$ be a polynomial of degree less than $\deg m$. Show that

$$\mathrm{ord}_\infty C_a(\tilde{\lambda}_m) = \deg m - \deg a - 1 - \frac{1}{q-1} \; .$$

   This important formula was proved in Chapter 12 using the Newton polygon. The analytic proof, sketched here, is due to D. Goss.

14

S-Units, S-Class Group, and the Corresponding L-Functions

Let K/F be an algebraic function field over the field of constants F. Throughout this book we have been emphasizing the analogy between the arithmetic of K and that of an algebraic number field. This analogy is particulary clear when we choose an element $x \in K$ which is not a constant. The ring $A = F[x] \subset k = F(x)$ then plays the role of the pair $\mathbb{Z} \subset \mathbb{Q}$ in number theory. K is an algebraic extension of $F(x)$ and the analogue of the ring of integers in an algebraic number field is the integral closure of A in K. Let's call this ring B. We will show that B is a Dedekind domain. We will investigate the unit group and the class group of B. We will also associate zeta and L-functions to B.

The ring B and its properties can be discussed in a slightly different, somewhat more intrinsic, way. Let ∞ denote the prime at infinity in the subfield $k = F(x)$ and denote by S the finitely many primes in K lying above ∞. We will show that B is the intersection of all the valuation rings O_P for $P \in \mathcal{S}_K - S$ (recall that $\mathcal{S}_K$ is the set of all primes of K). This being the case, let $S \subset \mathcal{S}_K$ be any finite set of primes. Define

$$O_S = \{a \in K \mid \mathrm{ord}_P(a) \geq 0, \ \forall P \notin S\},$$

the ring of S-integers. We will define S-units, S-divisors, S-class group, and even, an S-zeta function. After discussing these concepts and deriving their basic properties, we will show how all of this relates to the arithmetic properties of the field K.

Finally, we will discuss L-functions in a slightly more general situation. Namely, suppose that K/k is an abelian extension of global function fields and that S is a set of primes of k (not of K as in the above paragaph). We do not assume that k is a rational function field. Let A be the ring of S-integers of k and B the integral closure of A in K. We will introduce L-functions, $L_A(w, \chi)$ determined by the ring A and a character χ of the Galois group. Using properties of these functions, we will derive a very general class number formula relating the class number of B to the class number of A and certain finite character sums. In Chapter 16 we will look more closely into these formulas in the special cases of quadratic and cyclotomic function fields and find close analogues to a number of classical results in algebraic number theory.

Let us return to considering S as a finite set of primes of K. In addition to the definition of the ring of S-integers we will need a number of other definitions. The S-unit group is defined by

$$E(S) = \{a \in K^* \mid \mathrm{ord}_P(a) = 0, \ \forall P \notin S\}.$$

It is clear that $E(S) = O_S^*$, the units of the ring of S-integers. Moreover, $F^* \subseteq E(S)$. We will see that $E(S)/F^*$ is a finitely generated, free abelian group.

Since the field K will be fixed throughout the first part of our discussion, we denote by $\mathcal{D}$ its group of divisors, by $\mathcal{P}$ the subgroup of principal divisors, and by $Cl = \mathcal{D}/\mathcal{P}$ the group of divisor classes. The group of S-divisors, $\mathcal{D}_S$, is defined to be the subgroup of $\mathcal{D}$ generated by the primes in $\mathcal{S}_K - S$. Given an element $a \in K^*$, we define its S-divisor to be

$$(a)_S = \sum_{P \notin S} \mathrm{ord}_P(a)P.$$

A divisor which is of the form $(a)_S$ for some $a \in K^*$ is called a principal S-divisor. The principal S-divisors form a subgroup of $\mathcal{D}_S$, which is denoted by $\mathcal{P}_S$. The quotient group $Cl_S = \mathcal{D}_S/\mathcal{P}_S$ is called the S-class group. Later we will show that Cl_S is isomorphic to the ideal class group of the Dedekind domain O_S.

Finally, we define $\mathcal{D}(S)$ to be the subgroup of $\mathcal{D}$ generated by the primes in S and $\mathcal{P}(S) = \mathcal{P} \cap \mathcal{D}(S)$.

Consider the degree map deg : $\mathcal{D} \to \mathbb{Z}$. The image of this map is a principal ideal $i\mathbb{Z}$. The integer i is easily seen to be the greatest common divisor of all the elements of the set $\{\deg P \mid P \in \mathcal{S}_K\}$. When F is a finite field a theorem of F.K. Schmidt insures that $i = 1$. However, in the general case it is quite possible for i to be greater than 1. For example, consider the quotient field of the integral domain $\mathbb{R}[X, Y]/(X^2 + Y^2 + 1)$. This is a function field over the real numbers $\mathbb{R}$ as constant field. It is not hard to check that every prime has degree 2 and so we must have $i = 2$ for this example.

The image of $\mathcal{D}(S)$ under the degree map is also a principal ideal in $\mathbb{Z}$ which we denote by $d\mathbb{Z}$. The integer d is characterized as the greatest common divisor of the elements in $\{\deg P \mid P \in S\}$. Clearly, i divides d.

Proposition 14.1. *The following sequences are exact:*

(a)
$$(0) \to F^* \to E(S) \to \mathcal{P}(S) \to (0) \,,$$

(b)
$$(0) \to \mathcal{D}(S)^\circ/\mathcal{P}(S) \to Cl^\circ \to Cl_S \to C \to (0) \,,$$

where C is a cyclic group of order d/i.

Proof.   The map from $E(S)$ to $\mathcal{P}(S)$ is given by taking an S-unit to its divisor. This map is onto by the definition of $\mathcal{P}(S)$. If an S-unit e goes to the zero divisor, then $\mathrm{ord}_P(e) = 0$ for all $P \in S_K$ and so must be a constant. This proves the exactness of sequence (a).

To deal with the second exact sequence we first define a map $\tau : \mathcal{D} \to \mathcal{D}_S$ as follows:
$$\tau(D) = \sum_{P \notin S} \mathrm{ord}_P(D)P \,.$$

This map is an epimorphism with kernel $\mathcal{D}(S)$. The image of $\mathcal{P}$ under τ is $\mathcal{P}_S$. Thus, τ induces a homomorphism from $Cl \to Cl_S$ with kernel $(\mathcal{D}(S) + \mathcal{P})/\mathcal{P} \cong \mathcal{D}(S)/\mathcal{P}(S)$. From this we deduce the exactness of the sequence
$$(0) \to \mathcal{D}(S)^\circ/\mathcal{P}(S) \to Cl^\circ \to Cl_S \,,$$

and it remains to show that the cokernel of the last arrow is a cyclic group of order d/i.

To do this, we again use the fact that τ induces an isomorphism from $\mathcal{D}/(\mathcal{P}+\mathcal{D}(S))$ to Cl_S. The group we are interested in can also be described as the cokernel of the natural map from $\mathcal{D}^\circ/\mathcal{P}$ to $\mathcal{D}/(\mathcal{P} + \mathcal{D}(S))$. This cokernel is easily seen to be isomorphic to $\mathcal{D}/(\mathcal{D}^\circ + \mathcal{D}(S))$ (use the fact that $\mathcal{P} \subseteq \mathcal{D}^\circ$). The degree map provides an isomorphism of $\mathcal{D}/(\mathcal{D}^\circ+\mathcal{D}(S))$ with $i\mathbb{Z}/d\mathbb{Z} \cong \mathbb{Z}/(d/i)\mathbb{Z}$. This completes the proof.

This proof is due, in essence, to F.K. Schmidt. See his classic paper (Schmidt [1]).

Corollary 1. *The group $E(S)/F^*$ is a finitely generated free group of rank at most $|S| - 1$, where $|S|$ is the number of elements in S.*

Proof. By the exact sequence $a)$ we have $E(S)/F^* \cong \mathcal{P}(S)$, which is a subgroup of the free group $\mathcal{D}(S)^\circ$ on $|S| - 1$ generators. Thus, $\mathcal{P}(S)$ is free on at most $|S| - 1$ generators.

Corollary 2. *Cl_S is a finite group if Cl° is a finite group. Also, Cl_S is a torsion group if Cl° is a torsion group.*

Proof. Both statements are immediate consequences of exact sequence $b)$.

Proposition 14.2. *Let $K/\mathbb{F}$ be a function field over a finite field $\mathbb{F}$. Then, for all finite subsets $S \subset S_K$ we have that Cl_S is a finite group and $E(S)/\mathbb{F}^*$ is a free group on $|S| - 1$ generators.*

Proof. By Lemma 5.6, Cl^o is a finite group. By Corollary 2 to Proposition 14.1 we see that Cl_S is a finite group.

By exact sequence $b)$ of Proposition 14.1 we see that $\mathcal{D}(S)^o/\mathcal{P}(S)$ is finite. This shows that $\mathcal{P}(S)$ is free on $|S| - 1$ generators. We have already seen that $\mathcal{P}(S) \cong E(S)/F^*$.

It is fairly clear that the above results are analogues of finiteness of class number and the Dirichlet unit theorem in algebraic number theory.

For the rest of the chapter, we will assume that the constant field $F = \mathbb{F}$ is a finite field with q elements.

Our next task is to introduce the S-zeta function and investigate some of its properties. Recall the definition of $\zeta_K(w)$,

$$\zeta_K(w) = \prod_{P \in S_K} \left(1 - NP^{-w}\right)^{-1} .$$

If S is a finite set of primes, we define the S-zeta function to be

$$\zeta_S(w) = \prod_{P \notin S} \left(1 - NP^{-w}\right)^{-1} .$$

Two remarks are in order about the notation. Since we are not varying the field K in the discussion, we write $\zeta_S(w)$ rather than $\zeta_{K,S}(w)$. Secondly, we will use w as the variable instead of s, which we have used earlier. Among other reasons, this is because the notation $\zeta_S(s)$ is a bit confusing. Also, we want to reserve s to represent the number of elements in S, i.e., $s = |S|$.

It follows immediately from the definition that

$$\zeta_S(w) = \prod_{P \in S} \left(1 - NP^{-w}\right) \zeta_K(w) . \tag{1}$$

Since $\zeta_K(w)$ is a rational function of q^{-w}, the same is true for $\zeta_S(w)$. We will be interested in the power series expansion of $\zeta_S(w)$ about $w = 0$. By Theorem 5.9 we know that

$$\zeta_K(w) = \frac{L_K(q^{-w})}{(1 - q^{-w})(1 - q^{1-w})} ,$$

where $L_K(u) \in \mathbb{Z}[u]$ is a polynomial with the property that $L_K(1) = h_K$, the number of divisor classes of degree zero. It follows (as we have seen before) that

$$\lim_{w \to 0} w\zeta_K(w) = -\frac{h_K}{\ln(q)(q - 1)} .$$

Since $NP^{-w} = q^{-\deg P \, w} = e^{-\ln q \deg P \, w}$, we see that $1 - NP^{-w} = \ln q \deg P \, w + O(w^2)$. Using this information and substituting into Equation 1, we find

$$\zeta_S(w) = -(q - 1)^{-1} h_K \left(\prod_{P \in S} \deg P\right)(\ln q)^{s-1} w^{s-1} + O(w^s) . \tag{2}$$

From this we see that the order of vanishing of $\zeta_S(w)$ at $w = 0$ is $s - 1$, which is equal to the rank of the S-unit group by Proposition 14.2. We will now show how to rewrite the leading coefficient given in Equation 2 so that it becomes strikingly close to what it looks like in the number field case.

An ingredient in the calculation will be the S-regulator. To define this, we begin by choosing a set of S-units $\{e_1, e_2, \ldots, e_{s-1}\}$ whose projection to $E(S)/F^*$ is a basis. Consider the $(s-1) \times s$ matrix M whose ij-th entry is $\ln |e_i|_{\mathfrak{P}_j}$, where $S = \{\mathfrak{P}_1, \mathfrak{P}_2, \ldots, \mathfrak{P}_s\}$. We claim that the sum of the columns of this matrix is zero. To see this, note that for any $a \in K^*$ we have

$$-\sum_P \ln |a|_P = \sum_P \text{ord}_P(a) \, \deg P \, \ln q = \deg(a) \, \ln q = 0 \,.$$

For any S-unit, the only primes which occur in the sum are the primes in S. Our assertion follows.

It follows that the determinants of the $(s-1) \times (s-1)$ minors of M are all the same, up to sign. The absolute value of any of these determinants is then taken as the definition of the S-regulator. We denote the S-regulator by R_S. It is not hard to show that the S-regulator is independent of the choice of basis $\{e_1, e_2, \ldots, e_{s-1}\}$.

An associated regulator $R_S^{(q)}$ has the same definition as R_S except that throughout one uses $\log_q(x)$, the logarithm to the base q, instead of the natural logarithm, $\ln(x)$. The two regulators are related by the equation

$$(\ln q)^{s-1} R_S^{(q)} = R_S \,.$$

It is worthwhile to give a more direct definition of $R_S^{(q)}$, which has the advantage of showing that it is an ordinary integer. Simply notice that

$$\log_q(|e|_P) = \log_q(NP^{-\text{ord}_P(e)}) = -\deg P \, \text{ord}_P(e) \,.$$

Now, form the $(s-1) \times s$ matrix whose ij-th entry is $-\deg P_j \, \text{ord}_{P_j}(e_i)$. Then $R_S^{(q)}$ is the absolute value of the determinant of any $(s-1) \times (s-1)$ minor of this matrix.

Lemma 14.3.

$$[\mathcal{D}(S)^\circ : \mathcal{P}(S)] = \frac{dR_S^{(q)}}{\left(\prod_{P \in S} \deg P\right)} \,.$$

Proof. We begin by defining a map $l : \mathcal{D}(S) \to \mathbb{Z}^s$. If $D \in \mathcal{D}(S)$, we set $l(D) = (\ldots, -\text{ord}_P D \deg P, \ldots)$, where P varies over the set S. Note that l is a homomorphism and that if $a \in E(S)$, then $l((a)) = (\ldots, \log_q |a|_P, \ldots)$. Also, it is easy to see from the definition that $[\mathbb{Z}^s : l(\mathcal{D}(S))] = \prod_{P \in S} \deg P$.

Consider the elements of $\mathbb{Z}^s$ as row vectors and define $H^\circ \subset \mathbb{Z}^s$ as the subgroup consisting of row vectors the sum of whose coordinates is zero.

We have $l(\mathcal{P}(S)) \subseteq l(\mathcal{D}(S)^\circ) \subseteq H^\circ$. It is easy to check that l is one to one. It follows that

$$[\mathcal{D}(S)^\circ : \mathcal{P}(S)] = [l(\mathcal{D}(S)^\circ) : l(\mathcal{P}(S))] = \frac{[H^\circ : l(\mathcal{P}(S))]}{[H^\circ : l(\mathcal{D}(S)^\circ)]} . \tag{3}$$

We now calculate the numerator and denominator of this expression.

First we compute the index $[H^\circ : l(\mathcal{P}(S))]$. Let $\epsilon_s \in \mathbb{Z}^s$ be the vector with zeros everywhere except for a 1 at the s-th place. Then, $\mathbb{Z}^s$ is the direct sum of H° and $\mathbb{Z}\epsilon_s$. It follows that the index of $l(\mathcal{P}(S))$ in H° is the same as the index of $l(\mathcal{P}(S)) + \mathbb{Z}\epsilon_s$ in $\mathbb{Z}^s$. A free basis for this subgroup is $\{l((e_1)), \ldots, l((e_{s-1})), \epsilon_s\}$. Let $M^{(q)}$ be the $s - 1 \times s$ matrix whose i-th row is $l((e_i))$ and M' be the $s \times s$ matrix obtained from $M^{(q)}$ by adjoining ϵ_s as the bottom row. By a simple application of the elementary divisors theorem (see Lang [4], Theorem 7.8), the index we are looking for is the absolute value of the determinant of M'. Expanding this determinant in cofactors along the bottom row shows the index in question is $R_S^{(q)}$.

To compute $[H^\circ : l(\mathcal{D}(S)^\circ)]$, consider the exact sequence

$$(0) \to H^\circ / l(\mathcal{D}(S)^\circ) \to \mathbb{Z}^s / l(\mathcal{D}(S)) \to \mathbb{Z}/d\mathbb{Z} \to (0) .$$

The second arrow is induced by inclusion and the third arrow by the sum of coordinates map from $\mathbb{Z}^s \to \mathbb{Z}$. From this exact sequence, we deduce

$$[H^\circ : l(\mathcal{D}(S)^\circ)] = \frac{\prod_{P \in S} \deg P}{d} .$$

Substituting these results into Equation 3 completes the proof of the lemma.

Corollary 1. *Suppose all the primes in S have degree 1. Then, $[\mathcal{D}(S)^\circ : \mathcal{P}(S)] = R_S^{(q)}$.*

Corollary 2. *Both regulators R_S and $R_S^{(q)}$ are not zero.*

Theorem 14.4. *Let $K/\mathbb{F}$ be a function field over a finite field $\mathbb{F}$ with q elements. Let $S \subset S_K$ be a finite set of primes with s elements. Then*

$$\zeta_S(w) = -\frac{h_S R_S}{q - 1} w^{s-1} + O(w^s) .$$

Proof. Referring to Equation 2 we see that everything has already been proved except that in that equation the coefficient of w^{s-1} is given as

$$-(q - 1)^{-1} h_K (\prod_{P \in S} \deg P)(\ln q)^{s-1} . \tag{4}$$

Our task is to show that this number is the same as that given in the theorem.

By Proposition 14.1, part b), we see that $h_K d = h_S[\mathcal{D}(S)^\circ : \mathcal{P}(S)]$. From Lemma 14.3, we deduce $h_K \left(\prod_{P \in S} \deg P\right) = h_S R_S^{(q)}$. Since $R_S^{(q)} = (\ln q)^{-(s-1)} R_S$, we find that $h_K \left(\prod_{P \in S} \deg P\right) (\ln q)^{s-1} = h_S R_S$. Substituting this into Equation 4 we obtain

$$-\frac{h_S R_S}{q-1},$$

which proves the theorem.

The formula in this theorem is remarkably similar to the analogous formula for the S-zeta function in number fields. For a parallel "number field - function field" treatment see Tate [2].

The ring of S-integers can be characterized in other ways, as has already been suggested.

Theorem 14.5. *Let K/F be a function field with constant field F and let S be a non-empty, finite set of primes. There exist elements $x \in K$ such that the poles of x consist precisely of the elements of S. For any such element x, the integral closure of $F[x]$ in K is O_S. O_S is a Dedekind domain and there is a one-to-one correspondence between the non-zero prime ideals of O_S and the primes of K not in S. The S-units $E(S)$ are equal to the units of O_S and the class group of O_S, $Cl(O_S)$, is isomorphic to Cl_S.*

Proof. To begin with, let's label the primes in S, $S = \{P_1, P_2, \ldots, P_s\}$. For a large positive integer M consider the vector spaces $L(MP_i) = \{x \in K^* | (x) + MP_i \geq 0\}$. As soon as M is big enough (say, $M > 2g - 2$) we know from Corollary 4 to Theorem 5.4 that the dimension of this space is $M \deg P_i - g + 1$. It follows that $L(MP_i)$ is properly contained in $L((M+1)P_i)$. Pick an element x_i which is in the latter set, but not in the former set. Then x_i has a pole of order $M+1$ at P_i and no other poles. Now consider $x = x_1 x_2 \cdots x_s$. Then, x has each element of S as a pole and no other poles.

With x chosen to have poles at the elements of S, and nowhere else, let R be the integral closure of $F[x]$ in K. The ring R is a Dedekind domain. If $K/F(x)$ is a separable extension, this fact is well known and is proven in many places. As is shown in Chapter V, Theorem 19, of Samuel and Zariski [1], it remains true even if $K/F(x)$ is inseparable. If P is a prime of K not in S, then $x \in O_P$ and it follows that $R \subseteq O_P$. Thus,

$$R \subseteq \bigcap_{P \notin S} O_P = O_S.$$

We will show that $R = O_S$. Let $P \notin S$ be a prime of K and consider $P \cap R$. It cannot be that $P \cap R = (0)$ since otherwise the quotient field of R, namely, K, would inject into the residue class field O_P/P. However, O_P/P is finite over F. Thus, $P \cap R = \mathfrak{p}$ is a maximal ideal of R, and $R_{\mathfrak{p}} \subseteq O_P$.

This must be an equality because R_p is a discrete valuation ring and so is a maximal subring of K. On the other hand, if $\mathfrak{p}$ is a maximal ideal of R then $R_\mathfrak{p}$ is a discrete valuation ring and $(\mathfrak{p}R_\mathfrak{p}, R_\mathfrak{p})$ is a prime of K containing x. This shows that $\mathfrak{p} \to (\mathfrak{p}R_\mathfrak{p}, R_\mathfrak{p})$ is a one-to-one correspondence between the maximal ideals of R and the primes of K not in S. Again using the fact that R is a Dedekind domain, we find (see Jacobson [2], Section 10.4)

$$R = \bigcap_{\mathfrak{p} \subset R} R_\mathfrak{p} = \bigcap_{P \notin S} O_P = O_S \ .$$

We have shown that O_S is a Dedekind domain and that there is a one-to-one correspondence between the maximal ideals of O_S and the primes of K not in S. The remaining statements of the theorem are straightforward and we leave them as exercises for the reader.

It is of interest to see how these general ideas work out in particular cases. We will see how they apply in quadratic extensions of $\mathbb{F}(T)$ and in the cyclotomic functions fields which were defined and discussed in Chapter 12. We use the notation given there.

Let's assume that $q = |\mathbb{F}|$ is odd. Let $f(T) \in \mathbb{F}[T] = A$, be a square-free polynomial. Define $K = k(\sqrt{f(T)})$ (recall that $k = \mathbb{F}(T)$). One sees immediately that K/k is a Galois extension of degree 2 and that the non-trivial element, σ, of the Galois group is characterized by $\sigma\sqrt{f(T)} = -\sqrt{f(T)}$. A short calculation, completely analogous to what happens in quadratic number fields, shows that the integral closure of A in K, R, is equal to $A + A\sqrt{f(T)}$.

Recall that the prime at infinity, ∞, of k is defined by $\mathrm{ord}_\infty h = -\deg h$. Let $U = 1/T$. Then, $\mathrm{ord}_\infty U = 1$, i.e., U is a uniformizing parameter at infinity. Let $d = \deg f(T)$ and rewrite $f(T)$ in terms of U as follows:

$$f(T) = \sum_{I=0}^{d} a_i T^i = T^d \sum_{i=0}^{d} a_i T^{i-d} = U^{-d} \sum_{i=0}^{d} a_i U^{d-i} = U^{-d} f^*(U) \ .$$

Note that $f^*(U) \in \mathbb{F}[U]$ and that its constant term is $a_d \neq 0$, the leading term of $f(T)$.

Proposition 14.6. *Let $K = k(\sqrt{f(T)}\ )$, where $f(T) \in A = \mathbb{F}[T]$ is square-free. Let $d = \deg f(T)$ and a_d the leading coefficient of $f(T)$. If d is odd, then ∞ is ramified in K. If d is even, and a_d is a square in $\mathbb{F}^*$, then ∞ splits in K. Finally, if d is even and a_d is not a square in $\mathbb{F}^*$, then ∞ remains prime in K.*

Proof. Suppose d is odd. Since U is a uniformizing parameter at ∞ and $a_d \neq 0$ is the constant term of $f^*(U)$, we see that $f^*(U)$ is a unit at infinity. Suppose P_∞ is a prime of K lying above ∞. Then, setting e equal to the ramification index of P_∞ over ∞,

$$\mathrm{ord}_{P_\infty} \sqrt{f(T)} = \frac{1}{2}\mathrm{ord}_{P_\infty} f(T) = \frac{e}{2}\mathrm{ord}_\infty U^{-d} f^*(U) = -\frac{ed}{2} \ .$$

Since this number must be an integer and since d is assumed odd, it follows that $2|e$. Thus, $e = 2$ and K/k is ramified at ∞.

Now suppose that d is even. Then, K is generated over k by $\sqrt{f^*(U)}$. Since $f^*(U)$ is square-free as a polynomial in U, it follows that the integral closure, R', of $A' = \mathbb{F}[U]$ in K is $A' + A'\sqrt{f^*(U)}$. By Proposition 15 in Chapter 1 of Lang [5], the prime decomposition of ∞ follows from that of the irreducible polynomial $X^2 - f^*(U)$ reduced modulo U. The reduction is simply $X^2 - a_d \in \mathbb{F}[X]$. This either splits or is irreducible according to whether a_d is a square or not in $\mathbb{F}^*$. This completes the proof.

We have given a rather old-fashioned proof. A more modern proof can be given using the properties of the completion $\mathbb{F}((1/T))$ of k at ∞.

Following Emil Artin [1], we say that the quadratic function field $K = k(\sqrt{f(T)})$ is real if ∞ splits in K and is imaginary in the other two cases. This closely follows the terminology in the number field case.

Let $B = A + A\sqrt{f(T)}$ be the integral closure of A in K. We want to compare the class number of B, h_B, with h_K.

Proposition 14.7. *With the above notation, we have $h_B = h_K$ if ∞ is ramified, $h_B = 2h_K$ if ∞ is inert, and $h_B \log_q |e|_{P_\infty} = h_K$ if ∞ splits. In the latter case, e represents a fundamental unit in B, and P_∞ is the prime above ∞ at which e has negative ord.*

Proof. In the first two cases the set of primes above ∞ consists of one element P_∞. Thus, $s = 1$ and the rank of $\mathcal{D}(S)^\circ$ is zero. Also, in the first case the degree of P_∞ is 1 and in the second case it is 2. Thus, the first two assertions follow from Proposition 14.1, part (b), and Lemma 14.3.

In the third case, there are two primes above ∞, P_∞ and P'_∞. Thus, the unit group B^* has rank 1. Let e be a generator of B^* modulo torsion, i.e., $B^* = \mathbb{F}^* < e >$. If e' denotes the Galois conjugate of e, then $ee' \in \mathbb{F}^*$ which implies $\mathrm{ord}_{P_\infty} e + \mathrm{ord}_{P'_\infty} e = 0$. Thus we can chose P_∞ to be the prime over ∞ with $\mathrm{ord}_{P_\infty} e < 0$. Both primes above ∞ have degree 1, so by Lemma 14.3, $[\mathcal{D}(S)^\circ : \mathcal{P}(S)] = R_S^{(q)} = |\log_q |e|_{P_\infty}|$. By our choice of P_∞ we can remove the absolute value sign. Now, invoking Proposition 14.1 once again gives the result.

Remark. It is worth pointing out that the expression $\log_q |e|_{P_\infty}$ can be considerably simplified. Let $e = g + h\sqrt{f(T)}$, where $g, h \in \mathbb{F}[T]$. Then $e + e' = 2g$, which implies $\mathrm{ord}_\infty g = \mathrm{ord}_{P_\infty} g = \mathrm{ord}_{P_\infty}(e + e') = \mathrm{ord}_{P_\infty} e = -\log_q |e|_{P_\infty}$. Since $\mathrm{ord}_\infty g = -\deg g$, we arrive at the simple equation $h_K = h_B \deg g$.

Now let's consider briefly the cyclotomic function fields treated in Chapter 12. Recall that K_m is defined to be $k(\Lambda_m)$ where Λ_m are the m-torsion points on the Carlitz module. The ring $\mathcal{O}_m$ is the integral closure of A in K_m. Let S_m be the set of primes in K_m lying over ∞. Then, as we have seen, $\mathcal{O}_m$ is the ring of S_m-integers in K_m and its unit group is the group of S_m-units. What is the cardinality of S_m? The answer is implicitly given in Theorem 12.14. The fixed field of $\{\sigma_\alpha \mid \alpha \in \mathbb{F}^*\}$ is denoted by K_m^+. According to that theorem, ∞ splits completely in K_m^+ and each prime above ∞ in K_m^+ ramifies totally in K_m. Let S_m^+ denote the primes in K_m^+ lying above ∞. It follows that

$$|S_m^+| = |S_m| = [K_m^+ : k] = \frac{\Phi(m)}{q-1} .$$

From this information it also follows that each prime in K_m lying above ∞ has degree 1. Thus,

Proposition 14.8. *The groups $\mathcal{O}_m^*/\mathbb{F}^*$ and $\mathcal{O}_m^{+\,*}/\mathbb{F}^*$ are free of rank $\frac{\Phi(m)}{q-1} - 1$. Moreover, $h_{K_m} = h_{\mathcal{O}_m} R_{S_m}^{(q)}$ and $h_{K_m^+} = h_{\mathcal{O}_m^+} R_{S_m}^{(q)}$.*

Proof. With the information already provided, the proof is a straightforward application of Propositions 14.1, 14.2, and Lemma 14.3.

In Chapter 16 we will investigate the class numbers for quadratic and cyclotomic function fields in greater detail. A fundamental tool will be Artin L-functions and their properties in the special case where the Galois group is abelian. Some of this was already discussed in Chapter 9. We will provide a short review these ideas.

Let K/k be an Galois extension of global function fields. The number of elements in the constant field of k, $\mathbb{F}$, will be denoted by q (as usual). We will not suppose that k is a rational function field. Let G denote the Galois group of K/k. We suppose that G is abelian. If P is a prime of k and $\mathfrak{P}$ is a prime of K lying over P, then the decomposition and inertia groups, $Z(\mathfrak{P}/P)$ and $I(\mathfrak{P}/P)$, are independent of $\mathfrak{P}$ (because G is abelian). We denote them more simply by $Z(P)$ and $I(P)$. We recall that $|Z(P)| = e(P)f(P)$ and $|I(P)| = e(P)$. Here, $e(P) = e(\mathfrak{P}/P)$ and $f(P) = f(\mathfrak{P}/P)$ are the ramification index and relative degree of $\mathfrak{P}$ over P. We also know that $Z(P)/I(P)$ is cyclic, being isomorphic to the Galois group of the residue class field extension.

If $\mathfrak{P}/P$ is unramified, then the Artin automorphism $(P, K/k) \in G$ generates $Z(P)$ and is characterized by the congruence

$$(P, K/k)\omega \equiv \omega^{NP} \pmod{\mathfrak{P}} ,$$

where ω is any element of K integral at $\mathfrak{P}$.

Let P be any prime of k and $\chi \in \hat{G}$ a one-dimensional character of G. We want to define $\chi(P)$. If P is unramified in K, we set $\chi(P) = \chi((P, K/k))$. If P is ramified, suppose $\chi(I(P)) \neq 1$. In this case we say that χ is ramified at P and set $\chi(P) = 0$. If $\chi(I(P)) = 1$, then χ is a character on $G/I(P)$. Let the fixed field of $I(P)$ be denoted by M. Then, $\mathrm{Gal}(M/k) \cong G/I(P)$.

Under these conditions, set $\chi(P) = \chi((P, M/k))$. We have now defined $\chi(P)$ for any prime P and we define the Artin L-function of χ to be

$$L(w, \chi) = \prod_{P \in \mathcal{S}_k} (1 - \chi(P)NP^{-w})^{-1} .$$

When $\chi = \chi_o$, the trivial character, we see that $L(w, \chi_o) = \zeta_k(w)$, which has a simple pole at $w = 1$ but is analytic everywhere else. If $\chi \neq \chi_o$, then $L(s, \chi)$ is entire (as follows from the fact that it can be identified with a Hecke L-function; see Chapter 9). The following proposition is a special case of a more general result about Artin L-functions. Since we will use it so often, we provide the relatively simple proof.

Proposition 14.9. *With the above notations, we have*

$$\zeta_K(w) = \zeta_k(w) \prod_{\chi \neq \chi_o} L(w, \chi) . \tag{5}$$

Proof. By looking at the product decompositions on both sides we see that it is sufficient to prove the following "semi-local" identity for each prime P of k.

$$\prod_{\mathfrak{P}|P} (1 - N\mathfrak{P}^{-w}) = \prod_{\chi \in \hat{G}} (1 - \chi(P)NP^{-w}) . \tag{6}$$

Let $e = e(P), f = f(P)$, and $g = [K : k]/ef$. We see that g is the number of primes of K lying above P. The left-hand side of Equation 6 is thus

$$(1 - NP^{-fw})^g .$$

We want to show that the right-hand side of Equation 6 is equal to this same expression. Note first of all that if $\chi(I(P)) \neq 1$, then $1 - \chi(P)NP^{-w} = 1$. Thus, the right-hand side is equal to

$$\prod_{\chi \in \widehat{G/I(P)}} (1 - \chi(P)NP^{-w}) .$$

As before, let M be the fixed field of $I(P)$. Then, by definition, $\chi(P) = \chi((P, M/k))$ an f-th root of unity. Every f-th root of unity determines a unique character of the subgroup of $Z(P)/I(P)$ generated by $(P, M/k)$ and each such character will extend in $g = [G : Z(P)] = [G/I(P) : Z(P)/I(P)]$ ways to a character of $G/I(P)$. Thus,

$$\prod_{\chi \in \widehat{G/I(P)}} (1 - \chi(P)NP^{-w}) = \prod_{i=0}^{f-1} (1 - \zeta_f^i NP^{-w})^g = (1 - NP^{-fw})^g .$$

This concludes the proof.

The proof is a little easier to see when P is unramified in K, but it is important to include all the primes in the definition of $L(w, \chi)$.

Corollary 1. *Suppose K/k is abelian and geometric, i.e. that there is no constant field extension. Then*

$$h_K = h_k \prod_{\chi \neq \chi_o} L(0, \chi) \ .$$

Proof. By Theorem 5.9, we have

$$\zeta_K(w) = \frac{L_K(q^{-w})}{(1 - q^{-w})(1 - q^{1-w})} \ ,$$

where $L_K(u) \in \mathbb{Z}[u]$ is such that $L_K(1) = h_K$.

By the assumption that there is no constant field extension in K/k we can multiply both sides Equation 5 in the statement of the proposition by $(1 - q^{-w})(1 - q^{1-w})$ to derive

$$L_K(q^{-w}) = L_k(q^{-w}) \prod_{\chi \neq \chi_o} L(w, \chi) \ .$$

Now substitute $w = 0$ to get the result.

Corollary 2. *For $\chi \neq \chi_o$ we have $L(0, \chi) \neq 0$.*

Proof. This follows immediately from Corollary 1.

Remark. From Corollary 1, we can infer that $h_k \mid h_K$. In fact, h_K/h_k is a rational number equal to $\prod_{\chi \neq \chi_o} L(0, \chi)$, which is in $\mathbb{Z}[\zeta_n]$, where $n = [K : k]$. This is because $L(s, \chi)$ for $\chi \neq \chi_o$ is a polynomial in q^{-s} with coefficients in $\mathbb{Z}[\zeta_n]$. A rational number which is simultaneously an algebraic integer is a rational integer, which proves the assertion. By using formal properties of Artin L-functions, one can show in this way that for any finite extension K/k of global fields, the class number of k divides the class number of K. This fact was first shown by M. Madan [1] using cohomological methods. His proof is actually much more elementary than the analytic one we have just sketched.

For the remainder of the chapter, we will be concerned with finding a class number formula similar to that given in the above Corollary 1, but for the class number of the ring of S-integers rather than the group of divisor classes of degree zero.

Let K/k continue to denote a geometric, abelian extension of global function fields with Galois group G. Let S denote a finite set of primes of k and S' the set of primes of K lying above those in S. Let $A \subset k$ denote the ring of S-integers in k and $B \subset K$ denote the ring of S'-integers in K. By Theorem 14.5, both A and B are Dedekind domains. Using the method of proof of that theorem, it is not hard to see that B is the integral closure of A in K. We denote by h_A and h_B the class numbers of A and B, respectively. In the special case where $k = \mathbb{F}(T)$, $S = \{\infty\}$, and $A = \mathbb{F}[T]$ we have $h_A = 1$.

We have previously defined S-zeta functions. We now define S-L-functions in an analogous way. Namely, for $\chi \in \hat{G}$, define

$$L_S(w, \chi) = \prod_{P \notin S} (1 - \chi(P)NP^{-w})^{-1} \ .$$

By Theorem 14.5 there is a one-to-one correspondence between the primes in S_k not in S and the prime ideals of the ring A. Thus, it is natural to think of $L_S(w, \chi)$ as the L-function corresponding to the ring A. We set $L_S(w, \chi) = L_A(w, \chi)$ and work primarily with the latter notation.

In the following proposition we will need the definition of the Artin conductor $\mathcal{F}(\chi)$ of a character χ. Artin gave a definition in great generality. It applies even if the Galois group G is not abelian. In the abelian case, which is treated briefly in Chapter 9, $\mathcal{F}(\chi)$ is defined to be the minimal effective divisor $\mathcal{F}$ such that χ is trivial on the ray modulo $\mathcal{F}$, $\mathcal{P}^{\mathcal{F}}$. Recall that $\mathcal{P}^{\mathcal{F}}$ is the group of principal divisors generated by elements $a \in k^*$ such that $\mathrm{ord}_P(a - 1) \geq \mathrm{ord}_P\mathcal{F}$ for all primes P in the support of $\mathcal{F}$. That some effective divisor $\mathcal{F}$ exists with the property that χ vanishes on $\mathcal{P}^{\mathcal{F}}$ is part of the statement of the Artin reciprocity law, Theorem 9.23. It is then an exercise to show there is a unique minimal one with this property.

Proposition 14.10. *$L_A(w, \chi)$ is a polynomial in q^{-w} of degree $d(\chi)$ where*

$$d(\chi) = 2g - 2 + \deg \mathcal{F}(\chi) + \sum_{P \in S(\chi)} \deg P \ .$$

Here, g is the genus of k, $\mathcal{F}(\chi)$ is the Artin conductor of χ, and $S(\chi) \subseteq S$ is the set of primes in S at which χ is unramified (i.e., $\chi(I(P)) = 1$).

Proof. From the definition of $L_A(w, \chi)$ we have

$$L_A(w, \chi) = \prod_{P \in S} (1 - \chi(P)NP^{-w}) \, L(w, \chi) \ .$$

By a famous result of A. Weil [1], we know that $L(w, \chi)$ is a polynomial in q^{-w} of degree $2g - 2 + \deg \mathcal{F}(\chi)$. It remains to examine the factors of the product over the primes in S.

If χ is ramified at P we have $\chi(P) = 0$, so these terms do not contribute. If χ is not ramified at P, we have $\chi(P) \neq 0$ and so $1 - \chi(P)NP^{-w} = 1 - \chi(P)q^{-w \deg P}$, which is a polynomial of degree $\deg P$ in q^{-w}. The result follows from this.

Proposition 14.11. *We have*

$$\zeta_B(w) = \zeta_A(w) \prod_{\chi \neq \chi_o} L_A(w, \chi) \ . \tag{7}$$

Proof. This assertion follows immediately from the definitions and the method of proof of Theorem 14.9. The method there uses the semi-local identity given in Equation 6. We simply use that identity for all primes not in S, take the inverse of both sides, and then multiply over all primes not in S.

We want to use Equation 7 together with Theorem 14.4 to get a class number formula. An important first step is to find a formula for the order of vanishing of $L_A(w, \chi)$ at $w = 0$.

Proposition 14.12. *Suppose* $\chi \neq \chi_o$ *and let* $m(\chi)$ *denote the order of vanishing of* $L_A(w, \chi)$ *at* $w = 0$. *Then,*

$$m(\chi) = \#\{P \in S \mid \chi(Z(P)) = 1\} \ .$$

Proof. From the definition,

$$L_A(w, \chi) = \prod_{P \in S} (1 - \chi(P)NP^{-w})L(w, \chi) \ .$$

Since $L(0, \chi) \neq 0$ by Corollary 2 to Proposition 14.9, we see that $m(\chi)$ is just the number of $P \in S$ such that $\chi(P) = 1$. This only happens when χ is unramified and is trivial on $(P, M/k)$ (recall that M is the fixed field of $I(P)$). Since the Artin automorphism at P generates $Z(P)/I(P) \subset G/I(P)$ these conditions are equivalent to $\chi(Z(P)) = 1$.

We have now assembled all the background necessary to prove the main result of this chapter. However, we need one more piece of notation. For a character χ of G and a prime $P \in S_k$ we have defined $\chi(P)$. We now extend this definition to divisors $D \in \mathcal{D}_k$. If $D = \sum a(P)P \in \mathcal{D}_k$, set

$$\chi(D) = \prod_P \chi(P)^{a(P)} \ .$$

Theorem 14.13. (The Analytic Class Number Formula) *Let K/k be a geometric, abelian extension of global function fields. Let S be a finite set of primes of k, A the ring of S-integers and B the integral closure of A in K. Set $R_S^{(q)} = R_A^{(q)}$ and $R_{S'}^{(q)} = R_B^{(q)}$, where S' is the set of primes of K lying above those in S. Then*

$$h_B R_B^{(q)} = h_A R_A^{(q)} \prod_{\chi \neq \chi_o} C_\chi \ ,$$

where

$$C_\chi = \frac{(-1)^{m(\chi)}}{m(\chi)!} \sum_{\deg D \leq d(\chi)} \chi(D) \deg(D)^{m(\chi)} \ .$$

Here D runs over all effective divisors of k which are prime to S and of degree less than or equal to $d(\chi)$ (defined in the statement of Proposition 14.10). Alternatively, one can think of D as running through all integral ideals of A with $\dim_{\mathbb{F}}(A/D) \leq d(\chi)$. The number $m(\chi)$ is defined above.

Proof. By Proposition 14.12 we have

$$L_A(w, \chi) = c_\chi w^{m(\chi)} + O(w^{m(\chi)+1}) , \qquad (8)$$

where c_χ is a non-zero constant. Combining this with Equation 7 of Proposition 14.11 and the assertion of Theorem 14.4 yields the following identity:

$$h_B R_B = h_A R_A \prod_{\chi \neq \chi_0} c_\chi . \qquad (9)$$

We have set $R_{S'} = R_B$ and $R_S = R_A$. The same process shows the following fact (which can also be proved directly),

$$|S'| = |S| + \sum_{\chi \neq \chi_0} m(\chi) . \qquad (10)$$

From Equation 8 we see that c_χ is the $m(\chi)$-th derivative of $L_A(w, \chi)$ evaluated at $w = 0$ divided by $m(\chi)!$. By Proposition 14.10, we know that $L_A(w, \chi)$ is a polynomial in q^{-w} of degree $d(\chi)$. Thus,

$$L_A(w, \chi) = \sum_D \frac{\chi(D)}{ND^w} = \sum_D \chi(D) q^{-\deg D \, w} ,$$

where the sum is over effective divisors prime to S and of degree $\leq d(\chi)$. Thus, with the same restrictions on the sum we find

$$c_\chi = \frac{1}{m(\chi)!} \frac{d^{m(\chi)}}{dw^{m(\chi)}} (L_A(w, \chi)|_{w=0} = \frac{1}{m(\chi)!} \sum_D \chi(D)(-\deg D \, \ln \, q)^{m(\chi)} .$$

Notice that $c_\chi = (\ln \, q)^{m(\chi)} C_\chi$, where C_χ is defined in the statement of the theorem. Combining this remark with Equations 9 and 10 yields

$$h_B R_B = h_A R_A (\ln \, q)^{s'} \, {}^o \prod_{\chi \neq \chi_0} C_\chi .$$

where $s' = |S'|$ and $s = |S|$.

The result now follows from the fact (see the remarks preceding Lemma 14.3) that $(\ln \, q)^{s'-1} R_B^{(q)} = R_B$ and $(\ln \, q)^{s-1} R_A^{(q)} = R_A$.

In the number field case the situation is similar, but more complicated. If K/k is an abelian extension of number fields we can again choose a finite set S of primes of k and form S-units, S-class groups, S-L-functions, etc. Here it is standard to include in S at least the primes which ramify in K and (!) the archimedean primes. The local factors at the non-archimedean primes look exactly like their counterparts in the function field case, $1 - \chi(P) N P^{-s}$, and are handled similarly. On the other hand, the local factors at the archimedean primes involve the Γ-function, and this adds another level of complexity. The use of the Γ-function in "the local factors at infinity" is seen most clearly in the famous thesis of J. Tate [1]. An exposition is found in Chapter XIV of Lang [4].

Exercises

1. Let S be a finite set of primes in a global function field. In the definition of the S-regulator, R_S, we began by choosing a set of units $\{e_1, e_2, \ldots, e_{s-1}\}$ whose cosets in $E(S)/\mathbb{F}^*$ form a free $\mathbb{Z}$-basis. Show that R_S is independent of this choice.

2. In the proof of Lemma 14.3, we defined a map $l : \mathcal{D}(S) \to \mathbb{Z}^s$. Prove the assertion that $[\mathbb{Z} : l(\mathcal{D}(s))] = \prod_{P \in S} \deg P$.

3. Prove the last assertions of Theorem 14.5. Namely, prove that $E(S)$ is the group of units of $\mathcal{O}_S$ and that Cl_S is isomorphic to the ideal class group of $\mathcal{O}_S$.

4. Let $\mathbb{F}$ be a finite field of characteristic different from 2. Let $f(T) \in A = \mathbb{F}[T]$ be a square-free polynomial and let B be the integral closure of A in $\mathbb{F}(T)(\sqrt{f(T)})$. Show that $B = A + A\sqrt{f(T)}$.

5. Prove Proposition 14.6 by considering the completion of $k = \mathbb{F}(T)$ at ∞, k_∞, and the extension of k_∞ generated by the roots of $X^2 - f(T)$.

6. Let S be a finite set of primes in a global function field K. Suppose all the elements in S have degree 1. Show that $h_K = h_S R_S^{(q)}$.

7. Let $k = \mathbb{F}(T)$ and let $S = \{P_0, P_\infty\}$, the set consisiting of the prime at 0 and the prime at ∞. Show that $\mathcal{O}_S = \mathbb{F}[T, T^{-1}]$. What is $E(S)$ in this case?

8. (Continuation) Let $f(T) \in \mathbb{F}[T]$ be a square-free polynomial of even degree whose constant coefficient and leading coefficient are both squares in $\mathbb{F}^*$. Show that both P_0 and P_∞ split in $K = k(\sqrt{f(T)})$.

9. (Continuation) Let B be the integral closure of $\mathbb{F}[T, T^{-1}]$ in K. Show that

$$h_B R_B^{(q)} = \frac{1}{2} \sum_g \chi(g)(\deg g)^2 \, ,$$

where the sum is over all polynomials $g(T)$ with $\deg g \leq \deg f$ and $g(0) \neq 0$. Here, $\chi(g)$ means χ of the divisor $\sum_{P \notin S} \operatorname{ord}_P(g) P$.

10. Redo the last three exercises under the assumption that S constitutes all the primes of k of degree 1, i.e., $S = \{P_\alpha \mid \alpha \in \mathbb{F}\} \cup \{P_\infty\}$, where P_α is the prime corresponding to the localization of $\mathbb{F}[T]$ at $(T - \alpha)$.

15
The Brumer-Stark Conjecture

This chapter is devoted to the explanation and, in special cases, the proof of a conjecture which generalizes the famous theorem of Stickelberger about the structure of the class group of cyclotomic number fields. This important conjecture, due to A. Brumer and H. Stark, is unresolved in the number field case. The analogous conjecture in function fields is now a theorem due to the efforts of J. Tate and P. Deligne. A short time after Deligne completed Tate's work on this result, D. Hayes found a proof along completely different lines. We will give a proof for the cyclotomic function fields introduced in Chapter 12. We will do so by using a method of B. Gross which combines the approaches of Tate and Hayes as they apply in this relatively simple special case. The use of 1-motives, which is essential in Deligne's work, will not be needed here.

Before beginning, it will be useful to give an outline of this chapter. We start with some generalities about groups acting on abelian groups, the group ring and its properties, and a review of the orthogonality relations for group characters. After these preliminaries we will discuss Gauss sums and their prime decomposition in cyclotomic number fields. This culminates in the statement of Stickelberger's theorem. We then formulate the Brumer-Stark conjecture for both number fields and function fields (i.e., for all global fields). For abelian extensions of the rational numbers $\mathbb{Q}$ we show that the Brumer-Stark conjecture is a simple consequence of the theorem of Stickelberger. Finally, we come to the main result of this chapter, the proof of the Brumer-Stark conjecture for cyclotomic function fields. The proof is in two parts. The first is a general result due to Tate which asserts, roughly speaking, that the generalized Stickelberger element annihilates the group of divisor classes of degree zero (of a global function field). The second part involves determining the prime decomposition of a torsion point of the Carlitz module. Since this decomposition is implicit in the results of Chapter 12, we prove the second part first. We then conclude the chapter with a proof of Tate's result. This proof will be somewhat incomplete, because it relies heavily on work of A. Weil. Weil's results will be stated, but not proved, since this would require advanced methods of algebraic geometry. Accepting Weil's results as given, Tate's proof is very beautiful and ingenious. It can be described as a sophisticated application of the Cayley-Hamilton theorem of linear algebra.

Let V be an abelian group which is acted on by a finite group G. In other words, we are given a homomorphism $\rho : G \to \text{Aut}(V)$. Given this data, there is a canonical way to make V into a module over the group ring $\mathbb{Z}[G]$. Recall that the elements of $\mathbb{Z}[G]$ are formal linear combinations of group elements, $\sum_{\sigma \in G} a(\sigma)\sigma$, with coefficients $a(\sigma) \in \mathbb{Z}$. The addition of two such elements is done coordinate-wise. The product is given by the following formula:

$$\left(\sum_{\sigma \in G} a(\sigma)\sigma\right)\left(\sum_{\tau \in G} b(\tau)\tau\right) = \sum_{\gamma \in G}\left(\sum_{\substack{\sigma,\tau \in G \\ \sigma\tau = \gamma}} a(\sigma)b(\tau)\right)\gamma \ .$$

With these conventions, let $\sum_{\sigma \in G} a(\sigma)\sigma \in \mathbb{Z}[G]$ and $v \in V$. Then define

$$\left(\sum_{\sigma \in G} a(\sigma)\sigma\right)(v) = \sum_{\sigma \in G} a(\sigma)\rho(\sigma)(v) \ .$$

It is a simple matter to check that with this definition, V becomes a $\mathbb{Z}[G]$ module.

It is cumbersome to write $\rho(\sigma)(v)$. We often accept ρ as fixed and write more simply $\rho(\sigma)(v) = \sigma v$.

Another notational convention is worth mentioning. Suppose that the group operation in the abelian group V is written multiplicatively instead of additively. Then the group ring acts according to the following formula:

$$\left(\sum_{\sigma \in G} a(\sigma)\sigma\right)(v) = \prod_{\sigma \in G} (\sigma v)^{a(\sigma)} \ .$$

An example of when this notation is appropriate is the case where K/k is a Galois extension of number fields, G is the Galois group, and V is the ideal class group of K.

It is often useful to generalize these notions by assuming that V is not just an abelian group, but a module over a commutative ring R. In this case, we assume ρ maps G to $\text{Aut}_R(V)$, i.e., that the actions of G and of R on V commute with one another. In this case the action of G on V extends to an action of the group ring $R[G]$ on V in exactly the same manner as outlined above in the case $R = \mathbb{Z}$. Of course, the group ring $R[G]$ consists of formal R-linear combination of group elements. Addition and multiplication are given by the same formulas as in the case of the ring $\mathbb{Z}[G]$.

Let $f : G \to R^*$ be a homomorphism from G to the group of units of R. This is easily seen to extend to a homomorphism of rings from $R[G]$ to R by means of the formula

$$f\left(\sum_{\sigma \in G} \alpha(\sigma)\sigma\right) = \sum_{\sigma \in G} a(\sigma)f(\sigma).$$

Conversely, if such a homomorphism of rings (more precisely, R-algebras) is given, then, by restricting to G, one gets a homomorphism of groups $G \to R^*$.

Let's now specialize somewhat. We will assume that G is abelian. Set $|G| = n$ and suppose that n is a unit in R. Suppose further that R is an integral domain and that R^* contains an element of order n. These assumptions are satisfied if R is an algebraically closed field of characteristic zero. If R is an algebraically closed field of characteristic $p > 0$ and p does not divide n, then, once again, both assumptions hold.

Proposition 15.1. *The group $\hat{G} = \operatorname{Hom}(G, R^*)$ is isomorphic to G.*

Proof. (Sketch) The proof is very simple in the case that G is a finite cyclic group. The general case is handled by use of the theorem that a finite abelian group is isomorphic to a direct sum of cyclic groups. See Lang [4] for details.

Corollary. $|G| = |\hat{G}|$.

The elements of $\hat{G}$ are called characters of G and the groups $\hat{G}$ is called the character group of G or, sometimes, the dual group of G.

Lemma 15.2. *Let G be a finite abelian group and $\sigma \in G$, $\sigma \neq e$, the identity element of G. Then there is a $\chi \in \hat{G}$ such that $\chi(\sigma) \neq 1$.*

Proof. (Sketch) Suppose $\chi(\sigma) = 1$ for all $\chi \in \hat{G}$. There is a natural homomorphism from $\overline{G/\langle\sigma\rangle} \to \hat{G}$ which, under our assumptions, would be onto. This contradicts the corollary to Proposition 15.1.

Proposition 15.3. (The Orthogonality Relations) *Let G be a finite abelian group of order n. If $\sigma, \tau \in G$, then*

(a)
$$\frac{1}{n}\sum_{\chi \in \hat{G}} \chi(\sigma^{-1})\chi(\tau) = \delta(\sigma, \tau),$$

where $\delta(\sigma, \tau) = 1$ if $\sigma = \tau$ and is 0 otherwise. If $\chi, \psi \in \hat{G}$, then

(b)
$$\frac{1}{n}\sum_{\sigma\in G}\chi(\sigma^{-1})\psi(\sigma) = \delta(\chi,\psi) \; ,$$

where $\delta(\chi,\psi) = 1$ if $\chi = \psi$ and is 0 otherwise.

Proof. To prove the first relation, let $\gamma \in G$ and set $T(\gamma) = \sum_{\chi\in\hat{G}}\chi(\gamma)$. If $\gamma = e$, the identity of G, then clearly, $T(e) = n$. If $\gamma \neq e$ there is a $\psi \in \hat{G}$ such that $\psi(\gamma) \neq 1$ by Lemma 15.2. We have,

$$\psi(\gamma)T(\gamma) = \psi(\gamma)\sum_{\chi\in\hat{G}}\chi(\gamma) = \sum_{\chi\in\hat{G}}(\psi\chi)(\gamma) = T(\gamma) \; .$$

Thus, $(\psi(\gamma) - 1)T(\gamma) = 0$ and so, $T(\gamma) = 0$. In general, given $\sigma, \tau \in G$ set $\gamma = \sigma^{-1}\tau$ and note that for all characters χ, $\chi(\sigma^{-1}\tau) = \chi(\sigma^{-1})\chi(\tau)$. This proves the first relation.

The proof of the second relation is similar. Choose an element $\lambda \in \hat{G}$ and set $S(\lambda) = \sum_{\sigma\in G}\lambda(\sigma)$. If $\lambda = \chi_o$, the trivial character $(\chi_o(\sigma) = 1$ for all $\sigma \in G)$, then, clearly, $S(\chi_o) = n$. If $\lambda \neq \chi_o$, then there is a $\tau \in G$ such that $\lambda(\tau) \neq 1$. We have,

$$\lambda(\tau)S(\lambda) = \lambda(\tau)\sum_{\sigma\in G}\lambda(\sigma) = \sum_{\sigma\in G}\lambda(\tau\sigma) = S(\lambda) \; .$$

Thus, $(\lambda(\tau) - 1)S(\lambda) = 0$ and so $S(\lambda) = 0$. In general, if $\chi, \psi \in \hat{G}$, set $\lambda = \chi^{-1}\psi$. Then, $\lambda(\sigma) = (\chi^{-1}\psi)(\sigma) = \chi^{-1}(\sigma)\psi(\sigma) = \chi(\sigma^{-1})\psi(\sigma)$. The second relation follows immediately from this.

We have assumed that G is a finite, abelian group. For any finite group one can define irreducible characters and prove orthogonality relations which generalize those given in Proposition 15.3. We will have no need for this generalization in this chapter. We have discussed this situation in Chapter 9. The interested reader can find an elegant presentation of this topic in Serre [3].

Let V be an $R[G]$ module and $\chi \in \hat{G}$. Define $V(\chi) = \{v \in V \mid \sigma v = \chi(\sigma)v, \forall \sigma \in G\}$. The R-submodule $V(\chi)$ is called the χ-th isotypic component of V. Under the assumptions on G and R that we have made, we will show that V is the direct sum of the isotypic components $V(\chi)$, as χ varies over $\hat{G}$. This useful result is proved using certain idempotents in the group ring $R[G]$ which we will now define.

Let $\chi \in \hat{G}$ and define $\varepsilon(\chi) \in R[G]$ by the following formula:

$$\varepsilon(\chi) = \frac{1}{n}\sum_{\sigma\in G}\chi(\sigma^{-1})\sigma \; .$$

Since we are assuming that $n = |G|$ is a unit in R, the formula does indeed define an element of $R[G]$.

Lemma 15.4.

1) For all $\sigma \in G$, we have $\sigma\varepsilon(\chi) = \chi(\sigma)\varepsilon(\chi)$.

2) For all $\chi, \psi \in \hat{G}$, we have $\varepsilon(\chi)\varepsilon(\psi) = \delta(\chi, \psi)\varepsilon(\chi)$.

3) $\sum_{\chi \in \hat{G}} \varepsilon(\chi) = e$, the identity of G (and, also, of $R[G]$).

4) For $\chi, \psi \in \hat{G}$ we have $\chi(\varepsilon(\psi)) = \delta(\chi, \psi)$.

5) The set $\{\varepsilon(\chi) \mid \chi \in \hat{G}\}$ is a free R basis for the group ring $R[G]$.

Proof. To prove part 1, let $\tau \in G$ and calculate

$$\tau\varepsilon(\chi) = \tau\frac{1}{n}\sum_{\sigma \in G} \chi(\sigma^{-1})\sigma = \frac{1}{n}\sum_{\sigma \in G}\chi(\tau)\chi(\tau^{-1})\chi(\sigma^{-1})\tau\sigma =$$

$$\chi(\tau)\frac{1}{n}\sum_{\sigma \in G}\chi((\tau\sigma)^{-1})\tau\sigma = \chi(\tau)\varepsilon(\chi) .$$

To prove part 2, we use part 1 and the orthogonality relations as follows:

$$\varepsilon(\chi)\varepsilon(\psi) = \frac{1}{n}\sum_{\sigma \in G}\chi(\sigma^{-1})\sigma\varepsilon(\psi) =$$

$$\left(\frac{1}{n}\sum_{\sigma \in G}\chi(\sigma^{-1})\psi(\sigma)\right)\varepsilon(\psi) = \delta(\chi, \psi)\varepsilon(\psi) .$$

The proof of part 3 is another application of the orthogonality relations. We calculate again

$$\sum_{\chi \in \hat{G}}\varepsilon(\chi) = \sum_{\chi \in \hat{G}}\left(\frac{1}{n}\sum_{\sigma \in G}\chi(\sigma^{-1})\sigma\right)$$

$$= \sum_{\sigma \in G}\left(\frac{1}{n}\sum_{\chi \in \hat{G}}\chi(\sigma^{-1})\right)\sigma = \sum_{\sigma \in G}\delta(\sigma^{-1}, e)\sigma = e .$$

The property 4 is just a restatement of the second orthogonality relation. To see this, note that

$$\chi(\varepsilon(\psi)) = \chi\left(\frac{1}{n}\sum_{\sigma \in G}\psi(\sigma^{-1})\sigma\right) = \frac{1}{n}\sum_{\sigma \in G}\psi(\sigma^{-1})\chi(\sigma) = \delta(\chi, \psi).$$

Finally, to prove property 5 note first that by part 1, $R[G]\varepsilon(\chi) = R\varepsilon(\chi)$. From this and part 3 we see that the set $\{\varepsilon(\chi) \mid \chi \in \hat{G}\}$ spans $R[G]$ over R. The linear independence follows immediately from 2. .

For the sake of clarity, in the following proposition we restate the hypotheses under which we have been operating.

Proposition 15.5. *Let G be a finite abelian group of order n, R an integral domain whose units, R^*, contains an element of order n. Assume also that n is a unit in R. Let V be an $R[G]$ module. Then V is the direct sum of its isotypic components $V(\chi)$. In other words,*

$$V \cong \bigoplus_{\chi \in \hat{G}} V(\chi) .$$

Proof. To begin with, we claim that $V(\chi) = \varepsilon(\chi)V$. If $v \in V$, consider $\varepsilon(\chi)v$. By the first part of the above lemma, we see that $\sigma\varepsilon(\chi)v = \chi(\sigma)\varepsilon(\chi)v$ for all $\sigma \in G$. This shows that $\varepsilon(\chi)V \subseteq V(\chi)$. If $v \in V(\chi)$, then

$$\varepsilon(\chi)v = \frac{1}{n} \sum_{\sigma \in G} \left(\chi(\sigma^{-1})\sigma \right)v =$$

$$\frac{1}{n} \left(\sum_{\sigma \in G} \chi(\sigma^{-1})\chi(\sigma) \right)v = v.$$

This shows $V(\chi) \subseteq \varepsilon(\chi)V$, so our claim is proved. We have also shown that $\varepsilon(\chi)$ acts as the identity on $V(\chi)$, a fact which we will use shortly.

From the above Lemma, part 3, we see that for all $v \in V$, $v = \sum_{\chi \in \hat{G}} \varepsilon(\chi)v$. This shows that V is the sum of its isotypic components. It remains to show that the sum is direct. Suppose that for each $\chi \in \hat{G}$ we have an element $v_\chi \in V(\chi)$ and that $\sum_{\chi \in \hat{G}} v_\chi = 0$. Then, for each $\psi \in \hat{G}$ we have

$$0 = \varepsilon(\psi)(\sum_{\chi \in \hat{G}} v_\chi) = \sum_{\chi \in \hat{G}} \varepsilon(\psi)v_\chi =$$

$$\sum_{\chi \in \hat{G}} \varepsilon(\psi)\varepsilon(\chi)v_\chi = \varepsilon(\psi)v_\psi = v_\psi .$$

We have used part 2 of Lemma 15.4 and the fact that $\varepsilon(\chi)$ acts like the identity on $V(\chi)$. This completes the proof.

We have now presented all that we shall need from abstract algebra. Our next goal is to recall the relevant definitions and state the classical theorem of L. Stickelberger on the prime decomposition of Gauss sums. The details of this development and the proofs can be found in Ireland and Rosen [1]. Other sources are Lang [6] and Washington [1].

For every positive integer m let ζ_m denote the complex number $e^{\frac{2\pi i}{m}}$. Let $K_m = \mathbb{Q}(\zeta_m)$ and denote by D_m the ring of algebraic integers in K_m. D_m is generated, as a ring by ζ_m, i.e., $D_m = \mathbb{Z}[\zeta_m]$. We can assume that $m \not\equiv 2$ (mod 4), since if $m \equiv 2$ (mod 4) then $\zeta_{m/2} = \zeta_m^2$ and $\zeta_m = -\zeta_{m/2}^{\frac{m+2}{4}}$ and so $K_m = K_{m/2}$. With this convention, a prime $p \in \mathbb{Z}$ is ramified in K_m if and only if $p|m$.

Assume that $p \in \mathbb{Z}$ is a prime which does not divide m and that $P \subset D_m$ is a prime ideal lying above $p\mathbb{Z}$. D_m/P is a finite field with $NP = p^f$ elements, where f is the smallest positive integer such that $p^f \equiv 1$ (mod m). If $\alpha \notin P$ there is a unique integer i such that $0 \leq i < m$ and

$$\alpha^{\frac{NP-1}{m}} \equiv \zeta_m^i \pmod{P} .$$

We set $(\alpha/P)_m = \zeta_m^i$ and call $(\alpha/P)_m$ the m-th power residue symbol. If $\alpha \in P$, we set $(\alpha/P)_m = 0$. The m-th power residue symbol has a number of arithmetically interesting properties. For our purposes, the most important is that $\alpha \to (\alpha/P)_m$ induces a homomorphism from $(D_m/P)^* \to \langle \zeta_m \rangle$, i.e., a character of the multiplicative group of D_m/P. Let Tr_P be the trace map from D_m/P to $\mathbb{Z}/p\mathbb{Z}$ and define

$$g(P) = \sum_{\alpha \in (D_m/P)^*} (\alpha/P)_m^{-1} \zeta_p^{Tr_P(\alpha)} .$$

The quantity $g(P)$ is called the Gauss sum associated with the prime ideal P. We further define $\Phi(P) = g(P)^m$. These quantities possess the following properties

Proposition 15.6.

1) $g(P) \in \mathbb{Q}(\zeta_m, \zeta_p)$.

2) $\Phi(P) \in \mathbb{Q}(\zeta_m)$.

3) $|g(P)|^2 = NP$.

The proof of part 1 is immediate from the definition. Part 2 is somewhat surprising. The proof uses Galois theory. Part 3 is a standard property of Gauss sums. For details see Ireland and Rosen [1], Proposition 14.3.1.

The goal we are after is the prime decomposition of $\Phi(P)$ in D_m where P is any prime ideal not containing m. From part 3 of Proposition 15.6. we deduce that $\Phi(P)\overline{\Phi(P)} = NP^m = p^{fm}$. It follows that the primes which divide $(\Phi(P))$ are primes in D_m lying over $p\mathbb{Z}$. Since $\mathbb{Q}(\zeta_m)$ is a Galois extension, the primes above $p\mathbb{Z}$ are all conjugates of P. We thus take a moment to recall the explicit description of the Galois group of $\mathbb{Q}(\zeta_m)$ which we gave in Chapter 12.

If $t \in \mathbb{Z}$ is relatively prime to m, there is a unique automorphism σ_t in $G_m = \mathrm{Gal}(K_m/\mathbb{Q})$ with the property $\sigma_t(\zeta_m) = \zeta_m^t$. The map $t \to \sigma_t$ gives rise to an isomorphism $(\mathbb{Z}/m\mathbb{Z})^* \cong G_m$.

We can now state

Theorem 15.7. (L. Stickelberger):

$$(\Phi(P)) = \prod_{\substack{t=1 \\ (t,m)=1}}^{m-1} (\sigma_t^{-1} P)^t = \Big(\sum_{\substack{t=1 \\ (t,m)=1}}^{m-1} t\sigma_t^{-1} \Big) P .$$

This theorem dates from the 19-th century. In the case where m is a prime it was formulated and proved by E. Kummer (1847). Stickelberger's generalization came 43 years later (1890).

We have seen that the prime decomposition of $\Phi(P)$ involves only primes above pZ and that these are all conjugates of P under the action of the Galois group G_m. This is quite elementary. The remarkable feature of Stickelberger's theorem is that the *same* element of the group ring $\mathbb{Z}[G]$ describes the prime decomposition for all P not containing m. If we use the known fact that every element of the class group Cl_{K_m} contains infinitely many prime ideals, we derive the following important corollary.

Corollary. *The element*

$$\sum_{\substack{t=1 \\ (t,m)=1}}^{m-1} t\sigma_t^{-1} \in \mathbb{Z}[G_m]$$

annihilates the class group Cl_{K_m}.

For the proof of Stickelberger's theorem and some of the many important applications, see Ireland and Rosen [1], Lang [6], and/or Washington [1].

The goal of the Brumer-Stark conjecture is to generalize the above results to an arbitrary abelian extension of global fields K/k. If $G = \text{Gal}(K/k)$, we are looking for an element of $\mathbb{Z}[G]$ defined in some canonical way which annihilates the class group of K in the number field case and the divisor class group of K in the function field case. This canonical element should essentially be the one given in the above corollary when $K = K_m$ and $k = \mathbb{Q}$. Brumer (unpublished, but see Coates [1]) was the first to suggest a candidate for such an element. We now describe the background necessary to write this down.

Let K/k be a finite abelian extension of global fields of degree n, and G its Galois group. Let S be a non-empty finite set of primes of k which contains all the primes which ramify in K and, in the number field case, all the archimedean primes. If $\chi : G \to \mathbb{C}^*$ is a complex valued character on G we defined the S-L-function, $L_S(w,\chi)$, in Chapter 14 as follows:

$$L_S(w,\chi) = \prod_{P \notin S} (1-\chi(P)NP^{-w})^{-1} = \prod_{\substack{P \in S \\ P \text{ non-arch}}} (1-\chi(P)NP^{-w})\, L(w,\chi)\,,$$

where $L(w,\chi)$ is the complete Artin L-function attached to χ.

For the rest of this discussion the ring R will denote the ring of complex-valued meromorphic functions on the complex plane. It satisfies all the hypotheses we need; it is an integral domain, it contains n n-th roots of unity, and n is a unit in R.

Definition. The L-function evaluator $\theta_{K/k,S}(w) \in R[G]$ is defined as follows:

$$\theta(w) = \theta_{K/k,S}(w) = \sum_{\chi \in \hat{G}} L_S(w, \bar{\chi})\varepsilon(\chi) \ .$$

Proposition 15.9. *For all* $\chi \in \hat{G}$ *we have* $\chi(\theta(w)) = L_S(w, \bar{\chi})$.

Proof. This is an immediate consequence of the definition of $\theta(w)$ and Lemma 15.4, part 4.

This proposition explains the designation of $\theta(w)$ as the L-function evaluator.

One can rewrite the definition of $\theta(w)$ in terms of partial zeta functions. For $\sigma \in G$ the definition of the partial zeta function $\zeta_S(w, \sigma)$ is given by the sum

$$\zeta_S(w, \sigma) = \sum_{\substack{D,(D,S)=1 \\ (D,K/k)=\sigma}} ND^{-w} \ .$$

Here, in the function field case the sum is over all effective divisors whose support contains no prime in S and whose Artin symbol, $(D, K/k)$, is equal to σ. In the number field case the sum is over all integral ideals in the ring of integers of k which are prime to S and for which the Artin symbol, $(D, K/k)$, is equal to σ.

In all cases, the sum is absolutely convergent in the region $\Re(w) > 1$ and all these functions can be analytically continued to the whole complex plane with at most one simple pole at $w = 1$. The facts are reduced to known properties of zeta and L-functions by the following proposition.

Proposition 15.10. *With the above definitions and notations we have*

$$L_S(w, \chi) = \sum_{\sigma \in G} \chi(\sigma)\zeta_S(w, \sigma) \tag{1}$$

and

$$\zeta_S(w, \sigma) = \frac{1}{n} \sum_{\chi \in \hat{G}} \overline{\chi(\sigma)} L_S(w, \chi). \tag{2}$$

Proof. From the definition of $L_S(w, \chi)$, we find (summing over effective divisors or over integral ideals, prime to S)

$$L_S(w, \chi) = \sum_{(D,S)=1} \frac{\chi((D, K/k))}{ND^w} = \sum_{\sigma \in G} \chi(\sigma) \sum_{\substack{(D,S)=1 \\ (D,K/k)=\sigma}} \frac{1}{ND^w}$$

$$= \sum_{\sigma \in G} \chi(\sigma)\zeta_S(w, \sigma) \ .$$

This shows that the sum defining $\zeta_S(w, \sigma)$ is, essentially, a subsum of that defining $L_S(w, \chi)$. The latter sum is absolutely convergent in the region $\Re(w) > 1$ and so the sum defining $\zeta_S(w, \sigma)$ is absolutely convergent in this region as well.

To prove Equation 2 simply choose a $\tau \in G$, multiply Equation 1 on both sides by $\chi(\tau^{-1})/n = \overline{\chi(\tau)}/n$, sum over all $\chi \in \hat{G}$, and use the first orthogonality relation.

From Equation 2 we see that $\zeta_S(w, \sigma)$ can be analytically continued to the whole complex plane and is holomorphic everywhere except for a simple pole at $w = 1$ (corresponding to the simple pole of $L_S(w, \chi_o)$ at $w = 1$).

We can now give the promised alternate expression for the L-function evaluator $\theta(w)$.

Proposition 15.11.

$$\theta(w) = \theta_{K/k,S}(w) = \sum_{\sigma \in G} \zeta_S(w, \sigma)\sigma^{-1} .$$

Proof. Define $\tilde{\theta}(w) = \sum_{\sigma \in G} \zeta_S(w, \sigma)\sigma^{-1} \in R[G]$. By Equation 1 of the previous proposition, we find that $\chi(\tilde{\theta}(w)) = L_S(w, \bar{\chi})$ (we have used $\chi(\sigma^{-1}) = \overline{\chi(\sigma)}$). It follows that $\chi(\theta(w) - \tilde{\theta}(w)) = 0$ for all $\chi \in \hat{G}$. As we will see in a moment, this implies $\theta(w) = \tilde{\theta}(w)$.

Suppose $f \in R[G]$ has the property that $\chi(f) = 0$ for all $\chi \in \hat{G}$. Write $f = \sum r_\chi \epsilon(\chi)$ with $r_\chi \in R$ (that this is possible follows from Lemma 15.4, part 5. Let $\psi \in \hat{G}$ and apply ψ to both sides of this equation. We find $0 = r_\psi$ (by Lemma 15.4, part 4. Since this is true for all $\psi \in \hat{G}$ it follows that $f = 0$.

The values of the partial zeta functions $\zeta_S(w, \sigma)$ at $w = 0$ are especially important. It turns out that they are rational numbers and we have good control of their denominators. More precisely—

Theorem 15.12.

(a) $$\zeta_S(0, \sigma) \in \mathbb{Q} .$$

(b) $$W_K \zeta_S(0, \sigma) \in \mathbb{Z} .$$

where W_K denotes the number of roots of unity in K.

This theorem is quite deep. Part a was first proved, in the number field case, by C.L. Siegel [1] and part b was first proved, in this case, by P. Deligne and K. Ribet [1]. Other proofs of both results appeared soon thereafter, e.g., by D. Barsky and by P. Cassou-Noguès. Part a remains true when 0 is replaced by a negative integer $-n$ and part b remains true if we replace 0 by $-n$ and W_K by $W_K^{(n)}$, an integer also defined in terms of roots of unity. For details and applications see the instructive article of J. Coates [1].

We will give a proof later in the function field case using the fact that for non-trivial linear characters, χ, $L_S(w, \chi)$ is a polynomial in q^{-w}.

Definition. Define $\theta_{K/k,S} = \theta_{K/k,S}(0)$ and $\omega_{K/k,S} = W_K \theta_{K/k,S}$. The element $\omega_{K/k,S}$ is called Brumer element of K/k relative to S.

It follows from Proposition 15.11 and Theorem 15.12 that $\theta_{K/k,S} \in \mathbb{Q}[G]$ and that $\omega_{K/k,S} \in \mathbb{Z}[G]$. For the most part we will fix the abelian extension K/k and the non-empty set of primes S, so we will call these elements simply θ and ω.

We are now in a position to state the Brumer-Stark conjecture in both number fields and function fields.

The Brumer-Stark Conjecture (The Number Field Case). *We suppose that $|S| > 1$. Then, for every fractional ideal D of K we have $\omega D = (\alpha_D)$ where $\alpha_D \in K^*$ and α_D has absolute value 1 at all archimedean primes. Moreover, if λ_D is a W_K-th root of α_D, then $K(\lambda_D)/k$ is an abelian extension.*

Since the divisor of α_D is determined, α_D is determined up to a unit in O_K^*. The supplementary restrictions on α_D insure that it is well defined up to a root of unity in K.

The Brumer-Stark Conjecture (The Function Field Case). *Suppose first that $|S| > 1$. Then, for every divisor D of K, we have $\omega D = (\alpha_D)$ with $\alpha_D \in K^*$. If λ_D is a W_K-th root of α_D, then $K(\lambda_D)/k$ is an abelian extension. If $S = \{\mathfrak{P}\}$, then for every divisor D of K, there is an integer $n_D \in \mathbb{Z}$ and an element $\alpha_D \in K^*$ such that $\omega D = (\alpha_D) + n_D \sum_{\mathfrak{P}|\mathfrak{p}} \mathfrak{P}$. Once again, if λ_D is any W_K-th root of α_D, then $K(\lambda_D)/k$ is an abelian extension.*

In both the number field and the function field case, it is easy to see that the conditions imposed on α_D determine it up to multiplication by a root of unity. The same is true for λ_D. The question of whether $K(\lambda_D)/k$ is or is not abelian is not affected by this ambiguity.

In both versions, the conjecture that ω annihilates the class group is due to Brumer and the conjecture that $K(\lambda_D)/k$ is abelian is due to Stark.

We now show how Stickelberger's theorem implies the number field version of the Brumer-Stark conjecture for cyclotomic extensions of $\mathbb{Q}$.

Suppose m is a positive integer which is either odd or divisible by 4 and consider the cyclotomic field $K_m = \mathbb{Q}(\zeta_m)$. Let S be the set of primes dividing m together with the archimedean prime of $\mathbb{Q}$. The first task is to compute the element $\omega = \omega_{S,K_m/\mathbb{Q}}$.

Let $t \in \mathbb{Z}$ be relatively prime to m and $1 \le t < m$. Let σ_t be the corresponding element of $\mathrm{Gal}(K_m/\mathbb{Q})$. As is easily seen, if $n > 0$ is relatively prime to m, $\sigma_n = ((n), K_m/\mathbb{Q}) = \sigma_t$ if and only if $n \equiv t \pmod{m}$. Thus,

$$\zeta_S(w, \sigma_t) = \sum_{\substack{n=1 \\ n \equiv t \pmod{m}}}^{\infty} \frac{1}{n^w} = \sum_{h=0}^{\infty} \frac{1}{(t+hm)^w} .$$

For any real number b with $0 < b \le 1$, the Hurwitz zeta function is defined by the formula

$$\zeta(w, b) = \sum_{h=0}^{\infty} \frac{1}{(b+h)^w} .$$

It follows that $\zeta_S(w, \sigma_t) = m^{-w}\zeta(w, t/m)$. It is a well-known property of the Hurwitz zeta function that for every integer $n \ge 1$ we have $\zeta(1-n, b) = -B_n(b)/n$, where $B_n(b)$ is the n-th Bernoulli polynomial. A good source for this is Washington [1] or Lang [6]. For $n = 1$ we have $B_1(b) = b - \frac{1}{2}$. Putting all this together yields

$$\zeta_S(0, \sigma_t) = \frac{1}{2} - \frac{t}{m} ,$$

and so

$$\theta = \sum_{\substack{t=1 \\ (t,m)=1}}^{m-1} \left(\frac{1}{2} - \frac{t}{m}\right)\sigma_t^{-1} .$$

Assume first that m is odd. Then, $W_{K_m} = 2m$ and so

$$\omega = W_{K_m}\theta = \sum_{\substack{t=1 \\ (t,m)=1}}^{m-1} (m - 2t)\sigma_t^{-1} = mN - 2\sum_{\substack{t=1 \\ (t,m)=1}}^{m-1} t\sigma_t^{-1} .$$

Here, $N = \sum_{\sigma \in G} \sigma$ is the norm map.

Let P be a prime of K_m which is prime to m. Then, using the explicit expression we have just derived for ω and Stickelberger's theorem, we find

$$\omega P = \left(\frac{NP^m}{\Phi(P)^2}\right) = \left(\frac{NP^m}{g(P)^{2m}}\right) .$$

This verifies the first part of the Brumer-Stark conjecture when $D = P$ is a prime ideal which is prime to m with $\alpha_P = NP^m/g(P)^{2m}$. By Proposition 15.6, part 3, α_P has absolute value equal to 1. It is easily checked, by using the Galois properties of Gauss sums, that every Galois conjugate of α_P also has absolute value 1. This verifies the second condition of the conjecture. Finally, since $W_K = 2m$ in the case we are considering, we find that $\lambda_P = NP^{1/2}/g(P)$ so that $K_m(\lambda_P) \subseteq \mathbb{Q}(\zeta_m, \zeta_p, \sqrt{NP})$ which is abelian over $\mathbb{Q}$. If m is odd, the full Brumer-Stark conjecture for any divisor D prime to m follows from this.

When m is even and divisible by 4 we have $W_{K_m} = m$. The proof in this case differs insignificantly from the case we have considered. We leave the details as an exercise.

We now turn our attention to the function field case. To ease the exposition we will restrict our attention to abelian extensions K/k which are geometric; i.e., both K and k have the same constant field, $\mathbb{F}$, which is a finite field with q elements. Under this condition, group of roots of unity in K is just $\mathbb{F}^*$, so that $W_K = q - 1$.

Our first task is to consider more closely the L-function evaluator $\theta(w) = \theta_{K/k,S}(w)$ (we fix an abelian extension K/k of degree n and a finite set of primes S of k which contains all the ramified primes). In the function field case, all the S-L-functions which occur are rational functions of $u = q^{-w}$. We write $L_S(w, \chi) = \tilde{L}_S(u, \chi)$, $\zeta_S(w, \sigma) = \tilde{\zeta}_S(u, \sigma)$, and $\theta(w) = \tilde{\theta}(u)$. From Proposition 15.10, Equation 2, we find

$$\tilde{\zeta}_S(u, \sigma) = \frac{1}{n} \sum_{\chi \in \hat{G}} \overline{\chi(\sigma)} \, \tilde{L}_S(u, \chi) \, . \tag{3}$$

Let $E = \mathbb{Q}(\zeta_n)$. All the characters in $\hat{G}$ have values in E. It follows from Theorem 9.24, and the Artin reciprocity law (Artin L-functions can be identified with Hecke L-functions), that for χ non-trivial $\tilde{L}_S(u, \chi)$ is a polynomial in u with coefficients in E. If $\chi = \chi_o$, the trivial character, then

$$\tilde{L}_S(u, \chi_o) = \prod_{P \in S} (1 - u^{\deg P}) \, Z_k(u) = \prod_{P \in S} (1 - u^{\deg P}) \, \frac{L_k(u)}{(1 - u)(1 - qu)}.$$

Since S is non-empty by assumption, it follows that $(1 - qu)\tilde{L}_S(u, \chi_o) \in \mathbb{Z}[u]$.

It follows from all this and Equation 3, that $(1 - qu)\tilde{\theta}(u) \in E[u][G]$. We claim that it is actually in $\mathbb{Z}[u][G]$.

To see this, note that from the definition of the partial zeta functions we have

$$\zeta_S(w, \sigma) = \sum_{D, \, (D,K/k)=\sigma} \frac{1}{ND^w} = \sum_{D, \, (D,K/k)=\sigma} u^{\deg D} = \tilde{\zeta}_S(u, \sigma) \, .$$

It follows from this and Proposition 15.11 that $(1 - qu)\tilde{\theta}(u) \in \mathbb{Z}[[u]][G]$. Since $E[u] \cap \mathbb{Z}[[u]] = \mathbb{Z}[u]$, we have proved—

Theorem 15.13. Let $\theta(w) = \tilde{\theta}(u) = \tilde{\theta}_{K/k,S}(u)$ be the L-function evaluator. Then, $(1 - qu)\tilde{\theta}(u)$ is an element of $\mathbb{Z}[u][G]$. Evaluating at $u = 1$ we have $(q - 1)\theta = (q - 1)\tilde{\theta}(1) \in \mathbb{Z}[G]$.

The only point which perhaps needs some explanation is the last assertion. Recall that $u = q^{-w}$. It follows that $\theta = \theta(0) = \tilde{\theta}(1)$.

Since, as we have already pointed out, $W_K = q - 1$ under our hypothesis that K/k is a geometric extension, we see that Theorem 15.13 is a strong function field version of Theorem 15.12.

As before, set $\omega = (q - 1)\theta \in \mathbb{Z}[G]$. We can now state the theorem of J. Tate mentioned in the introduction to this chapter.

Theorem 15.14. *Let K/k be a geometric abelian extension of global function fields with Galois group G. Let $\omega = (q - 1)\theta \in \mathbb{Z}[G]$ be the Brumer element defined above. Then for every divisor D of K of degree zero, we have $\omega D = (\alpha_D)$, a principal divisor of K. In other words, ω annihilates the group of divisor classes of degree zero, Cl_K^0.*

This theorem proves a big piece of the Brumer-Stark conjecture in the general case. We will give the proof at the end of the chapter. Our next task is to use this result to prove the full Brumer-Stark conjecture for the cyclotomic function fields $K_m = k(\Lambda_m)$ and $K_m^+ = k(\Lambda_m)^+$ which were defined and investigated in Chapter 12. Note that K_m now denotes the cyclotomic function field generated by adding the m-torsion on the Carlitz module to the rational function field $k = \mathbb{F}(T)$. Here m is a non-constant monic polynomial of degree M in the ring $A = \mathbb{F}[T]$.

The sets S and S^+ corresponding to K_m/k and K_m^+/k will consist precisely of the ramified primes. Thus, $S = \{P \mid P|m\} \cup \{\infty\}$ and $S^+ = \{P \mid P|m\}$. We recall that ∞ is ramified in K_m and splits completely in K_m^+ (see Theorem 12.4). We wish to calculate $\theta = \theta_{K_m/k,S}$ and $\theta^+ = \theta_{K_m^+/k,S^+}$.

Proposition 15.15. *With the above definitions and notations we have*

(a)
$$\theta = \sum_{\substack{a \text{ monic} \\ \deg a < M,\ (a,m)=1}} \sigma_a^{-1} - \frac{1}{q-1} N .$$

(b)
$$\theta^+ = \sum_{\substack{a \text{ monic} \\ \deg a < M,\ (a,m)=1}} (M - \deg a - 1)\sigma_a^{-1} - \frac{1}{q-1} N^+ .$$

In the first equation, $N = \sum_{\sigma \in \text{Gal}(K_m/k)} \sigma$, and in the second, $N^+ = \sum_{\sigma \in \text{Gal}(K_m^+/k)} \sigma$, i.e., the norm maps.

Proof. Recall that $\text{Gal}(K_m/k) = \{\sigma_a \mid (a,m) = 1 \text{ and } \deg a < M\}$. Here σ_a is the unique automorphism with the property that $\sigma_a(\lambda) = C_a(\lambda)$ for all $\lambda \in \Lambda_m$. In fact, this condition defines σ_a for any $a \in A$ with $(a,m) = 1$. We have $\sigma_a = \sigma_b$ if and only if $a \equiv b \pmod{m}$. Moreover, $((a), K_m/k) = \sigma_a$ if and only if a is monic. For all this see Chapter 12.

Since S consists of the primes dividing m and ∞, in the definition of the partial zeta function we sum over effective divisors relatively prime to m with no component at ∞. This is the same as summing over ideals in A

which are prime to m. Every ideal D has a unique monic generator d and $ND = |d| = q^{\deg d}$. Thus, assuming a is monic, we have

$$\zeta_S(w, \sigma_a) = \sum_{\substack{(D,S)=1 \\ (D,K_m/k)=\sigma_a}} \frac{1}{ND^w} = \sum_{\substack{d \text{ monic, } (d,m)=1 \\ \sigma_d = \sigma_a}} |d|^{-w}$$

$$= |a|^{-w} + \sum_{\substack{h \in A \\ h \text{ monic}}} |a + hm|^{-w} = |a|^{-w} + |m|^{-w} \sum_{h \text{ monic}} |h|^{-w}$$

$$= |a|^{-w} + |m|^{-w} \frac{1}{1 - q^{1-w}}.$$

If a is not monic, the calculation is exactly the same except that the term $|a|^{-w}$ does not appear. Thus, $\zeta_S(0, \sigma_a) = 1 - (q - 1)^{-1}$ if a is monic and $\zeta_S(0, \sigma_a) = -(q - 1)^{-1}$ if a is not monic. The expression for θ given in part a of the proposition follows immediately from these results.

Recall that K_m^+ is the fixed field of $\{\sigma_\alpha \mid \alpha \in \mathbb{F}^*\}$. It follows that $\mathrm{Gal}(K_m^+/k) = \{\sigma_a \mid (a, m) = 1 \text{ and } \deg a < M \text{ and } a \text{ monic}\}$. Here we are identifying σ_a with its restriction to K_m^+. As automorphisms of K_m^+ we have $\sigma_d = \sigma_a$ if and only if $d \equiv \alpha a \pmod{m}$ for some $\alpha \in \mathbb{F}^*$.

Since S^+ consists only of primes dividing m, in the definition of the partial zeta we sum over all effective divisors of the form $D = D_f + i\infty$, where D_f is an effective divisor prime to m and ∞ and i is a non-negative integer. As before, D_f corresponds to an ideal of A with a monic generator d which is prime to m.

Since ∞ splits completely in K_m^+ we have $(\infty, K_M^+/k) = e$. Thus, for a monic we have

$$\zeta_{S^+}(w, \sigma_a) = \sum_{\substack{(D,S^+)=1 \\ (D,K_m^+/k)=\sigma_a}} \frac{1}{ND^w} = \sum_{i=0}^{\infty} \sum_{(D_f,K_m^+/k)=\sigma_a} \frac{1}{N(D_f + i\infty)^w}.$$

Now, $N(D_f + i\infty) = ND_f N(\infty)^i = |d|q^i$. Thus, we can rewrite this expression as

$$\zeta_{S^+}(w, \sigma_a) = \sum_{i=0}^{\infty} \sum_{\substack{d \text{ monic} \\ \sigma_d = \sigma_a}} |d|^{-w} q^{-iw} = \frac{1}{1 - q^{-w}} \sum_{\substack{d \text{ monic} \\ \sigma_d = \sigma_a}} |d|^{-w}.$$

Here, d runs over monic polynomials prime to m with $\sigma_d = \sigma_a$. As we have seen, the latter condition holds if and only if $d \equiv \alpha a \pmod{m}$ for some $\alpha \in \mathbb{F}^*$, which is equivalent to the condition $\alpha^{-1}d \equiv a \pmod{m}$. In other words, we can sum over all $d \in A$ (not just the monics) with $d \equiv a \pmod{m}$. Thus,

$$\sum_{\substack{d \text{ monic} \\ \sigma_d = \sigma_a}} |d|^{-w} = \sum_{\substack{d \in A \\ d \equiv a \pmod{m}}} |d|^{-w} = |a|^{-w} + \sum_{\substack{h \in A \\ h \neq 0}} |a + hm|^{-w}.$$

$$= |a|^{-w} + (q-1)|m|^{-w} \sum_{h \text{ monic}} |h|^{-w} = |a|^{-w} + \frac{q-1}{1-q^{1-w}}|m|^{-w} .$$

Putting all this together, we find

$$\zeta_{S^+}(w, \sigma_a) = (1 - q^{-w})^{-1}(|a|^{-w} + (q-1)(1 - q^{1-w})^{-1}|m|^{-w})$$

$$= \frac{(1 - qu)u^{\deg a} + (q-1)u^{\deg m}}{(1-u)(1-qu)} .$$

As usual, we have substituted $u = q^{-w}$ and simplified somewhat. We need the value of this function at $w = 0$ or what is the same, at $u = 1$. If we substitute $u = 1$ into the above expression, we find that both numerator and denominator vanish. Invoking L'Hôpital's rule, we differentiate both numerator and denominator and then substitute $u = 1$. The result is

$$\zeta_{S^+}(0, \sigma_a) = \tilde{\zeta}_{S^+}(1, \sigma_a) = \deg m - \deg a - 1 - \frac{1}{q-1} .$$

From this the proof of part b of the proposition is immediate.

Define

$$\eta = \sum_{\substack{a \text{ monic} \\ \deg a < M, \ (a,m)=1}} \sigma_a^{-1} \quad and \quad \eta^+ = \sum_{\substack{a \text{ monic} \\ \deg a < M, \ (a,m)=1}} (M - \deg a - 1)\sigma_a^{-1} .$$

We can now write $\theta = \eta - (q-1)^{-1}N$ and $\theta^+ = \eta^+ - (q-1)^{-1}N^+$. Also, for the Brumer elements we have $\omega = (q-1)\theta = (q-1)\eta - N$ and $\omega^+ = (q-1)\eta^+ - N^+$. This method of writing things will be of importance to us because of the following result of B. Gross [1].

Proposition 15.16. *The element η annihilates $Cl^o_{K_m}$ and the element η^+ annihilates $Cl^o_{K_m^+}$.*

We will prove this later as a corollary to the proof of Theorem 15.14.

The last ingredient we will need is the prime decomposition of a primitive m-torsion point on the Carlitz module. The miraculous thing that happens is that this decomposition is essentially given by the Brumer element ω^+.

Proposition 15.17. *Let $\mathfrak{P}_\infty$ be a prime of K_m lying over ∞ in k. There exists a primitive m-torsion point $\lambda \in \Lambda_m$ such that*

$$(\lambda) = ((q-1)\eta^+ - \eta)\mathfrak{P}_\infty + \mathfrak{P}_m .$$

The element λ^{q-1} is in K_m^+. As an element of K_m^+ its prime decomposition is given by

$$(\lambda^{q-1}) = \omega^+\mathfrak{P}_\infty^+ + \mathfrak{P}_m^+ .$$

Here, $\mathfrak{P}_m$ is the unique prime of K_m lying above P if $m = P^s$ is a prime power and is the zero divisor otherwise. $\mathfrak{P}_m^+$ is the prime of K_m^+ lying below $\mathfrak{P}_m$. Finally, $\mathfrak{P}_\infty^+$ is the prime of K_m^+ lying below $\mathfrak{P}_\infty$.

Proof. Let k_∞ be the completion of k at ∞ and let $\bar{k}_\infty$ be its algebraic closure. Let ord_∞ denote the normalized additive valuation of k_∞ extended to $\bar{k}_\infty$ in the usual way. Finally, let $\iota : K_m \to \bar{k}_\infty$ be an embedding and let $\mathfrak{P}_\infty$ be the corresponding prime of K_m.

Using the results of Chapter 12, in particular Proposition 12.13 and Theorem 12.14, we see there is a primitive m-torsion point for the Carlitz module, λ, such that $\mathrm{ord}_\infty(\iota\sigma_a\lambda) = M - \deg a - 1 - (q-1)^{-1}$ for any $a \in A$ relatively prime to m and with degree less than M. Since $\mathfrak{P}_\infty$ is ramified over k with ramification index $q-1$, we can write this as

$$\mathrm{ord}_{\sigma_a^{-1}\mathfrak{P}_\infty}(\lambda) = \mathrm{ord}_{\mathfrak{P}_\infty}(\sigma_a\lambda) = (q-1)(M - \deg a - 1) - 1 .$$

The decomposition group of $\mathfrak{P}_\infty$ is $\{\sigma_\alpha \mid \alpha \in \mathbb{F}^*\}$. It follows that the set of distinct primes above ∞ in K_m is $\{\sigma_a^{-1}\mathfrak{P}_\infty \mid a \text{ monic and } \deg a < M\}$. We recall Proposition 12.7, which shows that if $m = P^s$ is a prime power there is exactly one prime ideal above P in $\mathcal{O}_m \subset K_m$ and it is totally ramified and generated by λ. Otherwise, λ is a unit in $\mathcal{O}_m$ by the second part of Proposition 12.6. It follows that the prime decomposition of the divisor (λ) is given by

$$(\lambda) = \sum_{\substack{a \text{ monic} \\ \deg a < M, \ (a,m)=1}} ((q-1)(M - \deg a - 1) - 1)\sigma_a^{-1}\mathfrak{P}_\infty + \mathfrak{P}_m .$$

From this and the definitions of η and η^+ we get the first assertion.

All the primes $\{\sigma_a^{-1}\mathfrak{P}_\infty \mid a \text{ monic and } \deg a < M\}$ are totally and tamely ramified over K_m^+ of ramification index $q-1$. The same is true of $\mathfrak{P}_m$ when it is non-trivial. The second relation follows easily from these remarks, the first relation, and the fact that $\omega^+ = (q-1)\eta^+ - N^+$. It is also helpful to notice that η restricted to K_m^+ is N^+.

We have now assembled everything we need to prove the Brumer-Stark conjecture for K_m/k and K_m^+/k.

Theorem 15.18. *Let* $k = \mathbb{F}(T)$, $K_m = k(\Lambda_m)$, *and* $K_m^+ = k(\Lambda_m)^+$, *the maximal real subfield of* K_m. *The Brumer-Stark conjecture is valid for* K_m/k *and* K_m^+/k.

Proof. Let D be any divisor of K_m. Since $\mathfrak{P}_\infty$ has degree 1 we can write $D = D_0 + t\mathfrak{P}_\infty$ where $t = \deg D$ and D_0 has degree zero. Since the decomposition group of $\mathfrak{P}_\infty$ is $\{\sigma_\alpha \mid \alpha \in \mathbb{F}^*\}$ we see that $N\mathfrak{P}_\infty = (q-1)\eta\mathfrak{P}_\infty$. Thus,

$$\omega\mathfrak{P}_\infty = ((q-1)\eta - N)\mathfrak{P}_\infty = N\mathfrak{P}_\infty - N\mathfrak{P}_\infty = 0 .$$

From this and Theorem 15.14 we see that $\omega D = \omega D_0 = (\alpha_D)$ for some $\alpha_D \in K_m^*$. This proves the first part of the Brumer-Stark conjecture for K_m/k.

To prove the second part we make use of Gross's result, Proposition 15.16. From this we know that ηD_0 is already principal. Set $\eta D_0 = (\beta_D)$.

Notice, also, that $ND_0 = (d)$ where $d \in k^*$. This follows from the fact that Cl_k^0 is trivial. Therefore,

$$\omega D = \omega D_0 = (q-1)\eta D_0 - ND_0 = (\beta_D^{q-1}) - (d) = (\beta_D^{q-1}d^{-1}) .$$

We see that we can choose $\alpha_D = \beta_D^{q-1}d^{-1}$ and so the field generated by $\lambda_D = {}^{q-1}\!\!\sqrt{\alpha_D}$ over K_m is the same as the field generated over K_m by ${}^{q-1}\!\!\sqrt{d}$. Now, $k({}^{q-1}\!\!\sqrt{d})/k$ is a Kummer extension and consequently a cyclic extension of fields. Thus, $K_m(\lambda_D)$ is the composite of two abelian extensions of k, namely, K_m and $k({}^{q-1}\!\!\sqrt{d})$, and so is itself an abelian extension of k. This completes the proof for the case K_m/k.

Now consider the case K_m^+/k. Once again, any divisor D of K_m^+ can be written in the form $D_0 + t\mathfrak{P}_\infty^+$ where $t = \deg D$. By Theorem 15.14, we find $\omega^+ D_0 = (\alpha_{D_0})$ is principal. From Proposition 15.17, we have $\omega^+ \mathfrak{P}_\infty^+ = (\lambda^{q-1}) - \mathfrak{P}_m^+$. Thus,

$$\omega^+ D = (\alpha_{D_0}\lambda^{(q-1)t}) - t\mathfrak{P}_m^+ ,$$

which verifies the first part of the Brumer-Stark conjecture for K_m^+/k.

To prove the second part of the conjecture we use Proposition 15.15 once more to deduce that $\eta^+ D_0 = (\beta_{D_0})$ is principal. It follows that

$$\omega^+ D_0 = ((q-1)\eta^+ - N^+)D_0 = (\beta_{D_0}^{q-1}d^{-1}) ,$$

where $d \in k^*$ is such that $N^+D_0 = (d)$. Thus, we can choose $\alpha_{D_0} = \beta_{D_0}^{q-1}d^{-1}$ and so

$$\omega^+ D = (\beta_{D_0}^{q-1}d^{-1}\lambda^{(q-1)t}) - t\mathfrak{P}_m^+ .$$

We can set $\alpha_D = \beta_{D_0}^{q-1}d^{-1}\lambda^{(q-1)t}$. From this we see that λ_D^+ which is the $q-1$ root of α_D generates the same field over K_m as ${}^{q-1}\!\!\sqrt{d}$. Thus, $K_m^+(\lambda_D^+)$ is contained in $K_m({}^{q-1}\!\!\sqrt{d})$, which is abelian over k as we showed in the first part of the proof. This completes the proof for K_m^+/k.

Remarks.

1. We hope there is no confusion caused by the notation λ_D for the element appearing in the statement of the Brumer-Stark conjecture and the element λ, a primitive m-torsion point of the Carlitz module.

2. For the reader who is familiar with the classical situation there may be some surprise that the Brumer element for K_m^+ is non-trivial. The Brumer element for $\mathbb{Q}(\zeta_m)^+/\mathbb{Q}$ is zero. This is because in this case S^+ contains the archimedean prime of $\mathbb{Q}$ and this splits completely in $\mathbb{Q}(\zeta_m)^+$. It can be shown in general that a prime in S which splits completely in K forces the Brumer element $\omega_{K/k,S}$ to be zero. In the function field case, there are no archimedean primes. S^+ contains only those primes dividing the monic polynomial m, all of which ramify in K_m^+.

3. Using functorial properties of the Stickelberger element and Theorem 15.18 on can show that if k is the rational function field and $K \subseteq K_m$ for some monic $m \in A$, then the Brumer-Stark conjecture holds for K/k.

We now begin to describe the background necessary for the proof of Tate's theorem, Theorem 15.14.

As we have seen, in the function field case, when we describe everything in terms of $u = q^{-w}$, all the functions in question are rational in u with coefficients in $E = \mathbb{Q}(\zeta_n)$, where $n = [K : k]$. It will be necessary for us to work with characters whose values occur in the algebraic closure of $\mathbb{Q}_l$. Here, l is an arbitrarily chosen prime in $\mathbb{Z}$ different from p, the characteristic of $\mathbb{F}$. We write E_l for a finite extension of $\mathbb{Q}_l$ containing the n-th roots of unity. The same analysis given earlier shows that

$$\tilde{\theta}(u) = \sum_{\chi \in \hat{G}} \tilde{L}_S(u, \chi^{-1})\varepsilon(\chi) = \sum_{\sigma \in G} \tilde{\zeta}_S(u, \sigma)\sigma^{-1} \,.$$

Now, of course, $\hat{G} = \mathrm{Hom}(G, E_l^*)$, $\tilde{L}(u, \chi) \in E_l(u)$, and $\varepsilon(\chi) \in E_l[G]$. It is still the case that $(1 - qu)\theta \in \mathbb{Z}[u][G]$. The necessity for these changes will become apparent in a little while.

Let $\bar{\mathbb{F}}$ be the algebraic closure of $\mathbb{F}$, $\bar{k} = k\bar{\mathbb{F}}$, and $\bar{K} = K\bar{\mathbb{F}}$. Since K/k is a geometric extension, we have $K \cap \bar{k} = k$. It follows that the Galois group of $\bar{K}/k$ is the direct product of $\mathrm{Gal}(\bar{K}/\bar{k})$ and $\mathrm{Gal}(\bar{K}/K)$. The first group is naturally isomorphic to G, so we will now think of G as automorphisms of $\bar{K}$ which leave $\bar{k}$ fixed. Let ϕ be the automorphism of $\bar{K}/K$ which induces the automorphism "raising to the q-th power" on $\bar{\mathbb{F}}$. This is called the Frobenius automorphism of the extension. Note that ϕ commutes with the elements of G as automorphisms of $\bar{K}$.

In Chapter 11 we introduced the notation J for the divisor classes of degree zero of $\bar{K}$, i.e., $J = Cl^0_{\bar{K}}$. The corollary to Theorem 11.12 gives the algebraic structure of $J[N]$, the points of order dividing N on J. If $p \nmid N$, then

$$J[N] \cong \bigoplus_{i=1}^{2g} \mathbb{Z}/N\mathbb{Z} \,.$$

where g denotes the genus of K.

Choose and fix a rational prime $l \neq p$ and consider the groups $J[l^n]$. It is clear that for each positive integer n, multiplication by l maps $J[l^{n+1}]$ to $J[l^n]$. We define the Tate module, $T_l(J)$ as the inverse limit of the groups $J[l^n]$ under these maps. It is possible to give a very concrete interpretation of this group. Namely, the elements of $T_l(J)$ can be identified with infinite-tuples, $(a_1, a_2, a_3, \dots)$, where for all $n > 0$ we have $a_n \in J[l^n]$ and $la_{n+1} = a_n$. The Tate module is acted upon by the l-adic integers $\mathbb{Z}_l$ in the obvious way; if $\alpha \in \mathbb{Z}_l$ and $a = (a_1, a_2, a_3, \dots) \in T_l(J)$, then $\alpha a = (\alpha a_1, \alpha a_2, \alpha a_3, \dots)$. Similarly, since G and ϕ act on each $J[l^n]$, these actions can be extended diagonally to an action on $T_l(J)$. Thus, $T_l(J)$ is a

$\mathbb{Z}_l[G]$ module with an action by ϕ which commutes with the action of $\mathbb{Z}_l[G]$. Using the above structure theorem for $J[N]$, one can show that $T_l(J)$ is a free $\mathbb{Z}_l$ module of rank $2g$. We set $V_l = V_l(J) = \mathbb{Q}_l \otimes_{\mathbb{Z}_l} T_l(J)$. V_l is a vector space of dimension $2g$ over $\mathbb{Q}_l$ with a natural action by $\mathbb{Q}_l[G]$ and ϕ and these two actions commute.

We need enough roots of unity in our coefficient field. To this end define $V = E_l \otimes_{\mathbb{Q}_l} V_l$. By Proposition 15.5, we have the following decomposition:

$$V = \sum_{\chi \in \hat{G}} V(\chi) \ .$$

Since the actions of ϕ and G commute, it is easy to see that each E_l-vector space, $V(\chi)$, is mapped into itself by ϕ. We need the following two results.

Proposition 15.19. *If a polynomial in ϕ, $f(\phi) \in \mathbb{Z}[G][\phi]$, vanishes on V_l, then it vanishes on J.*

Theorem 15.20. *The determinant of $1 - \phi u$ acting on V_l is the numerator of the zeta function of the field K, i.e.,*

$$\det(1 - \phi u)|_{V_l} = L_K(u) \ .$$

Suppose $\chi \in \hat{G}$, $\chi \neq \chi_o$. Let $\phi(\chi)$ be the E_l-endomorphism of $V(\chi)$ induced by ϕ. Then

$$\det(I - \phi(\chi)u) = \tilde{L}(u, \chi^{-1}) \ .$$

For the trivial character, χ_o, we have

$$\det(I - \phi(\chi_o)u) = L_k(u) \ ,$$

where $L_k(u)$ is the numerator of the zeta function of k.

Proposition 15.19 is a consequence of a far more general result about geometric endomorphisms of abelian varieties. The point is that any such polynomial $f(\phi)$ can be thought of as an element of $\mathrm{End}_{\mathbb{F}}(J)$, regarding J as an abelian variety over $\mathbb{F}$. It is not a difficult result given the necessary background. Theorem 15.20, on the other hand, is a major theorem. It is due to Weil. The proof can be found in the original book of Weil [2]. A more modern exposition can be found in the article by J. Milne [1]. We will simply accept the result as true and deduce consequences.

We now have everything we need for the proof of Theorem 15.14. We know by Proposition 15.13 that

$$(1 - qu)\tilde{\theta}(u) = \sum_{\chi \in \hat{G}} (1 - qu)\tilde{L}_S(u, \chi^{-1})\varepsilon(\chi) \in \mathbb{Z}[u][G].$$

For each $\chi \neq \chi_0$, $\tilde{L}_S(u, \chi)$ is a polynomial in $E_l[u]$ which is divisible by the Artin L-function $\tilde{L}(u, \chi)$. For $\chi = \chi_o$ we know that $(1 - qu)\tilde{L}_S(u, \chi_o)$

is in $\mathbb{Z}[u]$ and is divisible by $L_k(u)$. From this, Theorem 15.20, and the Cayley-Hamilton theorem, we see that $(1 - q\phi^{-1})\tilde{\theta}(\phi^{-1})$ induces the zero endomorphism on V. Multiplying by a sufficiently large power of ϕ, ϕ^N say, we see that

$$f(\phi) = \phi^N(1 - q\phi^{-1})\tilde{\theta}(\phi^{-1}) \in \mathbb{Z}[\phi][G]$$

is a polynomial in ϕ with coefficients in $\mathbb{Z}[G]$. Since it vanishes on V, it vanishes on $V_l(J)$ and, by Proposition 15.19, it vanishes on J. Since $Cl_K^o = J(\mathbb{F}) \subset J$, we see $f(\phi)$ restricted to $J(\mathbb{F})$ is zero. However, ϕ restricted to $J(\mathbb{F})$ is the identity, so $f(\phi)$ restricted to $J(\mathbb{F})$ is $f(1) = (1 - q)\tilde{\theta}(1) = -\omega$. This shows that ω annihilates $J(\mathbb{F}) = Cl_K^o$, as asserted.

It remains to prove Gross's result, Proposition 15.16. This will follow from the proof, just given, of Tate's theorem. We need explicit expressions for the elements $\tilde{\theta}(u)$ and $\tilde{\theta}^+(u)$ associated to the cyclotomic function field extensions K_m/k and K_m^+/k, respectively. These were implicitly constructed in the course of the proof of Proposition 15.15.

For the extension K_m/k, we found that $\tilde{\zeta}(u, \sigma_a) = u^{\deg a} + (1 - qu)^{-1}u^{\deg m}$ if a is monic and $(1 - qu)^{-1}u^{\deg m}$ if a is not monic. Thus,

$$\tilde{\theta}(u) = \sum_{\substack{a \text{ monic} \\ \deg a < M, \, (a,m)=1}} u^{\deg a}\sigma_a^{-1} + \frac{u^M}{1 - qu}N \, .$$

We note that the norm map N induces the zero mapping on V_l, and thus on V, since $Cl_k^o = (0)$ (because k is the rational function field). As in the proof of Tate's theorem, substitute ϕ^{-1} into $(1 - qu)\tilde{\theta}(u)$ and multiply by ϕ^M to obtain a polynomial in $\mathbb{Z}[G][\phi]$ that annihilates V. Because the norm element annihilates V, we find that

$$(\phi - q) \sum_{\substack{a \text{ monic} \\ \deg a < M, \, (a,m)=1}} \phi^{M-1-\deg a}\sigma_a^{-1} \, ,$$

annihilates V. The endomorphism of V induced by $\phi - q$ is invertible since its determinant is given by

$$\det(\phi - q) = \det -q(1 - q^{-1}\phi) = q^{2g}L_K(q^{-1}) \, .$$

We have used the first part of Theorem 15.20. By the functional equation for the zeta function we find that the last quantity is a power of q times $L_K(1)$, which is the class number of K (see Theorem 5.9). Therefore $\det(\phi - q) \neq 0$. It follows that the element

$$\sum_{\substack{a \text{ monic} \\ \deg a < M, \, (a,m)=1}} \phi^{M-1-\deg a}\sigma_a^{-1} \in \mathbb{Z}[G][\phi]$$

annihilates V. By Proposition 15.19, it annihilates J and so its restriction to $J(\mathbb{F}) = Cl_K^o$ is the zero mapping. Since ϕ restricts to the identity, the restriction of this element is η. We have proven Proposition 15.16 in the case K_m/k.

The proof in the case K_m^+/k is similar, but a bit more complicated. We sketch the proof and leave it to the reader to check the details. Recall that in the proof of Proposition 15.15 we showed

$$\tilde{\zeta}(u, \sigma_a) = \frac{(1 - qu)u^{\deg a} + (q - 1)u^{\deg m}}{(1 - u)(1 - qu)}.$$

Using this, and a little algebraic manipulation, we deduce the following identity:

$$(1 - qu)\tilde{\theta}(u) = (1 - qu) \sum_{\substack{a \text{ monic} \\ \deg a < M, \ (a,m)=1}} \frac{u^{\deg a} - 1}{1 - u}\sigma_a^{-1} + \frac{1 - qu + (q - 1)u^M}{1 - u}N^+.$$

All the rational functions of u occurring as coefficients are actually polynomials since the numerators vanish at $u = 1$. Now, substituting $u = \phi^{-1}$, and following the same steps as in the first part of the proof leads to the conclusion that the following element annihilates $Cl^o_{K_m^+}$

$$\sum_{\substack{a \text{ monic} \\ \deg a < M, \ (a,m)=1}} -\deg a \ \sigma_a^{-1}.$$

However, this element differs from η^+ by an integer multiple of the norm map, N^+; so we find that η^+ annihilates $Cl^o_{K_m^+}$ and the proof is complete.

Having come this far, the reader who is interested in the proof of the Brumer-Stark conjecture in the general case for function fields has two directions to go. Learn the necessary background about 1-motives and read Deligne's proof as presented in Chapter V of Tate's monograph [1]. This proof does not involve Drinfeld modules at all. On the other hand, by learning more about the theory of Drinfeld modules one can build up enough background to read the paper Hayes [5], which gives an elegant proof involving no algebraic geometry beyond the Riemann-Roch theorem for curves. Hayes relies instead on the more advanced theory of Drinfeld modules. The "mixed" proof we have given here for the case of cyclotomic function fields should provide a good head start in either direction.

Exercises

In the following problems, K/k will denote a finite, geometric, abelian extension of global fields, G the Galois group of K/k, and S a finite set of primes of k which includes all those which ramify in K and, in the number field case, all the archimedean primes. We will often shorten the notation for the Stickelberger element $\theta_{K/k,S}$ to θ_S.

1. If $P \notin S$, show that $\theta_{S \cup \{P\}} = (1 - \sigma_P^{-1})\theta_S$ where $\sigma_P = (P, K/k)$.

2. For any prime P of k, let $N_P = \sum_{\sigma \in Z(P)} \sigma$, where the sum is over all the elements of the decomposition group of any prime in K lying over P. Suppose $P \in S$ and that $\#(S) \geq 2$. Show that $N_P \theta_S = 0$. Conclude that $\theta_S = 0$ if any prime in S splits completely in K. Hint: Consider $\chi(N_P \theta_S)$ for all $\chi \in \hat{G}$ and use Proposition 14.12.

3. Let K' be an intermediate extension between k and K. Let $G' = \mathrm{Gal}(K'/k)$ and $\pi : G \to G'$ the natural map given by restriction. Show that $\pi(\theta_{K/k,S}) = \theta_{K'/k,S}$.

4. Let $\varepsilon : \mathbb{Q}[G] \to \mathbb{Q}$ be the augmentation map defined by $\varepsilon(\sum_\sigma r(\sigma)\sigma) = \sum_\sigma r(\sigma)$. If $\#(S) \geq 2$, show that $\varepsilon(\theta_S) = 0$.

5. We showed in the text that $(q - 1)\theta_s \in \mathbb{Z}[G]$. Show that $(\sigma - 1)\theta_S \in \mathbb{Z}[G]$ for every $\sigma \in G, \sigma \neq 1$. Hint: If $\tilde{\theta}_S(u)$ is the L-function evaluator, show first that $(\sigma - 1)\tilde{\theta}_S(u)$ is a polynomial in u.

6. In Deligne-Ribet [1], the authors show that for primes P which are unramified in K and do not divide W_K we have $(\sigma_P - NP)\theta_S \in \mathbb{Z}[G]$, where $\sigma_P = (P, K/k)$. Let D be any divisor (ideal) of K and assume the Brumer-Stark conjecture is true. Assume also that $\#(S) \geq 2$. For $P \notin S$, $P \nmid W_K$, and $P \notin \mathrm{Supp}(D)$, show that $(\sigma_P - NP)\theta_S D = (\alpha_P)$ where $\alpha_P \in K^*$. Hint: By the Brumer-Stark conjecture, $W_K \theta_S D = (\alpha)$, where $\alpha \in K^*$ and $K(\lambda)/k$ is abelian where $\lambda^{W_K} = \alpha$. Let $\sigma'_P = (P, K(\lambda)/k)$. Show that $(\sigma'_P - NP)\theta_S D = ((\sigma'_P - NP)\lambda)$, where both sides are interpreted as divisors in $K(\lambda)$. Let $\alpha_P = (\sigma'_P - NP)\lambda$. Show that it suffice to prove that $\alpha_P \in K^*$ and then prove this using Galois theory and the fact that $\sigma_P - NP$ annihilates μ_K.

The next exercises are based on another conjecture of Stark, which is in turn a very special case of a broad class of conjectures on the value of Artin L-functions at zero. Let T be a finite set of primes in k such that all primes which ramify in K and all archimedean primes (if there are any) lie in T, $\#(T) \geq 2$, and at least one prime $P_o \in T$ splits completely in K. If $\#(T) \geq 3$, define $U^{(o)}$ to be the set of elements in K^* which are units except possibly at the primes lying above P_o. If $T = \{P_o, Q\}$ define $U^{(o)}$ to be all T-units u which satisfy $|u|_Q = |u|_{\sigma Q}$ for all $\sigma \in G$. Finally, let $\mathfrak{P}_o$ be a prime in K lying above P_o.

Conjecture A. If T satisfies the conditions just stated, there is an element $e_o \in U^{(o)}$ such that

$$L'(0,\chi) = -\frac{1}{W_K} \sum_{\sigma \in G} \chi(\sigma) \log |\sigma e_o|_{\mathfrak{P}_o},$$

for all $\chi \in \hat{G}$. Moreover, $K(^{W_K}\sqrt{e_o})/k$ is an abelian extension.

The next exercises indicate how one can deduce the Brumer-Stark conjecture from Conjecture A.

7. Let S be a non-empty, finite set of primes of k containing the primes which ramify in K and, in the number field case, the archimedean primes. Let P_o be a prime of k which splits in K and set $T = S \cup \{P_o\}$. Prove that $L'_T(0, \chi) = \log(NP_o)L_S(0, \chi)$.

8. Let P_o be a prime which splits completely in K and $\mathfrak{P}_o$ a prime of K lying above P_o. Prove that $\mathrm{ord}_{\mathfrak{P}_o} a = -\log |a|_{\mathfrak{P}_o} / \log(NP_o)$ for all $a \in K^*$.

9. Use Conjecture A, stated above, together with the last two exercises to deduce
$$L_S(0, \chi) = \frac{1}{W_K} \sum_{\sigma \in G} \chi(\sigma) \mathrm{ord}_{\mathfrak{P}_o}(\sigma e_o) .$$

10. Use Exercise 8 and Proposition 15.10 to deduce $W_K \zeta_S(0, \sigma) = \mathrm{ord}_{\sigma^{-1} \mathfrak{P}_o}(e_o)$.

11. Assuming $\#(S) \geq 2$ show that Conjecture A implies $W_K \theta_S \mathfrak{P}_o = (e_o)$, which verifies Brumer-Stark for the prime divisor $\mathfrak{P}_0$. Show Conjecture A also implies Brumer-Stark at $\mathfrak{P}_o$ in the remaining case where $\#(S) = 1$.

12. The result of Exercise 10 can be used to prove the full Brumer-Stark conjecture if one assumes Conjecture A. We sketch a proof and invite the reader to fill in the details. We have seen that Brumer-Stark is true for a prime $\mathfrak{P} \notin S$ provided that $\mathfrak{P}$ has relative degree 1 $(f(\mathfrak{P}/P) = 1)$. Choose one such prime $\mathfrak{P}_o$ and let S_o be the set consisting of $\mathfrak{P}_o$ alone. The S_o-class group (of K) is finite and every class in it is represented by infinitely many primes of relative degree one. This follows from considerations of L-functions associated to Cl_{S_o}. Using this, show that if D is any divisor of K we can write $D = \mathfrak{P} + (a) + m\mathfrak{P}_o$, where $\mathfrak{P} \notin S$ is a prime of relative degree one, $a \in K^*$, and m is an integer. The result now follows from the Exercise 11 and the fact, proved in Tate [3], that Brumer-Stark is true for principal divisors.

In the next set of exercises we sketch the proof of the Brumer-Stark conjecture in the case of relatively quadratic extensions of global function fields. Let K/k be a geometric extension of degree 2. Assume that the characteristic of k is not 2. Let the Galois group G of K/k be generated by τ. Let χ be the unique non-trivial character of G. Let S be a finite set of primes of k which include all those primes which ramify in K. Finally, let S' be the set of primes of K lying above those in S. We will assume $|S| \geq 2$ and that no prime in S splits in K (otherwise the Stickelberger element θ_S would be zero).

13. Prove that $\theta_S = 2^{-1} L_S(0, \chi)(1 - \tau)$.

14. Use the relation $\zeta_{K,S'}(w) = \zeta_{k,S}(w) L_S(w, \chi)$ to show (with obvious notation)

$$L_S(0, \chi) = \frac{h_{K,S'} R_{K,S'}}{h_{k,S} R_{k,S}} .$$

15. Let $s = |S|$, $U_{K,S'}$ be the S'-units of K, and $U_{k,S}$ be the S-units of k. Show

$$2^{s-1} R_{k,S} = [U_{K,S'} : U_{k,S}] R_{K,S'} .$$

Hint: Since no prime in S is ramified or split in K show that for every $u \in k^*$, $\log |u|_{\mathfrak{P}} = 2 \log |u|_P$ for $\mathfrak{P}|P \in S$.

16. Show that the kernel of the natural map from $Cl_{k,S} \to Cl_{K,S'}$ is isomorphic to $H^1(G, U_{K,S'}) = \{u \in U_{K,S'} \mid uu^\tau = 1\}/\{u/u^\tau \mid u \in U_{K,S'}\}$. The reader who does not know cohomology of groups may just want to accept this fact "on authority."

17. For all $u \in U_{K,S'}$ show that $u^\tau = \pm u$. Hint: First show $u^\tau/u \in \mathbb{F}^*$ by showing $|u^\tau/u|_{\mathfrak{P}} = 1$ for all primes $\mathfrak{P}$ of K.

18. The map $u \to u^\tau/u$ gives rise to an exact sequence $(1) \to U_{k,S} \to U_{K,S'} \to \langle \pm 1 \rangle$. Use this and the definition of $H^1(G, U_{K,S'})$ to prove that $[U_{K,S'} : U_{k,S}]|H^1(G, U_{K,S'})| = 2$. Hint: Consider individually the following two cases: the case where $u = u^\tau$ for all $u \in U_{K,S'}$ and the case where there is a $u_o \in U_{K,S'}$ such that $u_o^\tau = -u_o$.

19. Let M be the number of elements in the cokernel of the natural map from $Cl_{k,S} \to Cl_{K,S'}$. Use the last few exercises to give the following explicit description of the Brumer element:

$$\omega_S = (q-1)\theta_S = \frac{q-1}{2} 2^{s-2} M(1 - \tau) .$$

20. Use the result of Exercise 19 to verify the Brumer-Stark conjecture for the extension K/k and the set S. For all this in the case of algebraic number fields consult Tate [3].

16

The Class Number Formulas in Quadratic and Cyclotomic Function Fields

In this chapter we will discuss the analogues of some fascinating class number formulas which are well known in the case of quadratic and cyclotomic number fields. Some of these go back to the nineteenth century, e.g., the work of Dirichlet and Kummer. More recent contributions are associated with the names of Carlitz, Iwasawa, and Sinnott. We will review some of these results and then formulate and prove a number of analogues in the function field context.

Let's begin by reviewing the class number formulas for quadratic number fields (for details, see the classical text of E. Hecke [2]). We need the definition of the Kronecker symbol which is a mild generalization of the Jacobi symbol of elementary number theory. Suppose d is an integer congruent to either 0 or 1 modulo 4. If p is an odd prime, define (d/p) to be the usual Jacobi symbol. If $p = 2$, define $(d/2) = (d/-2)$ to be 0 if d is even, 1 if $d \equiv 1$ (mod 8), and -1 if $d \equiv 5$ (mod 8). Now define (d/m) for any non-zero integer m by multiplicativity. This new symbol is called the Kronecker symbol. It is useful in the theory of quadratic number fields, as we will see in a moment.

Let $d \in \mathbb{Z}$ be square-free and consider the field $K = \mathbb{Q}(\sqrt{d})$. The discriminant of $K/\mathbb{Q}$, δ_K, is d if $d \equiv 1$ (mod 4) and $4d$ if $d \equiv 2$ or 3 (mod 4). If χ_d is the non-trivial character of $\mathrm{Gal}(K/\mathbb{Q})$, then it can be shown that the Artin L-function $L(w, \chi_d)$ is given by

$$L(w, \chi_d) = \sum_{n=1}^{\infty} \frac{(\delta_K/n)}{n^w} .$$

By using the relation $\zeta_K(w) = \zeta_{\mathbb{Q}}(w)L(w, \chi_d)$ and comparing the residue of both sides at $w = 1$ we link up the class number of K with the value of $L(w, \chi_d)$ at $w = 1$. Pursuing these ideas leads to the following theorem which is due to Dirichlet.

Theorem 16.1. *The class number h of the quadratic number field $K = \mathbb{Q}(\sqrt{d})$ is given by*

(a)
$$h = -\frac{1}{|\delta_K|} \sum_{m=1}^{|\delta_K|-1} m(\delta_K/m) ,$$

if $\delta_K < -4$, and by

(b)
$$h = \frac{1}{2\log(\epsilon)} \log \frac{\prod_a \sin(\pi a/\delta_K)}{\prod_b \sin(\pi b/\delta_K)} ,$$

if $\delta_K > 1$. Here, $\epsilon > 1$ is the fundamental unit of K, and a varies over all integers between 1 and δ_K with $(\delta_K/a) = -1$, and b varies over all integers between 1 and δ_K with $(\delta_K/b) = 1$.

In the case that d is negative, K is called an imaginary quadratic number field. Part a of the theorem shows that the class number of such a field can always be computed in finitely many steps. It turns out that this is not the most efficient way to compute the class number, but the formula is remarkable nevertheless.

In the case where d is positive, K is said to be a real quadratic number field. The Dirichlet unit theorem tells us, in this case, that the unit group modulo $\langle \pm 1 \rangle$ is infinite cyclic. There is precisely one unit ϵ in K which is greater than 1 and projects on to a generator. This unit is called the fundamental unit in L. In part b of the theorem, let η be the quotient of the product of values of the sine-function which appears on the right hand side of the equation. We have

$$h = \frac{1}{2\log(\epsilon)} \log(\eta) ,$$

from which it follows that $\epsilon^{2h} = \eta$. This shows that η is a unit of K which can be explicitly constructed using special values of the sine-function. It is called a cyclotomic unit. It turns out to be a general phenomenon that for totally real abelian number fields, the class number is related to the index in the whole unit group of an explicitly constructed subgroup of cyclotomic units.

We next consider the cyclotomic fields $K_m = \mathbb{Q}(\zeta_m)$ and their maximal real subfields $K_m^+ = \mathbb{Q}(\zeta_m + \zeta_m^{-1})$. Let h_m denote the class number of K_m and h_m^+ denote the class number of K_m^+. It can be shown that $h_m^+ | h_m$ so that $h_m = h_m^+ h_m^-$, where h_m^- is an integer called the relative class number. We will state results about both h_m^+ and h_m^-.

For a prime to m let $\sigma_a \in \mathrm{Gal}(K_m/\mathbb{Q})$ be the automorphism which takes ζ_m to ζ_m^a. This induces an isomorphism $(\mathbb{Z}/m\mathbb{Z})^* \cong \mathrm{Gal}(K_m/\mathbb{Q})$. Note that σ_{-1} is complex conjugation. Any character of $\mathrm{Gal}(K_m/\mathbb{Q})$ can be thought of, via this isomorphism, as a character on $(\mathbb{Z}/m\mathbb{Z})^*$. We call χ an even character if $\chi(-1) = 1$ and an odd character if $\chi(-1) = -1$. Since -1 corresponds to complex conjugation, we see that the even characters are in one-to-one correspondence with the characters of $\mathrm{Gal}(K_m^+/\mathbb{Q})$.

For the sake of simplicity, we restrict ourselves, in the statement of the next two theorems, to the case where $m = p$, an odd prime.

Theorem 16.2. *Let h_p^- be the relative class number of $\mathbb{Q}(\zeta_p)$. Then,*

$$h_p^- = 2p \prod_{\chi \ \mathrm{odd}} \left(-\frac{1}{2} \sum_{a=1}^{p-1} \chi(a)\frac{a}{p} \right) ,$$

where the product is over all odd characters of $(\mathbb{Z}/p\mathbb{Z})^*$.

This beautiful result is due to Kummer. It shows that the relative class number can be computed in finitely many steps. It also turns out to be useful in deriving divisibility results about the class number. As we shall soon see, it is possible to rework this formula in such a way that the calculation of h_p^- involves nothing but elementary arithmetic in $\mathbb{Z}$.

Recall that in Chapter 12 we showed that the elements $\frac{\zeta_p^a - 1}{\zeta_p - 1}$ are units in the field K_p. Assuming that $K_p \subset \mathbb{C}$ we can choose $\zeta_p = e^{\frac{2\pi i}{p}}$. Then

$$\frac{\zeta_p^a - 1}{\zeta_p - 1} = e^{\frac{\pi i}{p}(a-1)} \frac{\sin(\pi a/p)}{\sin(\pi/p)} .$$

The element $e^{\frac{\pi i}{p}(a-1)}$ is a $2p$-th root of unity and so is a unit in K_p. Thus, the elements

$$\frac{\sin(\pi a/p)}{\sin(\pi/p)} \quad \text{for } a = 2, 3, \ldots, p-1$$

are units in K_p and, in fact, they are units in K_p^+. Note that the units corresponding to a and $p - a$ are the same. Kummer showed that the units corresponding to a in the interval $1 < a < \frac{p}{2}$ are independent. He actually showed much more.

Theorem 16.3. *Let C_p^+ be the subgroup of units in K_p^+ generated by the units*

$$\frac{\sin(\pi a/p)}{\sin(\pi/p)} \quad \textit{for} \quad 1 < a < \frac{p}{2} .$$

and by ± 1. Let E_p^+ be the full unit group of K_p^+. Then, $h_p^+ = [E_p^+ : C_p^+]$.

This result can easily be generalized to apply to K_{p^t}, i.e., to K_m where m is a prime power. For a proof see Lang [6] or Washington [1]. It can also be generalized to arbitrary positive integers m, but this is much more difficult than one might expect. It was accomplished by W. Sinnott in 1978 (see Sinnott [1]), over a hundred years after Kummer proved his result.

We will prove the analogue of Theorem 16.3 in the case of cyclotomic function fields K_m where $m = P$ is a monic irreducible polynomial in $\mathbb{F}[T]$. The result for prime powers was first proven in Galovich-Rosen [1] and generalized to arbitrary monic polynomials m in Galovich-Rosen [2]. In fact, much more general versions hold in the function field case. We will discuss these generalizations briefly after the proof of Theorem 16.12.

A tool which is useful both in number fields and function fields is the Dedekind determinant formula. This result was communicated by Dedekind in a letter to Frobenius in 1896.

Theorem 16.4. *Let G be a finite abelian group, f a function from G to $\mathbb{C}$, and $\hat{G} = \mathrm{Hom}(G, \mathbb{C}^*)$. Then*

(a)
$$\det_{\sigma, \tau} [f(\sigma^{-1}\tau)] = \prod_{\chi \in \hat{G}} \sum_{\sigma \in G} \chi(\sigma) f(\sigma) .$$

and

(b)
$$\det_{\sigma, \tau \neq e} [f(\sigma^{-1}\tau) - f(\sigma^{-1})] = \prod_{\chi \neq \chi_0} \sum_{\sigma \in G} \chi(\sigma) f(\sigma) .$$

Proof. Let V be the vector space of all complex-valued functions on G. A basis for this vector space is given by the functions $\delta_\sigma(x)$ defined by $\delta_\sigma(\sigma) = 1$ and $\delta_\sigma(\tau) = 0$ for $\tau \neq \sigma$. The proof is straightforward. This shows the dimension of V over $\mathbb{C}$ is $n := |G|$.

Another basis of V is given by $\{\chi \mid \chi \in \hat{G}\}$. The characters are linearly independent over $\mathbb{C}$, as can easily be seen from the orthogonality relation. Since $|\hat{G}| = |G|$ by the corollary to Proposition 15.1, it follows that the elements of $\hat{G}$ are a basis for V, as asserted.

Let G act on V by defining σf to be the function which takes x to $f(x\sigma)$. This extends to an action of the group ring $\mathbb{C}[G]$ on V as follows:

$$\left(\sum_{\sigma \in G} a_\sigma \sigma \right) f = \sum_{\sigma \in G} a_\sigma (\sigma f) .$$

Now, fix an element $f \in V$ and associate to it the group ring element $T = \sum_\sigma f(\sigma)\sigma$. The idea is to look at the matrix of T with respect to the two bases of V we have given and then take determinants.

First, note that $\sigma \delta_\tau = \delta_{\tau\sigma^{-1}}$. Thus,

$$T\delta_\tau = \sum_\sigma f(\sigma)\delta_{\tau\sigma^{-1}} = \sum_\sigma f(\sigma^{-1}\tau)\delta_\sigma .$$

It follows that the determinant of T is the determinant of the $n \times n$ matrix $[f(\sigma^{-1}\tau)]$.

Secondly, note that $(\sigma\chi)(x) = \chi(x\sigma) = \chi(x)\chi(\sigma) = \chi(\sigma)\chi(x)$, so that χ is an eigenvector for σ with eigenvalue $\chi(\sigma)$. It follows that χ is an eigenvector for T with eigenvalue $\sum_\sigma \chi(\sigma)f(\sigma)$. The determinant of T on V is just the product of its eigenvalues and this concludes the proof of part a.

To prove part b we restrict T to the subspace $V_o := \{f \in V \mid \sum_{\sigma \in G} f(\sigma) = 0\}$. It is easy to see that this subspace is mapped into itself by every element in G and thus by every element in $\mathbb{C}[G]$. We write down two bases for V_o. The first is $\{\delta_\tau - \delta_e \mid \tau \neq e\}$. The second is $\{\chi \in \hat{G} \mid \chi \neq \chi_o\}$. The proof that the first is a basis is a simple exercise. As for the second, we know that for $\chi \neq \chi_o$ we have $\sum_\sigma \chi(\sigma) = 0$. This shows the non-trivial characters are in V_o. There are $n-1$ of them and they are linearly independent. Since the dimension of V_o is $n-1$, we conclude that the non-trivial characters form a basis.

What is the matrix of T restricted to V_o with respect to the basis $\{\delta_\tau - \delta_e \mid \tau \neq e\}$? From our earlier computation we see

$$T(\delta_\tau - \delta_e) = T\delta_\tau - T\delta_e = \sum_{\sigma \in G}(f(\sigma^{-1}\tau) - f(\sigma^{-1}))\delta_\sigma .$$

For any $f \in V$ we have $0 = \sum_{\sigma \in G}(f(\sigma^{-1}\tau) - f(\sigma^{-1}))$. Multiply both sides of this equation on the right by δ_e and subtract the result from the last sum in displayed equation. We find

$$T(\delta_\tau - \delta_e) = \sum_{\sigma \in G, \sigma \neq e}(f(\sigma^{-1}\tau) - f(\sigma^{-1}))(\delta_\sigma - \delta_e) .$$

This shows the determinant of T restricted to V_o is the determinant of the $(n-1) \times (n-1)$ matrix $[f(\sigma^{-1}\tau) - f(\sigma^{-1})]$. The proof of part b now follows from considering the action of T on the basis $\{\chi \in \hat{G} \mid \chi \neq \chi_o\}$, exactly as in the proof of part a.

We remark that if σ is replaced by σ^{-1} in either determinant considered in the theorem, the effect is simply to permute the rows so that the determinant is multiplied by ± 1. Thus, $\det[f(\sigma^{-1}\tau)] = \pm \det[f(\sigma\tau)]$. We will use this remark shortly. It is a nice exercise to determine this sign change in terms of the structure of the group G.

As an illustration of the use of the Dedekind determinant formula we will state and prove the promised reformulation of Theorem 16.1.

First we recall a definition from elementary number theory. Let $r \in \mathbb{R}$ be any real number. Then there is a unique integer $n \in \mathbb{Z}$ such that $0 \leq r - n < 1$. We set $n = [r]$ and $r - n = \langle r \rangle$. The latter quantity is called the fractional part of r. Note that if $a, m \in \mathbb{Z}$ and $m \neq 0$, then $\langle \frac{a}{m} \rangle$ depends only on the residue class of a modulo m.

It will be convenient to write $G_m = (\mathbb{Z}/m\mathbb{Z})^*$ and $G_m^+ = G_m/\langle \pm 1 \rangle$.

Let χ be an odd character of $(\mathbb{Z}/p\mathbb{Z})^*$ and define the generalized Bernoulli number

$$B_{1,\chi} = \sum_{a=1}^{p-1} \chi(a)\langle\frac{a}{p}\rangle \ .$$

The reader will recognize these are the quantities that appear in the statement of Theorem 16.2. Since both $\chi(a)$ and $\langle\frac{a}{p}\rangle$ only depend on a modulo p we can rewrite this definition as

$$B_{1,\chi} = \sum_{a\in G_p} \chi(a)\langle\frac{a}{p}\rangle \ .$$

In this expression substitute $-a$ for a and use the fact that χ is odd to derive

$$B_{1,\chi} = \sum_{a\in G_p} -\chi(a)\langle\frac{-a}{p}\rangle \ .$$

Now add both expressions for $B_{1,\chi}$ and we find

$$B_{1,\chi} = \frac{1}{2}\sum_{a\in G_p} \chi(a)\left(\langle\frac{a}{p}\rangle - \langle\frac{-a}{p}\rangle\right) \ .$$

The summands are invariant under the substitution $a \to -a$, so we get our final expression for $B_{1,\chi}$,

$$B_{1,\chi} = \sum_{a\in G_p^+} \chi(a)\left(\langle\frac{a}{p}\rangle - \langle\frac{-a}{p}\rangle\right) \ .$$

Theorem 16.5. *Let h_p^- be the relative class number of $\mathbb{Q}(\zeta_p)$, where p is an odd prime number. Then*

$$h_p^- = \pm\frac{2p}{2^{\frac{p-1}{2}}} \det\left[\langle\frac{ab}{p}\rangle - \langle\frac{-ab}{p}\rangle\right] \ ,$$

where a and b are integers in the range $1 \le a, b \le \frac{p-1}{2}$.

Proof. From Theorem 16.2 and the expression we have derived for the generalized Bernoulli number $B_{1,\chi}$, we find

$$h_p^- = \pm\frac{2p}{2^{\frac{p-1}{2}}} \prod_{\chi \text{ odd}} \sum_{a\in G_p^+} \chi(a)\left(\langle\frac{a}{p}\rangle - \langle\frac{-a}{p}\rangle\right) \ .$$

From this, it is clear that all we have to prove is that the product in this expression is, up to sign, the determinant of the theorem.

Let ω be any odd character on G_p. The set of odd characters is easily seen to be the same as $\{\omega\chi' \mid \chi' \text{ even}\}$. Thus, the product in the last expression is the same as

$$\prod_{\chi' \text{ even}} \sum_{a\in G_p^+} \chi'(a)\omega(a)\left(\langle\tfrac{a}{p}\rangle - \langle\tfrac{-a}{p}\rangle\right) = \pm\det\left[\omega(ab)\left(\langle\tfrac{ab}{p}\rangle - \langle\tfrac{-ab}{p}\rangle\right)\right],$$

where we have invoked the Dedekind determinant formula as it applies to the group G_p^+ and the function $f(a) = \omega(a)\left(\langle\tfrac{a}{p}\rangle - \langle\tfrac{-a}{p}\rangle\right)$.

To complete the proof, simply notice that $\{a \in \mathbb{Z} \mid 1 \le a \le \tfrac{p-1}{2}\}$ is a set of representatives for G_p^+. Also, since ω is a character, $\omega(ab) = \omega(a)\omega(b)$. Thus, the terms involving ω in the determinant can be factored out to give the determinant of the theorem multiplied by $\prod_{a=1}^{\frac{p-1}{2}} \omega(a)^2$. This is easily seen to be $(-1)^{\frac{p+1}{2}}$, and that completes the proof.

This elegant result is was proved in Carlitz and Olson [1]. A discussion can also be found in Lang [6], Chapter 3. Later, we will give a function field analogue of Theorem 16.5. For now, we will concentrate on finding an analogue to Kummer's theorem, Theorem 16.3.

Let's begin by recalling some notation and results about cyclotomic function fields. Let $A = \mathbb{F}[T]$, $k = \mathbb{F}(T)$, $\Lambda_m = $ the m-torsion points on the Carlitz module ($m \in A$, a monic polynomial), $K_m = k(\Lambda_m)$, and $\mathcal{O}_m$, the integral closure of A in K_m.

We have an isomorphism $a \to \sigma_a$ from $(A/mA)^* \to G_m = \text{Gal}(K_m/k)$ where σ_a is characterized by $\sigma_a(\lambda) = C_a(\lambda)$ for all $\lambda \in \Lambda_m$.

Let $J = \{\sigma_\alpha \mid \alpha \in \mathbb{F}^*\}$. The fixed field of J is denoted by K_m^+ and by analogy with $\mathbb{Q}(\zeta_m)^+$ is called the maximal real subfield of K_m. We denote by $\mathcal{O}_m^+$ the integral closure of A in K_m^+. The prime ∞ of k splits completely in K_m^+. Each prime $\mathfrak{P}_\infty$ of K_m which lies above ∞ is totally and tamely ramified above K_m^+. The map $a \to \sigma_a$ gives rise to an isomorphism:

$$(A/mA)^*/J \cong G_m^+ = \text{Gal}(K_m^+/k) .$$

Let S_m be the set of primes of K_m lying over $S = \{\infty\}$ and S_m^+ the set of primes of K_m^+ lying over ∞. We have $|S_m^+| = |S_m| = \Phi(m)/(q-1)$ and, by Proposition 14.8,

$$h_{K_m} = h_{\mathcal{O}_m} R_{S_m}^{(q)} \quad \text{and} \quad h_{K_m^+} = h_{\mathcal{O}_m^+} R_{S_m^+}^{(q)} .$$

In these equations, $R_{S_m}^{(q)}$ is the q-regulator of the S_m-units and $R_{S_m^+}^{(q)}$ is the q-regulator of the S_m^+-units.

Our next goal is to give analytic formulas for the class numbers of $\mathcal{O}_m$ and $\mathcal{O}_m^+$. One approach would be to specialize Theorem 14.13 to the two abelian extensions K_m/k and K_m^+/k. However, that theorem involves some advanced material in its statement, for example the notion of the Artin conductor whose very definition depends on the Artin reciprocity law. We prefer to derive the necessary formula in the present circumstance in a much more elementary fashion.

Let's reconsider the Artin L-series $L_A(w, \chi)$, which appeared in Chapter 14,

$$L_A(w, \chi) = \sum_D \frac{\chi(D)}{ND^w} .$$

In this sum, D varies over all effective divisors of k prime to $S = \{\infty\}$. Recall $\chi(D) = \chi((D, K_m/k))$ if D is prime to the conductor of χ and $\chi(D) = 0$ otherwise (see the discussion of these points in Chapter 9). The effective divisors prime to ∞ are in one-to-one correspondence with ideals in A. We make this identification. The definition of the conductor of χ can be made very concrete in the present situation. Let $G_m(\chi) \subseteq G_m$ be the kernel of χ and let $K_m(\chi)$ be the corresponding subfield of K_m. Then the conductor of χ, or rather the part of the conductor which is prime to ∞, is given by the ideal $(m_\chi) \subset A$ where m_χ is the monic divisor of m of least degree such that $K_m(\chi) \subseteq K_{m_\chi} \subseteq K_m$. Since $\mathrm{Gal}(K_m/K_{m_\chi}) \subseteq G_m(\chi)$ we see that χ can be viewed as a character on $G_{m_\chi} \cong (A/m_\chi A)^*$.

Each ideal D in A has a unique monic generator, say, a. It follows immediately from Proposition 12.10 that if D is prime to (m_χ), then $(D, K_{m_\chi}/k) = \sigma_a$ as elements of G_{m_χ}. Thus, if we define, as we have been doing, $\chi(\sigma_a) = \chi(a)$, we can rewrite the L-series as follows:

$$L_A(w, \chi) = \sum_{\substack{a \in A \\ a \text{ monic}}} \frac{\chi(a)}{|a|^w} .$$

We have used $ND = N(aA) = \#(A/aA) = q^{\deg a} = |a|$. For emphasis, in this equation χ is being considered as a character on $(A/m_\chi A)^*$. From now on we make this convention: whenever a Dirichlet character modulo m occurs in an L-series, $L_A(w, \chi)$, we regard χ as a character modulo m_χ. With this convention, the Artin L-series and the Dirichlet L-series coincide.

Our Artin L-series (associated to K_m/k with $S = \{\infty\}$) have been revealed to be nothing more than the Dirichlet L-series that we treated in some detail in Chapter 4. Proposition 4.3 shows that when $\chi \neq \chi_o$, $L(w, \chi)$ is a polynomial in q^{-w} of degree at most $M_\chi - 1$, where $M_\chi = \deg m_\chi$. Setting $u = q^{-w}$ and $L_A(w, \chi) = \tilde{L}_A(u, \chi)$ we recast the content of Proposition 4.3 as follows:

$$\tilde{L}_A(u, \chi) = \sum_{\substack{a \text{ monic} \\ \deg a < M_\chi}} \chi(a) u^{\deg a} . \tag{1}$$

At this point we need to make a distinction between characters. For any monic polynomial m, we call a character of $(A/mA)^*$ even if $\chi(\alpha) = 1$ for all $\alpha \in \mathbb{F}^*$. Otherwise, χ is said to be an odd character. If we think of χ as a Galois character on G_m, then it is even if and only if it is trivial on $J = \text{Gal}(K_m/K_m^+)$. For this reason, one sometimes calls an even character a real character and an odd character an imaginary character. This is all done by analogy with the number field case.

Recall that the value $w = 0$ corresponds to the value $u = 1$. We will need the following result.

Lemma 16.6. *If $\chi \neq \chi_0$ is an even character, we have $\tilde{L}_A(1, \chi) = 0$.*

Proof. If χ is even, $\alpha \in \mathbb{F}^*$, and $a \in A$ is monic, then $\chi(\alpha a) = \chi(a)$. Thus,

$$\tilde{L}_A(1, \chi) = \sum_{\substack{a \text{ monic} \\ \deg a < M_\chi}} \chi(a) = (q - 1)^{-1} \sum_{\substack{a \neq 0 \\ \deg a < M_\chi}} \chi(a) = 0 .$$

The last equality is a consequence of the following facts: χ is not trivial, $\chi(a) = 0$ if $(a, m_\chi) \neq 1$, and the set $\{a \in A \mid (a, m_\chi) = 1, \ \deg a < M_\chi\}$ is a set of representatives for the group $(A/m_\chi A)^*$.

Proposition 16.7. *We have*

(a)
$$\zeta_{\mathcal{O}_m}(w) = \zeta_A(w) \prod_{\chi \neq \chi_0} L_A(w, \chi) .$$

(b)
$$\zeta_{\mathcal{O}_m^+}(w) = \prod_{\substack{\chi \neq \chi_0 \\ \chi \text{ even}}} L_A(w, \chi) .$$

The first product is over all non-trivial Dirichlet characters modulo m and the second is over all non-trivial even Dirichlet characters modulo m.

Proof. Both formulas are special cases of Proposition 14.11. To justify the second formula, note that even characters are the characters of $(A/m)^*/\mathbb{F}^* \cong G_m/J \cong G_m^+ = \text{Gal}(K_m^+/k)$.

It is possible to give a proof which avoids Artin L-series and just uses properties of Dirichlet L-series. One combines Lemma 4.4 with Proposition 12.10 and the definition of $\zeta_{\mathcal{O}_m}(w)$ to get the first equality. The second equality can also be done in a similar manner. This method is especially easy to carry out when m is irreducible. In the general case, there are technical difficulties introduced by having to consider conductors.

Theorem 16.8. *We have*

(a)
$$h_{K_m} = \prod_{\chi \text{ odd}} \left(\sum_{\substack{a \text{ monic} \\ \deg a < M_\chi}} \chi(a) \right) \prod_{\substack{\chi \text{ even} \\ \chi \neq \chi_0}} \left(\sum_{\substack{a \text{ monic} \\ \deg a < M_\chi}} -\deg a \, \chi(a) \right) .$$

(b)
$$h_{K_m^+} = \prod_{\substack{\chi \text{ even} \\ \chi \neq \chi_0}} \left(\sum_{\substack{a \text{ monic} \\ \deg a < M_\chi}} -\deg a \, \chi(a) \right) .$$

Proof. Recall that

$$\zeta_{\mathcal{O}_m}(w) = \prod_{\mathfrak{P} \in S_m} (1 - N\mathfrak{P}^{-w})\, \zeta_{K_m}(w)$$

and

$$\zeta_{K_m}(w) = \frac{L_{K_m}(q^{-w})}{(1 - q^{-w})(1 - q^{1-w})}\,,$$

where $L_{K_m}(u)$ is a polynomial whose value at $u = 1$ is h_{K_m}.

Since every prime in S_m has degree 1, we have $1 - N\mathfrak{P}^{-w} = 1 - q^{-w}$ for all $\mathfrak{P} \in S_m$. So, combining the last two equations and switching to the "u" language, we find

$$\zeta_{\mathcal{O}_m}(w) = (1 - u)^{\frac{\Phi(m)}{q-1} - 1}\, \frac{L_{K_m}(u)}{1 - qu}\,.$$

By Proposition 16.7, part a, we find

$$\zeta_{\mathcal{O}_m}(w) = \frac{1}{1 - qu} \prod_{\chi \neq \chi_0} \tilde{L}_A(u, \chi)\,.$$

Now, combining these last two equations and rewriting slightly, we find

$$L_{K_m}(u) = \prod_{\chi \text{ odd}} \tilde{L}_A(u, \chi) \prod_{\substack{\chi \text{ even} \\ \chi \neq \chi_0}} \frac{\tilde{L}_A(u, \chi)}{1 - u}\,. \tag{2}$$

We have used the fact that the number of non-trivial even characters is $\frac{\Phi(m)}{q-1} - 1$. This is because the set of even characters are in one-to-one correspondence with the characters of $(A/mA)^* / \mathbb{F}^*$, a group with $\frac{\Phi(m)}{q-1}$ elements.

We would like to just substitute $u = 1$ into Equation 2, but we must first deal with the expressions $\tilde{L}_A(u, \chi)/(1 - u)$ when χ is even and non-trivial. By Lemma 16.6, the numerator of these expressions are zero at $u = 1$. We can apply L'Hôpital's rule and Equation 1, which gives an explicit formula for the polynomial $\tilde{L}_A(u, \chi)$ to derive

$$\lim_{u \to 1} \frac{\tilde{L}_A(u, \chi)}{1 - u} = -\sum_{\substack{a \text{ monic} \\ \deg a < M_\chi}} \deg a\, \chi(a)\,,$$

whenever χ is even and non-trivial.

The proof of part a of the theorem now follows immediately by taking the limit as $u \to 1$ in Equation 2, and using Equation 1 once again.

The proof of part (b) follows along exactly the same lines using the fact that the even characters are in one to one correspondence with the characters of $(A/mA)^* / \mathbb{F}^* \cong G_m^+$, and the fact that $|S_m^+| = |S_m| = \frac{\Phi(m)}{q-1}$.

Before we can state and prove the main theorem of this chapter, we need three more preliminary results.

Lemma 16.9. *Let m be a monic polynomial and suppose that λ is a generator of Λ_m. If b is a polynomial prime to m, then $\sigma_b \lambda / \lambda \in K_m^+$.*

Proof. By definition of the automorphism σ_b we have $\sigma_b \lambda = C_b(\lambda)$. In particular, when $\alpha \in \mathbb{F}^*$, we have $\sigma_\alpha \lambda = C_\alpha(\lambda) = \alpha \lambda$. Thus,

$$\sigma_\alpha(\sigma_b \lambda / \lambda) = \sigma_b \sigma_\alpha \lambda / \sigma_\alpha \lambda = \sigma_b(\alpha \lambda)/\alpha \lambda = \sigma_b \lambda / \lambda \ .$$

It follows that $\sigma_b \lambda / \lambda$ is fixed for all elements in $\{\sigma_\alpha \mid \alpha \in \mathbb{F}^*\} = J$, the Galois group of K_m/K_m^+. The result follows.

It follows readily from the results of Chapter 12 that the elements $\sigma_b \lambda / \lambda$ are units. If m is not a prime power, then λ itself is a unit. If $m = P^e$ is a prime power, then λ generates the unique prime ideal in $\mathcal{O}_m$ lying above P. It follows easily that $\sigma_b \lambda$ is another such generator and therefore $\sigma_b \lambda / \lambda$ is a unit. Units of this type will play a key role in what follows.

Now that we know $\sigma_b \lambda / \lambda$ is a unit in $\mathcal{O}_m^+$ our next task is to determine its divisor as an element of K_m^+. In Chapter 15, using results developed in Chapter 12, we showed that there is a primitive m-torsion point $\lambda_m \in \Lambda$ and a prime $\mathfrak{P}_\infty$ of K_m lying over ∞ such that for all monics $a \in A$ with $(a, m) = 1$ and $\deg a < M$ we have

$$\operatorname{ord}_{\sigma_a^{-1} \mathfrak{P}_\infty}(\lambda_m) = (q-1)(M - \deg a - 1) - 1 \ . \tag{3}$$

See Proposition 15.17 and its proof.

We need to allow a to vary somewhat more freely. For $a \in A$ with $(a, m) = 1$, define $\langle a \rangle$ to be the unique polynomial c with $0 \leq \deg c < M$ and $a \equiv c \pmod{m}$. Define $f_m(a) = (q-1)(M - \deg\langle a \rangle - 1) - 1$. We can then rewrite Equation 3 as follows:

$$\operatorname{ord}_{\sigma_a^{-1} \mathfrak{P}_\infty}(\lambda_m) = f_m(a) \ . \tag{3'}$$

The advantage is that Equation 3' is valid for any a prime to m.

The $\Phi(m)/(q-1)$ primes in $S_m = \{\sigma_a^{-1} \mathfrak{P}_\infty \mid a \text{ monic}, \deg a < M, (a, m) = 1\}$ are all the primes in K_m lying over ∞. Let $\mathfrak{P}_\infty^+$ be the prime of K_m^+ lying below $\mathfrak{P}_\infty$. Then $S_m^+ = \{\sigma_a^{-1} \mathfrak{P}_\infty^+ \mid a \text{ monic}, \deg a < M, (a, m) = 1\}$ are all the primes of K_m^+ lying above ∞. Since $\sigma_b \lambda_m / \lambda_m$ is a unit in $\mathcal{O}_m^+$, the next proposition completely determines its divisor.

Proposition 16.10. *For $a, b \in A$ monic and prime to m we have*

$$\operatorname{ord}_{\sigma_a^{-1} \mathfrak{P}_\infty^+}(\sigma_b \lambda_m / \lambda_m) = \frac{f_m(ab) - f_m(a)}{q - 1} \ .$$

Proof. Using Equation 3' above, we find

$$\begin{aligned}
\operatorname{ord}_{\sigma_a^{-1} \mathfrak{P}_\infty}(\sigma_b \lambda_m) &= \operatorname{ord}_{\sigma_b^{-1} \sigma_a^{-1} \mathfrak{P}_\infty}(\lambda_m) \\
&= (q-1)(M - \deg\langle ab \rangle - 1) - 1 = f_m(ab) \ .
\end{aligned}$$

Combining Equations 3′ and 4, we find

$$\text{ord}_{\sigma_a^{-1}\mathfrak{P}_\infty}(\sigma_b\lambda_m/\lambda_m) = \text{ord}_{\sigma_a^{-1}\mathfrak{P}_\infty}(\sigma_b\lambda) - \text{ord}_{\sigma_a^{-1}\mathfrak{P}_\infty}(\lambda_m) = f_m(ab) - f_m(a)\,.$$

Finally, since $\sigma_a^{-1}\mathfrak{P}_\infty$ is totally and tamely ramified over $\sigma_a^{-1}\mathfrak{P}_\infty^+$ and $\sigma_b\lambda_m/\lambda_m \in K_m^+$ by Lemma 16.9 we have

$$\text{ord}_{\sigma_a^{-1}\mathfrak{P}_\infty^+}(\sigma_b\lambda_m/\lambda_m) = \frac{1}{q-1}\text{ord}_{\sigma_a^{-1}\mathfrak{P}_\infty}(\sigma_b\lambda_m/\lambda_m) = \frac{f_m(ab) - f_m(a)}{q-1}\,.$$

The units $\sigma_b\lambda/\lambda$ are similar to the cyclotomic units $\frac{\zeta_m^a - 1}{\zeta_m - 1}$ in the number field $\mathbb{Q}(\zeta_m)$. We now give the general definition of cyclotomic units in the function field case.

Definition. Let V_m be the subgroup of K_m^* generated by the non-zero elements of Λ_m and $\mathcal{E}_m = V_m \cap \mathcal{O}_m^*$. The group $\mathcal{E}_m$ is called the group of cyclotomic units in K_m.

Note that constants are cyclotomic units since if $\alpha \in \mathbb{F}^*$ we have $\alpha = \sigma_\alpha\lambda_m/\lambda_m$.

Lemma 16.11. *If $m = P^t$ is a power of an irreducible P, then the group of cyclotomic units, $\mathcal{E}_m$, is generated by $\mathbb{F}^*$ and the set*

$$T_m = \{\sigma_b\lambda_m/\lambda_m \mid b \text{ monic}, \ 0 < \deg b < \deg m, \ (b, m) = 1\}\,.$$

Proof. We will give the proof when $m = P$ is irreducible and leave the case $m = P^t, t > 1$, as an exercise.

Every non-zero element of Λ_P has the form $\sigma_a\lambda_P$, where a varies over the non-zero polynomials of degree less that $\deg P$. If u is a cyclotomic unit, then

$$u = \prod_a (\sigma_a\lambda_P)^{n_a}\,,$$

where the exponents n_a are in $\mathbb{Z}$. Rewrite this equation as

$$u = \prod_a (\sigma_a\lambda_P/\lambda_P)^{n_a} \times \lambda_P^{\Sigma n_a}\,.$$

Consider the fractional $\mathcal{O}_P$-ideal generated by both sides. We find, $\mathcal{O}_P = (\lambda_P)^{\Sigma n_a}$. Since (λ_P) is a prime ideal, this implies $\sum n_a = 0$. It remains to show that we can restrict our attention to a monic.

If a is not monic, $a = \alpha b$ with $\alpha \in \mathbb{F}^*$ and b monic. Then

$$\sigma_a\lambda_P/\lambda_P = \sigma_b\sigma_\alpha\lambda_P/\lambda_P = \alpha \ \sigma_b\lambda_P/\lambda_P\,.$$

The lemma now follows immediately.

Corollary. *If $m = P^t$, every cyclotomic unit is in $\mathcal{O}_m^{+\,*}$.*

Proof. It follows from Lemma 16.9 that the set T_m is contained in K_m^+. Since the constants, $\mathbb{F}^*$, are also in K_m^+, the result follows.

We remark that both the lemma and the corollary are false if m is not a prime power, since then λ_m is a cyclotomic unit and it is not in K_m^+.

Note that the set T_m has the same cardinality as the rank of $\mathcal{O}_m^*$. One might be tempted to think that it generates a subgroup of finite index. This is not always the case, but it is true if $m = P^t$ is a prime power. We will now show this and more when $m = P$ is itself a prime. The following theorem is the analogue of Kummer's theorem, Theorem 16.3.

Theorem 16.12. *Let $m = P$ be a prime. Then, the group of cyclotomic units, $\mathcal{E}_P$, is of finite index in $\mathcal{O}_P^{+*}$ and*

$$h_{\mathcal{O}_P^+} = [\mathcal{O}_P^{+*} : \mathcal{E}_P] .$$

Proof. We begin by specializing and reworking the analytic class number formula, Theorem 16.8, part b, in the case where $m = P$, a prime in $A = \mathbb{F}[T]$.

Since P has no monic divisors except 1 and itself, we see that any non-trivial character χ on $(A/PA)^*$ has P for its conductor. Let $d = \deg P$. Then

$$h_{K_P^+} = \prod_{\substack{\chi \text{ even} \\ \chi \neq \chi_o}} \left(\sum_{\substack{a \text{ monic} \\ \deg a < d}} -\deg a\, \chi(a) \right) .$$

Recall the definition, $f_P(a) = (q-1)(d - \deg\langle a \rangle - 1) - 1$. By Lemma 16.6, we see that for χ even and non-trivial, $\sum_{a \text{ monic, } \deg a < d} \chi(a) = 0$. It follows that

$$\sum_{\substack{a \text{ monic} \\ \deg a < d}} \chi(a) f_P(a) = (q-1) \sum_{\substack{a \text{ monic} \\ \deg a < d}} -\deg a\, \chi(a) .$$

Thus, our formula for $h_{K_m^+}$ can be rewritten

$$h_{K_P^+} = (q-1)^{1 - \frac{\Phi(P)}{q-1}} \prod_{\substack{\chi \text{ even} \\ \chi \neq \chi_o}} \left(\sum_{\substack{a \text{ monic} \\ \deg a < d}} \chi(a)\, f_P(a) \right) .$$

We want to apply the Dedekind determinant formula to rewrite the right-hand side of this equation. The group we are considering is $G_P^+ = \mathrm{Gal}(K_P^+/k)$. If $\sigma \in G_P^+$, then σ coincides with σ_a for some a representing an element of $(A/PA)^*/\mathbb{F}^*$. Define $f_P(\sigma) = f_P(a)$. It is easy to see that this definition is independent of the choice of a. Now, invoking Theorem 16.4, part (b), we find

$$h_{K_P^+} = \pm(q-1)^{1 - \frac{\phi(P)}{q-1}} \det[f_P(ab) - f_P(a)] = \pm\det\left[\frac{f_P(ab) - f_P(a)}{q-1} \right] . \quad (5)$$

In this matrix, a and b vary over all monic polynomials which are distinct from 1 and have degree less than $d = \deg P$.

By Proposition 16.10, the determinant on the right-hand side is the same as

$$\det\left[\operatorname{ord}_{\sigma_a^{-1}\mathfrak{P}_\infty^+}(\sigma_b\lambda_P/\lambda_P)\right].$$

Call the absolute value of this determinant $R_{\mathcal{E}_P}^{(q)}$. Using this remark, Equation 5, and the fact that $h_{K_P^+} = h_{\mathcal{O}_P^+}R_{S_P^+}^{(q)}$, (see Proposition 14.8), we derive

$$h_{\mathcal{O}_P^+} = \frac{R_{\mathcal{E}_P}^{(q)}}{R_{S_P^+}^{(q)}}. \tag{6}$$

It remains to show that the right-hand side of this equation is equal to the index of $\mathcal{E}_P$ in $\mathcal{O}_P^{+*}$. To do this we use the ideas that go into the proof of Proposition 14.3. Matters are even simpler in the present case, since all the primes above ∞ have degree 1.

Before proceeding, let's simplify the notation. Let U be a subgroup of $\mathcal{O}_P^{+*}$ and $S = S_P^+$. Let $s = |S| = \frac{\Phi(P)}{q-1}$. For each monic a with $\deg a < d$, let $\mathfrak{P}_a = \sigma_a^{-1}\mathfrak{P}_\infty^+$. Arrange the monics of degree less than d in some order with 1 being the first and label the coordinates of $\mathbb{Z}^s$ with these monics.

Define a map, l_q, from K_m^* to $\mathbb{Z}^s$, which takes an element x to the s-tuple whose a-th coordinate is $\operatorname{ord}_{\mathfrak{P}_a}(x)$. Let H° be the subgroup of elements in $\mathbb{Z}^s$ whose coordinates sum to zero. H° is a free group of rank $s - 1$. Using the fact that the primes in S have degree 1, it follows that if $u \in U$, we have $l_q(u) \in H^\circ$.

Let $T = \{u_1, u_2, \ldots, u_{s-1}\}$ be a set of elements in U. Consider the $(s - 1) \times (s - 1)$ matrix:

$$\mathcal{T} = [\operatorname{ord}_{\mathfrak{P}_a}(u_i)],$$

where $1 \leq i \leq s - 1$ and $a \neq 1$ varies over monics of degree less than d. We claim that the s-tuples $\{l_q(u_1), l_q(u_2), \ldots, l_q(u_{s-1})\}$ are linearly independent over $\mathbb{Z}$ if and only if $\det \mathcal{T} \neq 0$. Moreover, if $\det \mathcal{T} \neq 0$, then the group generated by $\{l_q(u_1), l_q(u_2), \ldots, l_q(u_{s-1})\}$ has index $|\det \mathcal{T}|$ in H°. To see this, let $e_1 \in \mathbb{Z}^s$ be the vector whose first coordinate is 1 and all of whose other coordinates are 0. Then, $\mathbb{Z}^s = \mathbb{Z}e_1 \oplus H^\circ$ and we are reduced to considering the $s \times s$ matrix whose first row is e_1 and whose i-th row, for $2 \leq i \leq s$, is $l_q(u_{i-1})$. The determinant of this matrix is the same as the determinant of $\mathcal{T}$ (consider the cofactor expansion along the first row), which proves the assertion.

Applying these general considerations to a set of fundamental units and to the generating set of cyclotomic units, we find that $R_{S_P^+}^{(q)} = [H^\circ : l_q(\mathcal{O}_P^{+*})]$ and $R_{\mathcal{E}_P}^{(q)} = [H^\circ : l_q(\mathcal{E}_P)]$. From this and Equation 6 we find

$$h_{\mathcal{O}_P^+} = [l_q(\mathcal{O}_P^{+*}) : l_q(\mathcal{E}_P)].$$

If U is any subgroup of the group of units containing $\mathbb{F}^*$ we have an exact sequence $(1) \to \mathbb{F}^* \to U \to l_q(U) \to (1)$. It follows that

$$h_{\mathcal{O}_P^+} = [\mathcal{O}_P^{+*} : \mathcal{E}_P] \, .$$

Remarks.

1. The main reason the case $m = P$, a prime, is so much easier than the general case is that all non-trivial Dirichlet characters have P as conductor. This makes it easy to use the Dedekind determinant formula in conjunction with the analytic class number formula. In other cases, the conductor depends on the character and one must take care. In the case $m = P^t$, a prime power, this is relatively easy to do (see the exercises). If m is not a prime power, keeping track of the conductor is quite difficult and requires rather sophisticated technique.

2. Theorem 16.12 first appeared in Galovich-Rosen [1]. Soon thereafter it was generalized to K_m^+ for arbitrary m (see Galovich-Rosen [2]) by following the methods of Sinnott [1]. These results were generalized in stages to the case of arbitrary global function field k as base and with ray class fields taking the place of K_m. See the work of Hayes [4], Shu [1], and Oukaba [1]. The most general case was handled by L. Yin [1].

We want to use Theorem 16.12 to help provide an analogue to Dirichlet's theorem, Theorem 16.1, part b. For simplicity we will assume $m = P$, a monic polynomial of even degree d.

It will be useful to define $\mathcal{M}$ to be the set of monic polynomials of degree less than d.

Lemma 16.13. *Assume $q = |\mathbb{F}|$ is odd and that P is a monic irreducible of even degree d. Then $k(\sqrt{P}) \subseteq K_P^+$.*

Proof. Recall the factorization of the Carlitz polynomial

$$\frac{C_P(u)}{u} = \prod_{\substack{\lambda \in \Lambda_P \\ \lambda \neq 0}} (u - \lambda) \, .$$

Comparing constant terms on both sides shows $(-1)^{q^d - 1} \prod \lambda = \prod \lambda = P$, where the product is over all elements in $\Lambda_P - \{0\}$.

The set of non-zero elements of Λ_P coincides with $\{\sigma_a \lambda_P \mid a \neq 0, \deg a < d\}$. Since every non-zero polynomial can be written uniquely as the product of a constant times a monic, we derive the following equation:

$$\left(\prod_{\alpha \in \mathbb{F}^*} \alpha \right)^{\frac{q^d - 1}{q - 1}} \prod_{a \in \mathcal{M}} (\sigma_a \lambda_P)^{q-1} = P \, .$$

The product of all the non-zero elements in a finite field is -1. Since q is assumed odd and d is even, $\frac{q^d - 1}{q - 1}$ is even. It follows that P is a $q - 1$-power in K_P, and so, a posteriori, a square. This shows that $k(\sqrt{P}) \subseteq K_P$.

To show $k(\sqrt{P}) \subseteq K_P^+$, note that G_P is cyclic. This shows there is a unique quadratic extension of k inside K_P. We have just seen that $\frac{q^d-1}{q-1} = |G_P^+|$ is even. It follows that k has a quadratic extension inside K_P^+. By the uniqueness, it must be $k(\sqrt{P})$.

We now want to define a cyclotomic unit in $k(\sqrt{P})$. Recall that $\sigma_a\lambda/\lambda \in K_P^+$ for all a not divisible by P. The Galois group G_P^+ is isomorphic to $(A/PA)^*/\mathbb{F}^*$. For the rest of the present discussion we are going to let a vary over $(A/P)^*/\mathbb{F}^*$. Define

$$\eta = \prod_{a\in(A/PA)^*/\mathbb{F}^*} \left(\frac{\sigma_a\lambda_P}{\text{sgn}(a)\lambda_P}\right)^{-(a/P)} , \tag{7}$$

where (a/P) is the quadratic character on $(A/PA)^*$. The factor sgn(a), the leading coefficient of a, is included so that the quotient $\sigma_a\lambda/\text{sgn}(a)\lambda$ is independent of the class of a in $(A/PA)^*/\mathbb{F}^*$. We also need to know that (a/P) is an even character, i.e., that it is equal to 1 on $\mathbb{F}^*$. This is true, since for $\alpha \in \mathbb{F}^*$,

$$(\alpha/P) = \alpha^{\frac{|P|-1}{2}} = \alpha^{\frac{q^d-1}{q-1}\frac{q-1}{2}} = 1 ,$$

because, under our assumptions, $\frac{q^d-1}{q-1}$ is an even integer.

Lemma 16.14. *The unit η is an element of $k(\sqrt{P})$.*

Proof. We have seen that $k(\sqrt{P}) \subseteq K_P^+$. An element $\sigma_b \in G_P^+$ is in $\text{Gal}(K_P^+/k(\sqrt{P}))$ if and only if b is a square in $(A/PA)^*/\mathbb{F}^*$. Notice that

$$\sigma_b\eta = \left(\prod_{a\in(A/PA)^*/\mathbb{F}^*} \left(\frac{\sigma_{ba}\lambda_P}{\text{sgn}(ab)\lambda_P}\right)^{-(a/P)}\right) \times \left(\frac{\sigma_b\lambda_P}{\text{sgn}(b)\lambda_P}\right)^{\sum_a(a/P)} .$$

The second factor is equal to 1, since $\sum_a(a/P) = 0$. If b is a square in $(A/PA)^*/\mathbb{F}^*$, then $(b/P) = 1$ so $(a/P) = (ba/P)$. Under these conditions the above equation shows $\sigma_b\eta = \eta$, which proves the lemma.

Theorem 16.15. *Let q be odd and P be a monic irreducible polynomial of even degree d. Let $K = k(\sqrt{P})$ and $\mathcal{O}_K$ the ring of integers of K (i.e., the integral closure of A in K). Let $\mathfrak{p}_\infty$ be a prime of K lying over ∞ and ϵ a fundamental unit of $\mathcal{O}_K^*$ such that $\text{ord}_{\mathfrak{p}_\infty}(\epsilon) < 0$. Finally, let η be the cyclotomic unit defined in Equation 7 (after an appropriate choice of λ_P). Then*

$$h_{\mathcal{O}_K} = \frac{\text{ord}_{\mathfrak{p}_\infty}(\eta)}{\text{ord}_{\mathfrak{p}_\infty}(\epsilon)} .$$

Moreover, $\epsilon^{h_{\mathcal{O}_K}} = \alpha\eta$ for some $\alpha \in \mathbb{F}^*$.

Proof. By Propositions 14.6 and 14.7, we know that ∞ splits in K and that

$$h_K = h_{\mathcal{O}_K} \log_q |\epsilon|_{\mathfrak{p}_\infty} = -h_{\mathcal{O}_K} \mathrm{ord}_{\mathfrak{p}_\infty}(\epsilon) .$$

The last equality uses the fact that $\mathfrak{p}_\infty$ has degree 1 .

The Artin character χ of $G_P \cong (A/PA)^*$ corresponding to K/k is even and of order 2. The only prime of K which is ramified over k is the prime above P, from which it is easy to see that $\chi(\sigma_a) = \chi(a) = (P/a)$ for all a not divisible by P. By the law of quadratic reciprocity (see Theorem 3.5), we find

$$(P/a) = (-1)^{\deg a \deg P \frac{q-1}{2}} \mathrm{sgn}_2(a)^{-\deg P}(a/P) = (a/P) ,$$

since $\deg P$ is even by hypothesis.

Using the same method of proof that led to Theorems 16.8, we can derive the following class number formula for h_K:

$$h_K = \sum_{\substack{a \text{ monic} \\ \deg a < d}} - \deg a \; \chi(a) = \sum_{\substack{a \text{ monic} \\ \deg a < d}} - \deg a \; (a/P) .$$

Recall the definition $f_P(a) = (q-1)(d - \deg\langle a\rangle - 1) - 1$. By Lemma 16.6, we rewrite this formula as

$$h_K = (q-1)^{-1} \sum_{\substack{a \text{ monic} \\ \deg a < d}} (a/P) \, f_P(a) . \tag{8}$$

Let $\mathfrak{P}_\infty$ be a prime of K_P lying over $\mathfrak{p}_\infty$ and let λ_P be a generator of Λ_P such that $\mathrm{ord}_{\mathfrak{P}_\infty} \lambda_P = (q - 1)(d - 1) - 1$. As we have seen, such a generator exists and for all a not divisible by P we have $\mathrm{ord}_{\mathfrak{P}_\infty}(\sigma_a \lambda) = f_P(a)$. Substituting this into Equation 8 and simplifying the result gives us

$$h_K = (q-1)^{-1} \mathrm{ord}_{\mathfrak{P}_\infty}(\eta^{-1}) .$$

By Lemma 16.14, $\eta \in K$. Also, $\mathfrak{P}_\infty$ is ramified over $\mathfrak{p}_\infty$ with ramification index $q - 1$. Thus,

$$h_K = -\mathrm{ord}_{\mathfrak{p}_\infty}(\eta) .$$

The first assertion of the theorem follows by combining this equation with $h_K = -h_{\mathcal{O}_K}\mathrm{ord}_{\mathfrak{p}_\infty}(\epsilon)$, which has already been demonstrated.

To prove the last assertion, consider $\alpha := \epsilon^{h_{\mathcal{O}_K}}\eta^{-1}$. This is a unit in $\mathcal{O}_K$ and the first part of the proof shows that the ord of α at $\mathfrak{p}_\infty$ is zero. Since the divisor of α has degree zero its ord at the other infinite prime must also be zero. We have shown the divisor of α is the zero divisor. Thus, α is a constant.

It may be useful to rewrite the unit η in a way that emphasizes the relationship of Theorem 16.15 to Theorem 16.1, part (b). Namely,

$$\eta = \prod_{\substack{(a/P)=-1 \\ a \text{ monic}}} \sigma_a \lambda_P \Big/ \prod_{\substack{(b/P)=1 \\ b \text{ monic}}} \sigma_b \lambda_P .$$

Using the analytic theory of Drinfeld modules which we sketched in Chapter 13 we can make the analogy even more precise. The Carlitz module is associated to a rank 1 lattice $A\tilde{\pi}$ and a Carlitz exponential function $e_{A\tilde{\pi}}(u)$. Set $e_C(u) = e_{A\tilde{\pi}}(u)$. In the exercises to Chapter 13 we point out that the set of P-torsion points for the Carlitz module which lie in $\mathbf{C}$ (the completion of the algebraic closure of k_∞) is given by

$$\{e_C(a\tilde{\pi}/P) \mid 0 \leq \deg a < d\} \cup \{0\} .$$

If we take λ_P to be $e_C(\tilde{\pi}/P)$, then $\sigma_a\lambda_P = C_a(\lambda_P) = C_a(e_C(\tilde{\pi}/P)) = e_C(a\tilde{\pi}/P)$. Hence, we can rewrite η once more (considering it as an element of $\mathbf{C}$) as follows:

$$\eta = \prod_{\substack{(a/P)=-1 \\ a \text{ monic}}} e_C(a\tilde{\pi}/P) \Big/ \prod_{\substack{(b/P)=1 \\ b \text{ monic}}} e_C(b\tilde{\pi}/P) .$$

The relationship of this unit with the expression after the logarithm in Theorem 16.1, part b, is now quite striking!

The final goal of this chapter is to produce a function field analogue of the Carlitz-Olsen Theorem, Theorem 16.5. We need a definition.

Definition. The relative class number, h_m^-, is defined to be $h_{K_m}/h_{K_m^+}$.

As it stands, h_m^- is a rational number, but it is actually an integer. This can be shown algebraically, by showing that the mapping from the divisor classes of K_m^+ to the divisor classes of K_m, induced by extension of divisors, is injective. It also follows from Theorem 16.8 since by parts a and b of that theorem and the definition of h_m^- we deduce

$$h_m^- = \prod_{\chi \text{ odd}} \left(\sum_{\substack{a \text{ monic} \\ \deg a < M_\chi}} \chi(a) \right) .$$

The right-hand side of this equation is an algebraic integer and the left-hand side is a rational number. This shows $h_m^- \in \mathbb{Z}$.

We will again assume $m = P$ a monic irreducible of degree d. As before, the advantage of assuming m is prime is that we don't have to worry about conductors. The last equation simplifies to

$$h_P^- = \prod_{\chi \text{ odd}} \left(\sum_{\substack{a \text{ monic} \\ \deg a < d}} \chi(a) \right) . \qquad (9)$$

To be precise, the product is over odd characters on $(A/PA)^*$. We will come back to this equation shortly.

Define $t = (q^d - 1)/(q - 1)$. Then t is the size of the set $\mathcal{M}$ of monic polynomials of degree less than d. For each character ψ of $\mathbb{F}^*$ we construct a $t \times t$ matrix $C(\psi)$ as follows:

$$C(\psi) = [\psi(\text{sgn}\langle ab \rangle)] .$$

More concretely, write $ab = cP + r$, where $r \in A$ and $\deg r < d$. The element r cannot be zero since neither a nor b is divisible by P. Then, $\operatorname{sgn}\langle ab \rangle = \operatorname{sgn}(r) =$ the leading coefficient of r.

The following theorem is the function field version of Theorem 16.5.

Theorem 16.16.

$$h_{\bar{P}}^- = \pm \prod_{\substack{\psi \in \widehat{\mathbb{F}}^* \\ \psi \neq \psi_o}} \det C(\psi) .$$

Proof. For every non-trivial character ψ of $\mathbb{F}^*$ define

$$h_\psi = \prod_{\chi|_{\mathbb{F}^*} = \psi} \left(\sum_{\substack{a \text{ monic} \\ \deg a < d}} \chi(a) \right) .$$

Then Equation 9 can be rewritten

$$h_{\bar{P}}^- = \prod_{\psi \neq \psi_o} h_\psi . \tag{10}$$

Fix a character ϕ on $(A/PA)^*$ whose restriction to $\mathbb{F}^*$ is ψ. Then every character with this property is of the form $\phi\chi'$ where χ' is an even character. We'll return to this in a moment.

If ψ is a non-trivial character on $\mathbb{F}^*$, define $\tilde\psi : A \to \mathbb{C}$ by $\tilde\psi(a) = \psi(\operatorname{sgn}(\langle a \rangle))^{-1}$ if $P \nmid a$ and 0 otherwise.

Using this definition and the previous remark we can write

$$h_\psi = \prod_{\chi' \text{ even}} \left(\sum_{\substack{a \text{ monic} \\ \deg a < d}} \chi'(a)\phi(a)\tilde\psi(a) \right) . \tag{11}$$

Of course, with a monic and $\deg a < d$ the term $\tilde\psi(a)$ is equal to 1. So why is it there? The point is that the product $\phi(a)\tilde\psi(a)$ defines a function on $(A/PA)^*$ since both terms depend only on the congruence class of a modulo P, and, in fact, it defines a function on $(A/PA)^*/\mathbb{F}^*$ since if $\alpha \in \mathbb{F}^*$ and a is monic with $\deg a < d$,

$$\phi(\alpha a)\tilde\psi(\alpha a) = \phi(\alpha)\phi(a)\psi(\alpha)^{-1} = \phi(a) = \phi(a)\tilde\psi(a) .$$

The upshot is that the sums in Equation 11 over all monic a with $\deg a < d$ can be replaced with sums over $(A/PA)^*/\mathbb{F}^*$. We then apply the Dedekind determinant formula to Equation 11 and deduce

$$h_\psi = \pm \det[\phi(ab)\tilde\psi(ab)] = \det[\phi(a)\phi(b)\tilde\psi(ab)] .$$

By elementary properties of the determinant we can factor out $\phi(b)$ from the b-th row and $\phi(a)$ from the a-th column. The result is that the determinant is multiplied by $\phi(\prod_a a)^2$, where the product is over all

$a \in (A/PA)^*/\mathbb{F}^*$. This product is the product of the elements of order two in this group and so its square is the identity. Thus,

$$h_\psi = \pm \det[\tilde{\psi}(ab)] = \pm \det[\psi(\text{sgn}\langle ab \rangle)^{-1}] = \pm \det C(\psi^{-1}) .$$

The proof is concluded by substituting this result into Equation 10 and noting that ψ^{-1} runs through the non-trivial characters of $\mathbb{F}^*$ as ψ does.

Theorem 16.16 is taken from Rosen [4], where other similar results are proven and some applications to the size of h_P^- are given. As far as I know, no one has published anything generalizing Theorem 16.16 although L. Shu has a preprint which addresses this problem.

There is an elaborate theory of the so-called Stickelberger ideal and its relation to the relative class number. We have not discussed this circle of ideas. The interested reader may wish to consult Iwasawa [1] and Sinnott [1] for the case of cyclotomic fields. The (very general) function field case has been dealt with in Yin [3]. In a paper which is to appear, Yin [4], gives a new definition of the Stickelberger ideal in the function field case which enables him to deal with the class number itself and not just the relative class number.

Exercises

1. Prove Lemma 16.11 in the general case. More specifically, if $m = P^t$ is a prime power, show that the group of cyclotomic units in K_{P^t} is generated by $\mathbb{F}^*$ and the set $\{\sigma_a \lambda_{P^t}/\lambda_{P^t} \mid a \text{ monic}, (a, P) = 1, \deg a < t \deg P\}$.

2. Generalize the statement of Theorem 16.12 to the case where $m = P^t$ is a prime power and prove it. Hint: Prove first that a character of $\text{Gal}(K_{P^t}/k)$ has conductor P^s if s is the smallest power of P such that $\chi(\sigma_a) = 1$ for all $a \equiv 1 \pmod{P^s}$. The reader may wish to consult the proof of the classical case. See, for example, Chapter 8 of Washington [1].

3. Let P be a monic irreducible of even degree. In the proof of Theorem 16.15 it is claimed that

$$h_K = \sum_{\substack{a \text{ monic} \\ \deg a < \deg P}} - \deg a \, (a/P) .$$

Prove this formula.

4. There is a natural map from $Cl_{K_m^+} \to Cl_{K_m}$ induced by extension of divisors. Show this map is one to one, thereby giving a new proof that $h_{K_m^+}$ divides h_{K_m}. Hint: Assume D is a divisor of K_m^+ and that $i_{K_m/K_m^+} D = (\alpha)$ for some $\alpha \in K_m^*$. Show $\alpha^{q-1} = a \in K_m^+$ and deduce $K_m^+(\alpha)/K_m^+$ is unramified everywhere.

5. Let P be a monic irreducible of degree d in $\mathbb{F}[T]$. Since $\mathrm{Gal}(K_P/k)$ is cyclic of degree $q^d - 1$, there is a unique subfield $L \subset K_P$ such that $[L : k] = q - 1$. Show that $L = k(\sqrt[q-1]{P})$ if d is even, and $L = k(\sqrt[q-1]{-P})$ if d is odd.

6. (Continued) Show that $L \subseteq K_P^+$ if and only if $q - 1 \mid d$.

7. (Continued) More generally, show $[L \cap K_P^+ : k] = (d, q - 1)$.

8. (Continued) Let $L^+ = L \cap K_P^+$. Derive analytic class number formulas for h_{L^+} and h_L and show h_{L^+} divides h_L.

17
Average Value Theorems
in Function Fields

In Chapter 2 we touched upon the subject of average value theorems in $A = \mathbb{F}[T]$. The technique which we used goes back to Carlitz who associated certain Dirichlet series with some of the basic number-theoretic functions and then expressed these Dirichlet series in terms of $\zeta_A(s)$. The zeta function is so simple in the case of the polynomial ring that it was possible to arrive at very precise results for the average values in question. For example, for $n \in A$ define $d(n)$ to be the number of monic divisors of n. Then we showed

$$\sum_{\substack{n \text{ monic} \\ \deg n = N}} d(n) = q^N (N + 1) \,.$$

The corresponding classical result goes as follows. For $n \in \mathbb{Z}$, let $d(n)$ denote the number of positive divisors of n. Then

$$\sum_{1 \leq n \leq x} d(n) = x \log x + (2\gamma - 1)x + O(\sqrt{x}) \,.$$

The constant γ is Euler's constant.

The relation of the two results is made clearer if one recalls that the size of a non-zero polynomial n of degree N was defined to be $|n| = q^N$. Setting $x = q^N$, the first equation can be rewritten

$$\sum_{\substack{n \text{ monic} \\ |n| = x}} d(n) = x \log_q(x) + x \,,$$

which makes the analogy much clearer.

In the first part of this chapter we consider average values of the generalizations of some elementary number-theoretic functions to global function fields. Everything becomes a little more complicated, as will be seen. The work is made a lot easier through the use of a function-field version of the famous Wiener-Ikehara Tauberian theorem. The proof of the function-field version of this theorem is relatively simple, being an application of the Cauchy integral formula. The idea behind this is due to Jeff Hoffstein.

In the second part of the chapter we work over $A = \mathbb{F}[T]$ once again, but we consider average values of a "not so elementary" number-theoretic function. Namely, to each non-square polynomial $m \in A$ we consider the order $\mathcal{O}_m = A + A\sqrt{m} \subset k(\sqrt{m})$ and its class number $h_m = |\text{Pic}(\mathcal{O}_m)|$. We will average these class numbers in various ways, thereby obtaining analogies to two famous conjectures of Gauss. We will discuss this connection as well as possible variants and generalizations.

Let $K/\mathbb{F}$ be an algebraic function field with field of constants $\mathbb{F}$ with $|\mathbb{F}| = q$. We could set aside a few prime divisors, S, the primes "at infinity," and work with the ring A of functions whose only poles lie in S. Our functions would then be defined on the ideals of A. Instead, we will work with functions on the semigroup of all effective divisors. Everything we do can be extended to the former situation without much difficulty.

Let $\mathcal{D}_K$ be the group of divisors of K and $\mathcal{D}_K^+$ be the sub-semigroup of effective divisors. We explicitly include the zero divisor as an element of $\mathcal{D}_K^+$. Let $f : \mathcal{D}_K^+ \to \mathbb{C}$ be a function and define

$$\zeta_f(s) = \sum_{D \in \mathcal{D}_K^+} \frac{f(D)}{ND^s} , \qquad (1)$$

the Dirichlet series associated to f.

Since the use of the variable s will cause no confusion in this chapter, we go back to using s for the variable instead of w. Also, when we use D as a summation variable, it will be assumed that the sum is over D in $\mathcal{D}_K^+$ with, perhaps, some other restrictions.

For $N \geq 0$ an integer, define $F(N) = \sum_{\deg D = N} f(D)$. Equation 1 can be rewritten

$$\zeta_f(s) = \sum_{N=0}^{\infty} F(N) q^{-Ns} .$$

Finally, define $Z_f(u)$ as the function for which $Z_f(q^{-s}) = \zeta_f(s)$. Then

$$Z_f(u) = \sum_{N=0}^{\infty} F(N) u^N . \qquad (2)$$

In Chapter 5 we investigated the function $b_N(K)$, the number of effective divisors of K with degree N. We showed that if $N > 2g - 2$ (where g is the genus of K)

$$b_N(K) = h_K \frac{q^{N-g+1} - 1}{q-1} .$$

Definition. Let $f : \mathcal{D}_K^+ \to \mathbb{C}$ be a function. The average value of f is defined to be

$$\text{Ave}(f) = \lim_{N \to \infty} \frac{\sum_{\deg D = N} f(D)}{\sum_{\deg D = N} 1} = \lim_{N \to \infty} \frac{F(N)}{b_N(K)} ,$$

provided the limit exists.

This definition is certainly the right one in the current context. In many interesting cases, the limit doesn't exist. In that case the task is to find some simple formula for $F(N)$, or a simple formula plus an error term. We will give a number of examples after proving the next theorem which is the function field version of the Wiener-Ikehara Tauberian theorem. The original theorem is much more difficult to prove. See Lang [5]. A little later we will give a function field version of greater generality.

Before stating the theorem we have to establish a convention which will be used throughout the remainder of this chapter. The function q^{-s} is easily seen to be periodic with period $2\pi i / \log(q)$. The same therefore applies to all functions of q^{-s} such as our functions $\zeta_f(s)$. For this reason, nothing is lost by confining our attention to the region

$$B = \left\{ s \in \mathbb{C} \mid -\frac{\pi i}{\log(q)} \leq \Im(s) < \frac{\pi i}{\log(q)} \right\} .$$

In what follows, we will always suppose that s is confined to the region B. This makes life a lot easier. For example, $\zeta_K(s)$ has two simple poles, one at $s = 1$ and one at $s = 0$ if s is confined to B, but it has infinitely many poles on the lines $\Re(s) = 1$ and $\Re(s) = 0$ if s is not so confined.

Theorem 17.1. *Let $f : \mathcal{D}_K^+ \to \mathbb{C}$ be given and suppose $\zeta_f(s)$ converges absolutely for $\Re(s) > 1$ and is holomorphic on $\{s \in B \mid \Re(s) = 1\}$ except for a simple pole at $s = 1$ with residue α. Then, there is a $\delta < 1$ such that*

$$\sum_{\deg D = N} f(D) = \alpha \log(q) q^N + O(q^{\delta N}) .$$

If $\zeta_f(s) - \frac{\alpha}{s-1}$ is holomorphic in $\Re(s) \geq \delta'$, then the error term can be replaced with $O(q^{\delta' N})$.

Proof. The hypothesis implies that $Z_f(u)$ is holomorphic on the disk $\{u \in \mathbb{C} \mid |u| \leq q^{-1}\}$ with the exception of a simple pole at $u = q^{-1}$ (just use the transformation $s \to u = q^{-s}$). What is the residue of $Z_f(u)$ at $u = q^{-1}$? The answer is given by

$$\lim_{u \to q^{-1}} (u - q^{-1}) Z_f(u) = \lim_{s \to 1} \frac{q^{-s} - q^{-1}}{s - 1} (s - 1)\zeta_f(s) = -\frac{\log(q)}{q} \alpha .$$

Next, notice that since the circle $\{u \in \mathbb{C} \mid |u| = q^{-1}\}$ is compact, there is a $\delta < 1$ such that $Z_f(u)$ is holomorphic on the disk $\{u \in \mathbb{C} \mid |u| \leq q^{-\delta}\}$ except for the simple pole at $u = q^{-1}$. Let C be the boundary of this disk oriented counterclockwise and let C_ϵ be a small disc about the origin of radius $\epsilon < q^{-1}$. Orient C_ϵ clockwise, and consider the integral

$$\frac{1}{2\pi i} \oint_{C_\epsilon + C} \frac{Z_f(u)}{u^{N+1}} \, du .$$

By the Cauchy integral formula, this equals to sum of the residues of $Z_f(u)u^{-N-1}$ between the two circles. There is only one pole at $u = q^{-1}$ and the residue there is

$$-\frac{\log(q)}{q} \alpha \, q^{N+1} = -\alpha \log(q) q^N .$$

On the other hand, using the power series expansion of $Z_f(u)$ about $u = 0$, we see

$$\frac{1}{2\pi i} \oint_{C_\epsilon} \frac{Z_f(u)}{u^{N+1}} \, du = -F(N) .$$

It follows that

$$F(N) = \alpha \log(q) q^N + \frac{1}{2\pi i} \oint_C \frac{Z_f(u)}{u^{N+1}} \, du .$$

Let M be the maximum value of $|Z_f(u)|$ on the circle C. The integral in the last formula is bounded by $Mq^{\delta N}$, which completes the proof of the first assertion of the theorem.

To prove the last part, we may assume $\delta' < 1$ since otherwise the error term would be the same size or bigger than the main term. If $\zeta_f(s) - \alpha/(s-1)$ is holomorphic for $\Re(s) \geq \delta'$, then $Z_f(u)$ is holomorphic on the disc $\{u \in \mathbb{C} \mid |u| \leq q^{-\delta'}\}$ except for a simple pole at $u = q^{-1}$. In that case we can repeat the above proof with the role of the circle C being replaced by the circle $C' = \{u \in \mathbb{C} \mid |u| = q^{-\delta'}\}$. The result follows.

We illustrate the use of this theorem by investigating the generalization of the question: what is the probability that a polynomial is square-free? In Chapter 2 we showed, after making the question more precise, that the answer is $1/\zeta_A(2)$.

What would it mean for a divisor to be square-free? This is initially confusing, but only because we write the group law for divisors additively. A moment's reflection shows that the following to be the right definition. An effective divisor D is square-free if and only if $\mathrm{ord}_P D$ is either 0 or 1 for all prime divisors P, i.e., if and only if D is a sum of distinct prime divisors.

Proposition 17.2. *Let $f : \mathcal{D}_K^+ \to \mathbb{C}$ be the characteristic function of the square-free effective divisors. Then $F(N) = \sum_{deg D = N} f(D)$ is the number of square-free effective divisors of degree N. Given $\epsilon > 0$, we have*

$$F(N) = \frac{1}{\zeta_K(2)} \frac{h_K}{q^{g-1}(q-1)} \, q^N + O_\epsilon(q^{(\frac{1}{4}+\epsilon)N}) \, .$$

Moreover, $\mathrm{Ave}(f) = 1/\zeta_K(2)$.

Proof. Recall that for divisors C and D we have $N(C+D) = NC \, ND$. From this we calculate

$$\zeta_f(s) = \sum_D \frac{f(D)}{ND^s} = \sum_{D \text{ square-free}} \frac{1}{ND^s} = \prod_P \left(1 + \frac{1}{NP^s}\right) = \frac{\zeta_K(s)}{\zeta_K(2s)} \, .$$

By the function-field Riemann Hypothesis we know that all the zeros of $\zeta_K(s)$ are on the line $\Re(s) = \frac{1}{2}$. Thus $1/\zeta_K(2s)$ has no poles in the region $\Re(s) > \frac{1}{4}$. On the other hand, we know that in this region $\zeta_K(s)$ is holomorphic except for a simple pole at $s = 1$.

Choose an $\epsilon > 0$ and set $\delta' = \frac{1}{4} + \epsilon$. Then all the hypotheses of Theorem 17.1 apply to $\zeta_f(s)$ and we find

$$F(N) = \alpha \log(q) q^N + O_\epsilon(q^{(\frac{1}{4}+\epsilon)N}) \, , \tag{3}$$

where α is the residue of $\zeta_K(s)/\zeta_K(2s)$ at $s = 1$. We have seen in Chapter 5 that the residue of $\zeta_K(s)$ at $s = 1$ is

$$\rho_K = \frac{h_K}{q^{g-1}(q-1)\log(q)} \, . \tag{4}$$

It follows that $\alpha = \rho_K/\zeta_K(2)$. Substituting this information into Equation 3 completes the proof of the first assertion of the proposition.

To prove the second assertion recall that $\mathrm{Ave}(f) = \lim_{N\to\infty} F(N)/b_N(K)$ and that for all $N > 2g - 2$, $b_N(K) = h_K(q^{N-g+1} - 1)/(q-1)$. By the first part of the proposition we find, for N in this range,

$$\frac{F(N)}{b_N(K)} = \frac{1}{\zeta_K(2)} \frac{q^{N-g+1}}{q^{N-g+1} - 1} + O_\epsilon(q^{(-\frac{3}{4}+\epsilon)N}) \, .$$

Now, simply pass to the limit as N tends to ∞.

As a second example, we generalize the function $\sigma(n)$, the sum of the divisors of n. If D is an effective divisor, what is a divisor of D ? A little thought leads to the following definition; C is a divisor of D if and only if $D - C \geq 0$. D has only finitely many effective divisors in this sense and we define $\sigma(D)$ to be

$$\sigma(D) = \sum_{O \leq C \leq D} NC \, .$$

The reader should not have trouble being convinced that this is a sensible generalization of the usual "sum of the size of divisors" function.

The Dirichlet series $\zeta_\sigma(s)$ is equal to $\zeta_K(s)\zeta_K(s-1)$, as can be verified by the following calculation:

$$\zeta_K(s)\zeta_K(s-1) = \left(\sum_B \frac{1}{NB^s}\right)\left(\sum_C \frac{NC}{NC^s}\right) = \sum_D \left(\sum_{B+C=D} NC\right) ND^{-s} .$$

Proposition 17.3. *Let $\sigma : \mathcal{D}_K^+ \to \mathbb{C}$ be the sum of norms of divisors function defined above. Given an $\epsilon > 0$, we have*

$$\sum_{\deg D=N} \sigma(D) = \zeta_K(2)\frac{h_K}{q^{g-1}(q-1)} q^{2N} + O_\epsilon(q^{(1+\epsilon)N}) .$$

Proof. Since $\zeta_\sigma(s) = \zeta_K(s)\zeta_K(s-1)$, it has a pole at $s = 2$, a double pole at $s = 1$, and a pole at $s = 0$. The conditions of Theorem 17.1 do not hold! However, we can make progress by substituting $s + 1$ for s. This yields $\zeta_\sigma(s+1) = \zeta_K(s+1)\zeta_K(s)$. This function has a simple pole at $s = 1$ and is otherwise holomorphic on the region $\Re(s) > 0$. Choose an $\epsilon > 0$ and set $\delta' = \epsilon$ in Theorem 17.1. We have $\zeta_\sigma(s+1)$ is holomorphic on the region $\Re(s) \geq \epsilon$ except for a simple pole at $s = 1$ with residue $\zeta_K(2)\rho_K$.

We are all set to apply Theorem 17.1, except that we need the expansion of $\zeta_\sigma(s+1)$ as a power series in $q^{-s} = u$. This is easy,

$$\zeta_\sigma(s+1) = \sum_D \frac{\sigma(D)}{ND^{s+1}} = \sum_{N=0}^\infty \left(\sum_{\deg D=N} \frac{\sigma(D)}{ND}\right)q^{-Ns}$$

$$= \sum_{N=0}^\infty \left(q^{-N} \sum_{\deg D=N} \sigma(D)\right)u^N .$$

It follows that

$$q^{-N} \sum_{\deg D=N} \sigma(D) = \zeta_K(2)\rho_K \log(q)q^N + O_\epsilon(q^{\epsilon N}) .$$

Multiply both sides of this equation by q^N and use the explicit expression for ρ_K given by Equation 4. This finishes the proof.

It is amusing to carry matters a step further. Divide both sides of the equation in the proposition by $b_N(K)$ and use the reasoning at the end of the proof of Proposition 17.2. We find that the average of $\sigma(D)$ among all effective divisors of degree N is approximately $\zeta_K(2)q^N$.

As a final application of these methods we want to investigate the function $d(D)$, the number of effective divisors of D. More precisely, $d(D) = \#\{C \in \mathcal{D}_K^+ \mid O \leq C \leq D\}$.

It is relatively easy to check that $\zeta_d(s) = \zeta_K(s)^2$. This function has a double pole at $s = 1$ so Theorem 17.1 doesn't immediately apply. Moreover, it is hard to imagine any simple trick reducing us to the conditions of that Theorem. What is needed is a generalization. This is provided by the next result.

Theorem 17.4. *Let $f : \mathcal{D}_K^+ \to \mathbb{C}$ and let $\zeta_f(s)$ be the corresponding Dirichlet series. Suppose this series converges absolutely in the region $\Re(s) > 1$ and is holomorphic in the region $\{s \in B \mid \Re(s) = 1\}$ except for a pole of order r at $s = 1$. Let $\alpha = \lim_{s \to 1}(s-1)^r \zeta_f(s)$. Then, there is a $\delta < 1$ and constants c_{-i} with $1 \le i \le r$ such that*

$$F(N) = \sum_{\deg D = N} f(D) = q^N \left(\sum_{i=1}^{r} c_{-i} \binom{N+i-1}{i-1} (-q)^i \right) + O(q^{\delta N}) .$$

The sum in parenthesis is a polynomial in N of degree $r - 1$ with leading term

$$\frac{\log(q)^r}{(r-1)!} \, \alpha \, N^{r-1} .$$

Proof. As in the proof of Theorem 17.1, we can find a $\delta < 1$ such that $Z_f(u)$ is holomorphic on the disc $\{u \in \mathbb{C} \mid |u| \le q^{-\delta}\}$. We again let C be the boundary of this disc oriented counterclockwise and C_ϵ a small circle about $s = 0$ oriented clockwise. By the Cauchy integral theorem, the integral

$$\frac{1}{2\pi i} \oint_{C_\epsilon + C} \frac{Z_f(u)}{u^{N+1}} \, du$$

is equal to the sum of the residues of the function $Z_f(u)u^{-N-1}$ in the region between the two circles. There is only one pole in this region. It is located at $u = q^{-1}$. To find the residue there, we expand both $Z_f(u)$ and u^{-N-1} in Laurent series about $u = q^{-1}$, multiply the results together, and pick out the coefficient of $(u - q^{-1})^{-1}$.

By using the Taylor series formula or the general binomial expansion theorem we find

$$u^{-N-1} = q^{N+1} \sum_{j=0}^{\infty} \binom{-N-1}{j} q^j (u - q^{-1})^j .$$

The Laurent series for $Z_f(u)$ has the form

$$Z_f(u) = \sum_{i=-r}^{\infty} c_i (u - q^{-1})^i ,$$

with $c_{-r} \ne 0$.

Multiplying these two series together and isolating the coefficient of $(u - q^{-1})^{-1}$ in the result yields

$$\operatorname{Res}_{u=q^{-1}} Z_f(u) u^{-N-1} = q^{N+1} \sum_{i=-r}^{-1} c_i \binom{-N-1}{-i-1} q^{-i-1}$$

$$= q^N \sum_{i=1}^{r} c_{-i} \binom{-N-1}{i-1} q^i .$$

To get the last equality we simply transformed i to $-i$ and redistributed one factor of q.

It is easy to see that $\binom{-N-1}{k} = (-1)^k \binom{N+k}{k}$, so the residue can be rewritten as

$$-q^N \sum_{i=1}^{r} c_{-i} \binom{N+i-1}{i-1} (-q)^i .$$

As in the proof of Theorem 17.1, it now follows that

$$F(N) = q^N \left(\sum_{i=1}^{r} c_{-i} \binom{N+i-1}{i-1} (-q)^i \right) + O(q^{\delta N}) .$$

Finally, we must prove the assertions about the term in parenthesis. First of all, it is clear that when $k \geq 0$, $\binom{N+k}{k}$ is a polynomial in N of degree k, and that its leading term is $k!^{-1} N^k$. Thus the sum in parenthesis is a polynomial in N of degree $r-1$ and its leading term is

$$\frac{c_{-r}}{(r-1)!} (-q)^r N^{r-1} .$$

It remains to relate $\alpha = \lim_{s \to 1} (s-1)^r \zeta_f(s)$ to c_{-r}. This relationship follows from the calculation

$$c_{-r} = \lim_{u \to q^{-1}} (u - q^{-1})^r Z_f(u)$$

$$= \lim_{s \to 1} \left(\frac{q^{-s} - q^{-1}}{s-1} \right)^r (s-1)^r \zeta_f(s) = \left(-\frac{\log(q)}{q} \right)^r \alpha .$$

Substitute this expression for c_{-r} into the previous expression for the leading term of the sum in parentheses and we arrive at

$$\frac{\log(q)^r}{(r-1)!} \alpha \, N^{r-1}$$

for the leading term. This completes the proof.

Corollary. *With the assumptions and notation of the theorem, we have, as $N \to \infty$,*

$$F(N) \sim \frac{\log(q)^r}{(r-1)!} \alpha \, q^N N^{r-1} .$$

(Here, "$\sim$" means "is asymptotic to").

Proof. This is immediate from the theorem.

In order not to clutter the statement of Theorem 17.4, we did not include a refinement similar to the last part of the statement of Theorem 17.1. However, it is easy to establish the following generalization in the present case. Suppose the hypotheses of Theorem 17.4 hold and that $P(s) = \sum_{i=1}^{r} a_{-i}(s-1)^{-i}$ is the polar part of the Laurent expansion of $\zeta_f(s)$ about $s = 1$. If $\zeta_f(s) - P(s)$ is holomorphic in $\Re(s) \geq \delta'$, then the error term can be replaced with $O(q^{\delta'N})$. The proof is the same as in the earlier situation.

For general Dirichlet series there is a generalization of the Wiener-Ikehara Tauberian theorem, which is analogous to Theorem 17.4. It is due to H. Delange [1]. A statement of the theorem is given in Appendix II of Narkiewicz [1].

We now want to apply Theorem 17.4 to the divisor function $d(D)$ on $\mathcal{D}_K^+$.

Proposition 17.5. *Let $K/\mathbb{F}$ be a global function field and $d(D)$ the divisor function on the effective divisors. Then, there exist constants μ_K and λ_K such that for fixed $\epsilon > 0$ we have*

$$\sum_{\deg D = N} d(D) = q^N \left(\lambda_K N + \mu_K \right) + O_\epsilon(q^{\epsilon N}) .$$

More explicitly, $\lambda_K = h_K^2 q^{2-2g}(q-1)^{-2}$.

Proof. We have already seen that $\zeta_d(s) = \zeta_K(s)^2$, a function which has a double pole at $s = 1$ and is otherwise holomorphic for $\Re(s) > 0$. Choose $\epsilon > 0$. Notice that $\lim_{s \to 1}(s-1)^2\zeta_K(s)^2 = \rho_K^2$. Applying Theorem 17.4 and the remarks given after that theorem we find there are constants λ_K and μ_K such that

$$\sum_{\deg D = N} d(D) = q^N \left(\lambda_K N + \mu_K \right) + O_\epsilon(q^{\epsilon N}) .$$

Applying the formula for the leading term of the polynomial in parenthesis given in the statement of Theorem 17.4, we find

$$\lambda_K = \frac{\log(q)^r}{(r-1)!}\alpha = \frac{\log(q)^2}{1!}\rho_K^2 = \frac{h_K^2}{q^{2g-2}(q-1)^2} .$$

This finishes the proof.

Another interesting fact is that the average value of $d(D)$ over the effective divisors of degree N is asymptotic to

$$\frac{h_K}{q^{g-1}(q-1)} N .$$

This is easy to establish on the basis of the proposition.

We have merely touched on the fringes of a large subject. The reader who wishes to explore this area further should consult the book of J. Knopfmacher [1]. We will not pursue these matters here. Instead, we turn to another topic in the area of average values of arithmetic functions.

In his famous work *Disquisitiones Mathematicae*, C.F. Gauss considered at length the arithmetic of binary quadratic forms $ax^2 + 2bxy + cy^2$ defined over the integers $\mathbb{Z}$. The discriminant of such a form is by definition $D = 4b^2 - 4ac$ (because of the restriction that the coefficient of xy be even, Gauss considered only even discriminants). He defined an equivalence between such forms and showed that equivalent forms have the same discriminant. Moreover, he showed that the number of equivalence classes of forms with the same discriminant is finite. Call that number h_D. Based on extensive numerical evidence he made two conjectures about the average value of these class numbers h_D. Slightly reformulated, they read as follows.

1. Let $D = -4k$ vary over all negative even discriminants with $1 \leq k \leq N$. Then
$$\sum_{1 \leq k \leq N} h_D \sim \frac{4\pi}{21\zeta(3)} \, N^{\frac{3}{2}} \, .$$

2. Let $D = 4k$ vary over all positive even discriminants such that $1 \leq k \leq N$. Then
$$\sum_{1 \leq k \leq N} h_D R_D \sim \frac{4\pi^2}{21\zeta(3)} \, N^{\frac{3}{2}} \, .$$

The number R_D in the second conjecture is closely related to the regulator of the real quadratic number field $\mathbb{Q}(\sqrt{D})$. In fact, the both conjectures can be reformulated in terms of orders $\mathcal{O}$ in quadratic number fields where the class numbers h are interpreted in terms of the size of the Picard group of $\mathcal{O}$, $\mathrm{Pic}(\mathcal{O})$, i.e., invertible fractional ideals of $\mathcal{O}$ modulo principal ideals.

Both of these conjectures have been proven. There is a long history. The interested reader can find a brief review of all this in Hoffstein-Rosen [1].

We will consider the function field analogue of Gauss's conjectures. As usual, instead of $\mathbb{Z}$ and $\mathbb{Q}$ we consider the pair $A = \mathbb{F}[T]$ and $k = \mathbb{F}(T)$. For the remainder of the chapter, we assume that the characteristic of $\mathbb{F}$ is odd. Let $m \in A$ be any non-square polynomial, and consider the quadratic function field $K = k(\sqrt{m})$. Write $m = m_0 m_1^2$, where m_0 is square-free. The polynomial m_0 is well defined up to the square of a constant. Define $\mathcal{O}_m$ to be the ring $A + A\sqrt{m} \subset K$. It is an A-order in K, i.e., it is a ring, finitely generated as an A-module, and its quotient field is K.

Proposition 17.6. *With the notations introduced above, the integral closure of A in K is $\mathcal{O}_{m_0}$. The ring $\mathcal{O}_m$ is a subring of $\mathcal{O}_{m_0}$ and the polynomial m_1 is a generator of the annihilator of the A-module $\mathcal{O}_{m_0}/\mathcal{O}_m$. Finally, if $\mathcal{O}$ is any A-order in K, then $\mathcal{O} = \mathcal{O}_m$ for some $m \in A$.*

Proof. Since the characteristic of $\mathbb{F}$ is odd, it is easy to see that K/k is a Galois extension. Let σ generate the Galois group.

Clearly, $K = k(\sqrt{m}) = k(\sqrt{m_0})$. Every element in K has the form $r + s\sqrt{m_0}$ for suitable $r, s \in k$. The automorphism σ takes $\sqrt{m_0}$ to $-\sqrt{m_0}$.

Suppose $r + s\sqrt{m_0}$ is integral over A. Applying σ we see that $r - s\sqrt{m_0}$ is also integral over A. Thus, so is the sum and product of these two elements, i.e. $2r$ and $r^2 - m_0 s^2$ are integral over A. Since A is integrally closed, we have $2r \in A$ and $r^2 - m_0 s^2 \in A$. We may divide by 2, so $r \in A$ and it follows that $m_0 s^2 \in A$. Since m_0 is square-free, we must have $s \in A$. We have proved that if $r + s\sqrt{m_0}$ is integral over A, then $r + s\sqrt{m_0} \in A + A\sqrt{m_0}$. The converse is clear, so our first assertion is established.

From the definitions, $\mathcal{O}_m = A + A\sqrt{m} = A + Am_1\sqrt{m_0} \subseteq A + A\sqrt{m_0} = \mathcal{O}_{m_0}$. It is then immediate that as A-modules

$$\mathcal{O}_{m_0}/\mathcal{O}_m \cong A/m_1 A \,,$$

which proves the second assertion.

Let $\mathcal{O}$ be any A-order in K. One can easily show that every element of $\mathcal{O}$ is integral over A. Since $1 \in \mathcal{O}$ by definition, we have $A \subset \mathcal{O}$. Since K is the quotient field of $\mathcal{O}$ there is an element $\alpha \in \mathcal{O}$ such that $\alpha \notin A$. By the first part of the proof we can write $\alpha = a + b\sqrt{m_0}$ with $a, b \in A$ and $b \neq 0$. It follows that $b\sqrt{m_0} \in \mathcal{O}$ with $b \in A - \{0\}$. Choose $m_1 \in A - \{0\}$ to be a non-zero polynomial of least degree such that $m_1\sqrt{m_0} \in \mathcal{O}$. Set $m = m_0 m_1^2$. We claim that $\mathcal{O} = \mathcal{O}_m$. It is clear that $\mathcal{O}_m \subseteq \mathcal{O}$, so we must show the reverse inclusion. Suppose $a + b\sqrt{m_0} \in \mathcal{O}$ with $a, b \in A$. By the division algorithm in A we can write $b = cm_1 + r$, where $c, r \in A$ and either $r = 0$ or $\deg r < \deg m_1$. Multiply this relation on the right by $\sqrt{m_0}$ and we can deduce that $r\sqrt{m_0} \in \mathcal{O}$. Since m_1 is a non-zero polynomial of least degree with this property, we conclude that $r = 0$. Thus, $\alpha = a + cm_1\sqrt{m_0} \in \mathcal{O}_m$ and we are done.

Definition. Let $m \in A$, m not a square, and let $\mathcal{O}_m \subset k(\sqrt{m})$ be the A-order described above. $\mathrm{Pic}(\mathcal{O}_m)$, the Picard group of $\mathcal{O}_m$, is the group of invertible fractional ideals of $\mathcal{O}_m$ modulo the subgroup of principal fractional ideals. The class number h_m is defined to be the cardinality of this group (we will see shortly that h_m is finite).

If m_0 is square-free, then as we have just seen, $\mathcal{O}_{m_0}$ is the integral closure of A in $K = k(\sqrt{m_0})$. In this case, $\mathcal{O}_{m_0}$ is a Dedekind domain, $\mathrm{Pic}(\mathcal{O}_{m_0}) = Cl(\mathcal{O}_{m_0})$, the class group of $\mathcal{O}_{m_0}$, and h_{m_0} is the usual class number. Moreover, h_{m_0} is finite by Proposition 14.2 (take S to be the primes of K lying above ∞).

Before going further with this analysis, we need another definition. If $m \in A$, m a non-square, define $\chi_m(a)$ as follows:

$$\chi_m(a) = \left(\frac{m}{a}\right)_2 \,.$$

Recall that if P is irreducible, then $\chi_m(P) = 0$ if $P|m$, and if $P \nmid m$ then $\chi_m(P) = 1$ if m is a square modulo P and -1 otherwise. If a is a product of irreducibles, one extends $\chi_m(P)$ by multiplicativity; i.e., if $a = \prod_{i=1}^{t} P_i$, then $\chi_m(a) = \prod_{i=1}^{t} \chi_m(P_i)$.

If $m = m_0 m_1^2$ we have $\chi_m(a) = \chi_{m_0}(a)$ whenever $(a, m) = 1$. However, if P is an irreducible such that $P|m_1$ and $P \nmid m_0$, then we have $\chi_m(P) = 0$, whereas $\chi_{m_0}(P) \neq 0$.

Define $L(s, \chi_m)$ as follows:

$$L(s, \chi_m) = \sum_{n \text{ monic}} \frac{\chi_m(n)}{|n|^s} = \prod_{P \nmid m} \left(1 - \frac{\chi_m(P)}{|P|^s}\right)^{-1}.$$

Notice that if $m = m_0 m_1^2$, we have

$$L(s, \chi_m) = \prod_{P|m_1} \left(1 - \frac{\chi_{m_0}(P)}{|P|^s}\right) L(s, \chi_{m_0}).$$

When m is square-free, the next proposition shows that $L(s, \chi_m)$ is closely related to the Artin L-function associated to the abelian extension $k(\sqrt{m})/k$.

Proposition 17.7. *Suppose m is square-free. Consider the quadratic extension $K = k(\sqrt{m})$ of k. Let $L_\infty(s, \chi_m)$ be 1 if ∞ is ramified in K, $(1 - q^{-s})^{-1}$ if ∞ splits in K, and $(1 + q^{-s})^{-1}$ if ∞ is inert in K. Then*

$$L_\infty(s, \chi_m) L(s, \chi_m)$$

is the Artin L-function associated to the unique non-trivial character of $\mathrm{Gal}(K/k)$.

Proof. We have seen that $A + A\sqrt{m}$ is the integral closure of A in K. The discriminant of this ring over A is $4m$. Since 4 is a non-zero constant, a prime P of A is ramified if and only if it divides m.

Let $L(s, \chi)$ be the Artin L-function associated to the unique non-trivial character χ of $\mathrm{Gal}(K/k)$. If P is a finite prime, $\chi(P) = 0$ if P is ramified and $\chi(P) = \chi((P, K/k))$ if P is unramified. By the definition of the Artin symbol, $(P, K/k)$ is e if P splits, and σ if P is inert (σ being the non-trivial element of the Galois group). Thus, $\chi(P) = 1$ if P splits and $\chi(P) = -1$ if P is inert. By the decomposition law in quadratic extensions (take $l = 2$ in Proposition 10.5), P splits in K if and only if $\chi_m(P) = 1$. Thus, for finite primes $\chi(P) = \chi_m(P)$. At ∞ we know that $|\infty| = q$ so $(1 - \chi(\infty)q^{-s})^{-1}$ is 1 if ∞ is ramified, $(1 - q^{-s})^{-1}$ if ∞ splits, and $(1 + q^{-s})^{-1}$ if ∞ is inert. Thus, $(1 - \chi(\infty)|\infty|^{-s})^{-1} = L_\infty(s, \chi_m)$. We have shown that $L(s, \chi)$ and $L_\infty(s, \chi_m) L(s, \chi_m)$ have the same Euler factors for all primes. Thus, they are equal.

We are now in a position to state the connection between $L(1, \chi_m)$ and class numbers. We begin with the case of m square-free. This relation is proven in the thesis of E. Artin (see Artin [1]). We will show, in a little while, how to generalize this result to the case of non-square polynomials m.

If $m \in A$, recall the definition of $\text{sgn}_2(m)$. This is 1 if the leading coefficient of m is a square in $\mathbb{F}^*$ and is -1 if it is not.

Theorem 17.8A. *Let $m \in A$ be a square-free polynomial of degree M. Then,*

1) If M is odd, $L(1, \chi_m) = \dfrac{\sqrt{q}}{\sqrt{|m|}} h_m$.

2) If M is even and $\text{sgn}_2(m) = -1$, $L(1, \chi_m) = \dfrac{q+1}{2\sqrt{|m|}} h_m$.

3) If M is even and $\text{sgn}_2(m) = 1$, $L(1, \chi_m) = \dfrac{q-1}{\sqrt{|m|}} h_m R_m$.

Here, R_m is the regulator of the ring $\mathcal{O}_m$.

Proof. Set $K = k(\sqrt{m})$. From Proposition 17.7 and Proposition 14.9 we derive

$$\zeta_K(s) = \zeta_k(s) L_\infty(s, \chi_m) L(s, \chi_m) .$$

Multiply both sides of this equation by $s - 1$ and take the limit as $s \to 1$. One finds

$$\frac{h_K}{q^{g-1}(q-1)\log(q)} = \frac{1}{q^{-1}(q-1)\log(q)} L_\infty(1, \chi_m) L(1, \chi_m) .$$

Simplifying, we obtain

$$h_K q^{-g} = L_\infty(1, \chi_m) L(1, \chi_m) . \tag{5}$$

Proposition 10.4 and the following remarks show that the genus, g, of K is $\frac{M-1}{2}$ in case 1 and $\frac{M}{2} - 1$ in cases 2 and 3.

Proposition 14.6 shows that in case 1, ∞ is ramified, in case 2, ∞ is inert, and in case 3, ∞ splits. By Proposition 14.7, we find $h_m = h_K$, $h_m = 2h_K$, and $h_m R_m = h_K$ in cases 1, 2, and 3, respectively.

Let's consider case 1. We have $g = \frac{M-1}{2}$ and $L_\infty(1, \chi_m) = 1$. Also, $h_m = h_K$. Substituting this information into Equation 5, and noting $\sqrt{|m|} = q^{\frac{M}{2}}$, we find

$$\frac{h_m \sqrt{q}}{\sqrt{|m|}} = L(1, \chi_m) .$$

This proves case 1.

To deal with case 2 we note $g = \frac{M}{2} - 1$, $L_\infty(1, \chi_m) = (1 + q^{-1})^{-1}$, and $h_m = 2h_K$. Substituting into Equation 5 once again, we find

$$\frac{h_m}{2} \frac{q}{\sqrt{|m|}} = \left(1 + \frac{1}{q}\right)^{-1} L(1, \chi_m) \, .$$

Case 2 of the Proposition is immediate from this.

The last case, case 3, is done in exactly the same way. Here, $g = \frac{M}{2} - 1$, $L_\infty(1, \chi_m) = (1 - q^{-1})^{-1}$, and $h_K = h_m R_m$. Substituting into Equation 5 one more time yields

$$\frac{h_m R_m q}{\sqrt{|m|}} = \left(1 - \frac{1}{q}\right)^{-1} L(1, \chi_m) \, .$$

Case 3 follows easily, and this concludes the proof.

Let's now return to the general case. Let m be a non-square polynomial and write $m = m_0 m_1^2$ with m_0 square-free. We will need the relation between h_m and h_{m_0}. While this is not a very difficult relationship to find and prove, it does take a rather detailed investigation which is off to the side of our main purpose. For this reason, we will merely state the result and refer to Lang [1], Chapter 8, Theorem 7, where the corresponding result for quadratic number fields is proven. The function field version, stated below, is proved in exactly the same way. For a general result (for number fields) along the same lines the reader may wish to look at Neukirch [1], Chapter 1, Theorem 12.12.

Proposition 17.9. *Let $m \in A$ be a non-square and write $m = m_0 m_1^2$ with m_0 square-free. Then,*

$$h_m = h_{m_0} \frac{|m_1|}{[\mathcal{O}_{m_0}^* : \mathcal{O}_m^*]} \prod_{P | m_1} (1 - \chi_{m_0}(P)|P|^{-1}) \, .$$

Implicit in this result is that the index $[\mathcal{O}_{m_0}^* : \mathcal{O}_m^*]$ is finite. If ∞ either ramifies or is inert, both groups are equal to $\mathbb{F}^*$ and the index is 1. If ∞ splits, then both groups have $\mathbb{Z}$-rank 1 and one can show without much difficulty that the index is the same as the quotient of regulators R_m / R_{m_0}. We set $R_m = R_{m_0} = 1$ in the first two cases. Then the relationship given by Proposition 17.9 can be rewritten

$$\frac{h_m R_m}{\sqrt{|m|}} = \frac{h_{m_0} R_{m_0}}{\sqrt{|m_0|}} \prod_{P | m_1} (1 - \chi_{m_0}(P)|P|^{-1}) \, . \tag{6}$$

Theorem 17.8B. *All the assertions of Theorem 17.8A remain valid if $m \in A$ is a non-square polynomial.*

Proof. Suppose $m = m_0 m_1^2$ with m_0 square-free. From the definitions,

$$L(s, \chi_m) = L(s, \chi_{m_0}) \prod_{P|m_1} (1 - \chi_{m_0}(P)|P|^{-s}) .$$

It follows that Equation 6 can be rewritten as

$$\frac{h_m R_m}{\sqrt{|m|}} \frac{1}{L(1, \chi_m)} = \frac{h_{m_0} R_{m_0}}{\sqrt{|m_0|}} \frac{1}{L(1, \chi_{m_0})} .$$

With the help of this equation, we see that Theorem 17.8B follows from Theorem 17.8A.

By Theorem 17.8B, we see that the task of averaging class numbers reduces to the task of averaging the numbers $L(1, \chi_m)$. It turns out that it is no harder to average $L(s, \chi_m)$ for any value of s. This is what we shall do.

To begin with, notice that

$$L(s, \chi_m) = \sum_{n \text{ monic}} \frac{\chi_m(n)}{|n|^s} = \sum_{d=0}^{\infty} \Big(\sum_{\substack{n \text{ monic} \\ \deg(n)=d}} \chi_m(n) \Big) q^{-ds} .$$

Definition. For $d \in \mathbb{Z}$, $d \geq 0$, define

$$S_d(\chi_m) = \sum_{\substack{n \text{ monic} \\ \deg(n)=d}} \chi_m(n)$$

Using this definition, we can rewrite $L(s, \chi_m)$ as $\sum_{d=0}^{\infty} S_d(\chi_m) q^{-ds}$. This sum is actually finite as we see from the following Lemma.

Lemma 17.10. *If $m \notin \mathbb{F}^*$ is not a square, $S_d(\chi_m) = 0$ for $d \geq M = \deg(m)$.*

Proof. By the reciprocity law, Theorem 3.5, we have

$$\left(\frac{m}{n}\right)\left(\frac{n}{m}\right) = (-1)^{\frac{q-1}{2} Md} \operatorname{sgn}(m)^d .$$

Call the quantity on the right of this equation c_d. Then, we have $\chi_m(n) = c_d(n/m)$. Thus, if $d \geq M$,

$$S_d(\chi_m) = c_d \sum_{\substack{n \text{ monic} \\ \deg(n)=d}} \left(\frac{n}{m}\right) = 0 ,$$

by the proof of Proposition 4.3.

Corollary. *If $m \notin \mathbb{F}^*$ is not a square, then*

$$L(s, \chi_m) = \sum_{d=0}^{M-1} S_d(\chi_m) q^{-ds} ,$$

a polynomial of degree at most $M-1$ in q^{-s}.

Proof. This is immediate from the lemma and the previous remarks.

Our goal is to understand the sums $\sum_{\deg(m)=M} L(s, \chi_m)$ or the same sums restricted to monic polynomials m of degree M. By the corollary we are reduced to considering the sums $\sum_{\deg(m)=M} S_d(\chi_m)$ where $d < M$.

We will need the following definition and the subsequent proposition.

Definition. Let M and N be non-negative integers and n a monic polynomial of degree N. Define $\Phi_n(M)$ to be the number of monic polynomials m of degree M such that $\gcd(n, m) = 1$. Define $\Phi(N, M)$ to be the number of pairs (n, m) of monic polynomials such that $\deg(n) = N$, $\deg(m) = M$, and $\gcd(n, m) = 1$.

Note that

$$\sum_{\substack{n \text{ monic} \\ \deg(n)=N}} \Phi_n(M) = \Phi(N, M) .$$

Also, it is obvious that $\Phi(N, M) = \Phi(M, N)$.

Proposition 17.11. $\Phi(0, M) = q^M$ *and if* $M, N \geq 1$, *then*

$$\Phi(N, M) = q^{M+N} \left(1 - \frac{1}{q} \right) .$$

Proof. From the definition, $\Phi(0, M)$ is equal to the number of monic polynomials of degree M which we know is q^M. This proves the first assertion. To prove the second assertion, call two pairs (n, m) and (n', m') equivalent if $\gcd(n, m) = \gcd(n', m')$. Breaking the set $\{(n, m) \mid \deg(n) = N, \deg(m) = M\}$ into equivalence classes and counting leads to the identity

$$q^{N+M} = \sum_{d=0}^{\min(N,M)} q^d \, \Phi(N - d, M - d) .$$

Suppose $M, N \geq 1$. The proof now proceeds by induction on the number $M + N$. The smallest value this number can have is 2, in which case the formula yields $q^2 = \Phi(1, 1) + q\Phi(0, 0)$, or $\Phi(1, 1) = q^2 - q = q^2(1 - q^{-1})$.

Now suppose the formula is correct for all pairs $N', M' \geq 1$ with $N' + M' < N + M$. We may also suppose, by symmetry, that $N \leq M$. Then

$$q^{M+N} = \Phi(N, M) + \sum_{d=1}^{N-1} q^d \Phi(N - d, M - d) + q^N \Phi(0, M - N) .$$

For $1 \leq d \leq N-1$ we have $\Phi(N-d, M-d) = q^{M+N-2d}(1-q^{-1})$ whereas by the first part of the proof, $\Phi(0, M - N) = q^{M-N}$. Substituting into the above formula and simplifying slightly,

$$q^{M+N} = \Phi(N, M) + q^{M+N} \sum_{d=1}^{N-1} q^{-d}(1 - q^{-1}) + q^M = \Phi(N, M) + q^{M+N-1} .$$

The second assertion now follows immediately.

This elegant proof is due to David Hayes.

It is convenient to extend the definition of $\Phi(N, M)$ to half integers by defining $\Phi(N/2, M) = 0$ if N is odd.

Proposition 17.12. *Suppose $1 \leq d \leq M - 1$. Then*

$$\sum_{\substack{m \text{ monic} \\ \deg(m)=M}} S_d(\chi_m) = (q - 1)^{-1} \sum_{\deg(m)=M} S_d(\chi_m) = \Phi(d/2, M) .$$

Proof. To begin with assume all sums are over monics. Then

$$\sum_{\deg(m)=M} S_d(\chi_m) = \sum_{\deg(m)=M} \sum_{\deg(n)=d} \left(\frac{m}{n}\right) = \sum_{\deg(n)=d} \sum_{\deg(m)=M} \left(\frac{m}{n}\right) .$$

If n is not a square, $(*/n)$ is a non-trivial character modulo n. Thus, in this case, since $M > \deg n = d$,

$$\sum_{\deg(m)=M} \left(\frac{m}{n}\right) = 0 ,$$

by the proof of Proposition 4.3.

Now, suppose that $n = n_1^2$ is a square. Then $(m/n) = (m/n_1)^2 = 1$ whenever $\gcd(m, n_1) = 1$ and $(m/n_1)^2 = 0$ otherwise. It follows that

$$\sum_{\deg(m)=M} \left(\frac{m}{n}\right) = \sum_{\deg(m)=M} \left(\frac{m}{n_1}\right)^2 = \Phi_{n_1}(M) .$$

Thus

$$\sum_{\deg(m)=M} S_d(\chi_m) = \sum_{\deg(n_1)=d/2} \Phi_{n_1}(M) = \Phi(d/2, M) .$$

To do the general case, let $\alpha \in \mathbb{F}^*$ and sum over all αm as m runs through the monics of degree M. The above calculation shows the answer is again equal to $\Phi(d/2, M)$. It follows that if we sum over all polynomials of degree d the answer is $(q - 1)\Phi(d/2, M)$. This completes the proof.

We now have all the information we need to state our main results about averages of L-functions. We begin with the easiest case, averaging over all monics of fixed odd degree.

Theorem 17.13. *Let M be odd and positive. We have, for all $s \in B$ with $s \neq \frac{1}{2}$,*

$$q^{-M} \sum_{\substack{m \text{ monic} \\ \deg(m)=M}} L(s, \chi_m) = \frac{\zeta_A(2s)}{\zeta_A(2s+1)} - \left(1 - \frac{1}{q}\right)(q^{1-2s})^{\frac{M+1}{2}} \zeta_A(2s) .$$

For $s = \frac{1}{2}$, we have

$$q^{-M} \sum_{\substack{m \text{ monic} \\ \deg(m)=M}} L(1/2, \chi_m) = 1 + \left(1 - \frac{1}{q}\right)\left(\frac{M-1}{2}\right) .$$

Proof. By the corollary to Lemma 17.10, $L(s, \chi_m) = \sum_{d=0}^{M-1} S_d(\chi_m) q^{-ds}$. From this, Proposition 17.11 and Proposition 17.12, we find

$$\sum_{\substack{m \text{ monic} \\ \deg(m)=M}} L(s, \chi_m) = \sum_{d=0}^{M-1}\left(\sum_{\substack{m \text{ monic} \\ \deg(m)=M}} S_d(\chi_m)\right) q^{-ds}$$

$$= q^M + \Phi(1, M)q^{-2s} + \Phi(2, M)q^{-4s} + \cdots + \Phi((M-1)/2, M)q^{-(M-1)s}$$

$$= q^M \left(1 + \left(1 - \frac{1}{q}\right)\left[q^{1-2s} + (q^{1-2s})^2 + \cdots + (q^{1-2s})^{\frac{M-1}{2}}\right]\right) .$$

The result for $s = \frac{1}{2}$ follows from this by substitution. For $s \neq \frac{1}{2}$ we sum the geometric series to derive

$$q^{-M} \sum_{\substack{m \text{ monic} \\ \deg(m)=M}} L(s, \chi_m) = 1 + \left(1 - \frac{1}{q}\right) q^{1-2s} \frac{1 - (q^{1-2s})^{\frac{M-1}{2}}}{1 - q^{1-2s}}$$

$$= 1 + \left(1 - \frac{1}{q}\right) \frac{q^{1-2s}}{1 - q^{1-2s}} - \left(1 - \frac{1}{q}\right)(q^{1-2s})^{\frac{M+1}{2}} \zeta_A(2s) .$$

We have used the fact that $\zeta_A(s) = (1 - q^{1-s})^{-1}$, a fact we will use again almost immediately.

A close look at the last line shows that it only remains to identify the sum of the first two terms with a quotient of zeta values. This follows from the calculation

$$1 + \left(1 - \frac{1}{q}\right) \frac{q^{1-2s}}{1 - q^{1-2s}} = 1 + \frac{q^{1-2s} - q^{-2s}}{1 - q^{1-2s}} = \frac{1 - q^{-2s}}{1 - q^{1-2s}} = \frac{\zeta_A(2s)}{\zeta_A(2s+1)} .$$

Corollary 1. *If $\Re(s) > \frac{1}{2}$, then*

$$q^{-M} \sum_{\substack{m \text{ monic} \\ \deg(m)=M}} L(s, \chi_m) \to \frac{\zeta_A(2s)}{\zeta_A(2s+1)},$$

as $M \to \infty$ through odd values.

Proof. This follows immediately from the theorem together with the observation that if $\Re(s) > \frac{1}{2}$ then $|q^{1-2s}| < 1$.

Corollary 2. *If M is odd and positive, then*

$$q^{-M} \sum_{\substack{m \text{ monic} \\ \deg(m)=M}} h_m = \frac{\zeta_A(2)}{\zeta_A(3)} q^{\frac{M-1}{2}} - q^{-1}.$$

Proof. We begin by substituting $s = 1$ into the identity given in the theorem. We find

$$q^{-M} \sum_{\substack{m \text{ monic} \\ \deg(m)=M}} L(1, \chi_m) = \frac{\zeta_A(2)}{\zeta_A(3)} - \left(1 - \frac{1}{q}\right) q^{-\frac{M+1}{2}} \zeta_A(2) = \frac{\zeta_A(2)}{\zeta_A(3)} - q^{-\frac{M+1}{2}}.$$

The last equality follows from $\zeta_A(2) = (1 - q^{-1})^{-1}$.

By Theorem 17.8A and Theorem 17.8B, we see that $L(1, \chi_m) = h_m \frac{\sqrt{q}}{\sqrt{|m|}} = h_m q^{-\frac{M-1}{2}}$. Substituting this information into the last equation yields the result.

We remark that in Theorem 17.13 and the corollaries we could have averaged over all polynomials of odd degree M instead of the monics of that degree and the result would be the same. This is implied by Proposition 17.12. We leave the details to the reader.

We are left with consideration of the two cases where $\deg(m) = M$ is even and the leading coefficient of m is either a square in $\mathbb{F}^*$ or a non-square. These cases are complicated by the possibility of m being itself a square or a constant times a square. In these cases, $k(\sqrt{m})$ is either equal to k or is a constant field extension. We wish to exclude both possibilities. This can be done without much difficulty, but the calculations are more involved. We will be content with stating the results in these cases and referring the reader to Hoffstein-Rosen [1], Section 1, for the proofs.

Theorem 17.14. *Let M be even and positive. The following sums are over all non-square monic polynomials of degree M.*

(1) Suppose $s \neq \frac{1}{2}$ or 1. Then

$$q^{-M} \sum L(s, \chi_m) = \frac{\zeta_A(2s)}{\zeta_A(2s+1)} - \left(1 - \frac{1}{q}\right) (q^{1-2s})^{\frac{M}{2}} \zeta_A(2s)$$

$$-q^{-\frac{M}{2}}\left(\frac{\zeta_A(2s)}{\zeta_A(2s+1)} - \left(1 - \frac{1}{q}\right)(q^{1-s})^M \zeta_A(s)\right) .$$

(2) For $s = 1$ we have

$$q^{-M}\sum L(1, \chi_m) = \frac{\zeta_A(2)}{\zeta_A(3)} - q^{-\frac{M}{2}}\left(2 + \left(1 - \frac{1}{q}\right)(M-1)\right) .$$

Corollary 1. If $Re(s) > \frac{1}{2}$, then as $M \to \infty$ though even integers,

$$q^{-M}\sum L(s, \chi_m) \to \frac{\zeta_A(2s)}{\zeta_A(2s+1)} .$$

Corollary 2. With the hypotheses of the theorem, we have

$$q^{-M}\sum h_m R_m = (q-1)^{-1}\left(\frac{\zeta_A(2)}{\zeta_A(3)}q^{\frac{M}{2}} - \left(2 + \left(1 - \frac{1}{q}\right)(M-1)\right)\right) .$$

Nothing mysterious happens at $s = \frac{1}{2}$. We leave the evaluation of these averages at $s = \frac{1}{2}$ to the exercises.

We now state the result in the remaining case.

Theorem 17.15. Let M be positive and even, and let $\gamma \in \mathbb{F}^*$ be a non-square constant. The following sum is over all non-square monic polynomials of degree M. For $s \neq \frac{1}{2}$ we have

$$q^{-M}\sum L(s, \chi_{\gamma m}) =$$

$$\frac{\zeta_A(2s)}{\zeta_A(2s+1)} - \left(1 - \frac{1}{q}\right)(q^{1-2s})^{\frac{M}{2}}\zeta_A(2s) - q^{-\frac{M}{2}}\left(\frac{1+q^{-s}}{1+q^{1-s}} - \left(1 - \frac{1}{q}\right)\frac{(q^{1-s})^M}{1+q^{1-s}}\right) .$$

Corollary 1. If $\Re(s) > \frac{1}{2}$, then as $M \to \infty$ through even integers,

$$q^{-M}\sum L(s, \chi_{\gamma m}) \to \frac{\zeta_A(2s)}{\zeta_A(2s+1)} .$$

Corollary 2. With the hypotheses of the theorem, we have

$$q^{-M}\sum h_{\gamma m} = 2(q+1)^{-1}\left(\frac{\zeta_A(2)}{\zeta_A(3)}q^{\frac{M}{2}} - 1 - q^{-1}\right) .$$

There is a question that could be asked about all the occurrences of expressions involving the zeta function, $\zeta_A(s)$, which occur in these last few theorems. After all, all these expressions are simple rational functions of q^{-s} that can be written down explicitly. We have maintained the zeta function notation for two reasons. First, it makes the analogy with average value results in the number field case more striking. Secondly, consider the following research project. Fix a global function field k other than $\mathbb{F}(T)$ as base field and fix a ring of S-integers A in k. Investigate average value of class numbers of the integral closure of A in quadratic extensions of k. This would all have to be formulated more exactly, but after making everything precise it is fairly clear that special values of $\zeta_A(s)$ will appear in the answer. In this general situation, the zeta function $\zeta_A(s)$ is no longer as simple as in the rational function field case, nevertheless it will play a similar structural role. Thus, it seems reasonable to phrase the more elementary results in the way we have done it.

We want to conclude this chapter by mentioning a number of refinements and generalizations of the above results on class numbers.

The first refinement is to consider only polynomials m that are square-free. In this case, $\mathcal{O}_m$ is the integral closure of $A = \mathbb{F}[T]$ in $K = k(\sqrt{m})$. Thus the class numbers h_m are similar to the class numbers associated to quadratic number fields. In the language of binary quadratic forms, we would be restricting consideration to forms with fundamental discriminants. Averaging in this case is surprisingly difficult. In Hoffstein-Rosen [1], the task is accomplished with the help of functions defined on the metaplectic two-fold cover of $\mathrm{GL}(2, k_\infty)$, where k_∞ is the completion of k at the prime at infinity.

Definition. For $s \in \mathbb{C}$, $\Re(s) \geq \frac{1}{2}$, define

$$c(s) = \prod_P (1 - |P|^{-2} - |P|^{-(2s+1)} + |P|^{-(2s+2)}) .$$

It is easy to see the product converges uniformly and absolutely in the region under consideration.

For simplicity we state the next theorem for the region $\Re(s) \geq 1$. The full result concerns the region $\Re(s) \geq \frac{1}{2}$.

Theorem 17.16. *Let $\epsilon > 0$ be given and assume $s \in \mathbb{C}$ with $\Re(s) \geq 1$.*

(1) If $M = 2n + 1$ is odd, then

$$(q-1)^{-1}(q^M - q^{M-1})^{-1} \sum_m L(s, \chi_m) = \zeta_A(2)\zeta_A(2s)c(s) + O(q^{-n(1-\epsilon)}) ,$$

where the sum is over all square-free m such that $\deg(m) = M$.

(2) If $M = 2n$ is even, then

$$2^{-1}(q-1)^{-1}(q^M - q^{M-1})^{-1} \sum_m L(s, \chi_m) = \zeta_A(2)\zeta_A(2s)c(s) + O(q^{-n(1-\epsilon)}) ,$$

where the sum is over all square-free m such that $\deg(m) = M$ and $\mathrm{sgn}_2(m) = 1$, or over all square-free m with $\deg(m) = M$ and $\mathrm{sgn}_2(m) = -1$.

The problem of working out the generalization of these results to the case of arbitrary global function fields as base field has been solved by B. Fisher and S. Friedberg. They use a new technique of "double Dirichlet series." Their paper should appear soon.

Motivated by questions about rank 2 Drinfeld modules on $A = \mathbb{F}[T]$ with complex multiplication by orders in quadratic extensions of $k = \mathbb{F}(T)$, D. Hayes has formulated average value results about degrees of "minimal ideals" and j-invariants which refine Theorem 17.13. Results and conjectures in the case of all discriminants have been published in Hayes [7]. Using Theorem 17.16, he and his former student, Z. Chen, have also treated the case of averaging over fundamental discriminants. This has not yet appeared.

Finally, we point out that one can move beyond the consideration of quadratic extensions. Let l be a prime dividing $q - 1$. Then, $\mathbb{F}^*$ contains a primitive l-th root of unity. It follows that every cyclic extension of degree l of $k = \mathbb{F}(T)$ is obtained by adjoining an l-th root of a polynomial in A. One can develop a theory of orders in such an extension and try to build a theory of average values of class numbers similar to what we have seen in the case of quadratic extensions. This was done in Rosen [3]. Even in the case where all discriminants are under consideration, the averaging process becomes more difficult. In this paper the case $l = 3$ is treated in detail. Later, in her Brown University Ph.D. thesis, G. Menochi was able to handle the case $l = 5$. The complication increases at each step. To get a completely general result looks out of reach without new methods. At this time, no one has attempted the task of averaging class numbers over fundamental discriminants when $l > 2$.

Exercises

1. Suppose $f : \mathcal{D}_K^+ \to \mathbb{C}$ and that $\mathrm{Ave}(f)$ exists in the sense defined at the beginning of this Chapter. Show that

$$\mathrm{Ave}(f) = \lim_{N \to \infty} \frac{\sum_{\deg D \leq N} f(D)}{\sum_{\deg D \leq N} 1} \ .$$

2. Let $D = \sum a(P)P$ be an effective divisor of K. Let $m > 0$ be a positive integer We say that D is m-th power free if for each P, $a(P) \neq 0$ implies $m \nmid a(P)$. Let f_m be the characteristic function of the m-th power free divisors. Show $\mathrm{Ave}(f_m) = \zeta_K(m)^{-1}$.

3. Let $m > 0$ be a positive integer and D an effective divisor. Define

$$\sigma_m(D) = \sum_{0 \leq C \leq D} NC^m \ .$$

Find an asymptotic formula for $S_m(N) = \sum_{\deg D = N} \sigma_m(D)$.

4. Let $D = \sum a(P)P$ be an effective divisor of K. Define $\mu(D) = 1$ if D is the zero divisor, $\mu(D) = (-1)^t$ if D is square-free and exactly t of the coefficients $a(P)$ are not zero, and $\mu(D) = 0$ otherwise. For every fixed $\epsilon > 0$, show

$$\sum_{\deg D = N} \mu(D) = O(q^{(\frac{1}{2}+\epsilon)N}) \, .$$

5. Let D be an effective divisor of K. Define $d_m(D)$ to be the number of $m+1$-tuples of effective divisors $(C_1, C_2, \ldots, C_m, C_{m+1})$ such that $\sum_{i=1}^{m+1} C_i = D$. Note that $d_1(D) = d(D)$ is equal to the number of divisors of D. Show that $\zeta_{d_m}(s) = \zeta_K(s)^{m+1}$ and use Theorem 17.4 to derive an asymptotic formula for $\Delta_m(N) = \sum_{\deg D = N} d_m(D)$.

6. Prove Theorems 17.14 and 17.15.

7. In the situations of Theorem 17.14 and 17.15 find a formula for

$$q^{-M} \sum_{\substack{m \text{ monic} \\ \deg m = M}} L(1/2, \chi_m) \, .$$

Appendix

A Proof of the Function Field Riemann Hypothesis

In this Appendix we will give a detailed exposition of E. Bombieri's proof of the Riemann Hypothesis for function fields over finite fields or, in other language, for curves over finite fields. For the statement, see Theorem 5.10 of Chapter 5 or Theorem A7 below.

For hyperelliptic curves this result was first conjectured, in the 1920s, by E. Artin. In the 1930s, H. Hasse made the first substantial contribution by proving it in the case of function fields of genus one (the case of elliptic curves). In the late 1940s, A. Weil found a way to prove the general result. In fact, he gave two proofs; one involved intersection theory on algebraic surfaces, the second involved l-adic representations and abelian varieties. Both proofs used sophisticated algebraic geometry. In fact Weil had to redo the foundations of algebraic geometry to provide the necessary background for his proofs. It was a surprise then when S.A. Stepanov, in the late 1960s, found a proof, albeit in special cases, which involved nothing deeper than the Riemann-Roch theorem. Soon thereafter, W. Schmidt was able ot use Stepanov's ideas to prove the general result. Finally, E. Bombieri found a substantial simplification of the proof of Stepanov and Schmidt, see Bombieri [1], both for his original treatment and for references to this history.

Although the proof we are about to give is extremely ingenious and "elementary" it has to be admitted that Weil's original method, especially the approach involving algebraic surfaces, is much more natural. However, this intersection-theoretic proof requires extensive background whereas Bombieri's proof uses nothing more than material developed in this book.

It is possible to give Bombieri's proof purely in the context of function fields without mentioning algebraic curves. The resulting treatment is logically coherent, but feels very artificial. As a compromise, we will assume that the reader is familiar with the beginnings of the theory of algebraic curves as is presented, for example, in Fulton [1], and sketch the connection between the algebraic-geometric language and the language we have used in this book. Having done that, we will switch back and forth as convenient.

Let C be a complete, non-singular algebraic curve defined over a finite field $\mathbb{F}$. We assume that C is embedded in projective N-space, $\mathbb{P}^N(\bar{\mathbb{F}})$, where $\bar{\mathbb{F}}$ denotes a fixed algebraic closure of $\mathbb{F}$. Let $K = \mathbb{F}(C)$ be the function field of C over $\mathbb{F}$. Recall that a typical element of $f \in K$ is represented by a quotient F/G where F and G are homogeneous polynomials of the same degree in the ring $\mathbb{F}[X_0, X_1, \ldots, X_N]$, and where G does not vanish identically on C. In this circumstance, G only vanishes at finitely many points of C, and f defines an actual function on the complement of this set to $\bar{\mathbb{F}}$. K is an algebraic function field in one variable over $\mathbb{F}$ and one can show $\mathbb{F}$ is algebraically closed in K. Now, let $\bar{K} = \bar{\mathbb{F}}(C)$. Then $\bar{K} = K\bar{\mathbb{F}} = \bigcup_n K\mathbb{F}_n = \bigcup_n K_n$.

There is a one to one correspondence between points on $C(\bar{\mathbb{F}})$ and primes in $\bar{K}$. If $\alpha = [\alpha_0, \alpha_1, \ldots, \alpha_N] \in C(\bar{\mathbb{F}})$, let O_α be the set of elements $f \in \bar{K}$ represented by F/G where $G(\alpha) \neq 0$ (F and G are homogeneous polynomials with coefficients in $\bar{\mathbb{F}}$). This is easily seen to be a ring. Let $\mathcal{P}_\alpha \subset O_\alpha$ be the set of $f \in O_\alpha$ such that $f(\alpha) = 0$. Then, O_α is a discrete valuation ring and $\mathcal{P}_\alpha$ is its maximal ideal. The fact that O_α is a dvr follows from the assumption that every point on C, in particular α, is non-singular. One can show that the map $\alpha \to (O_\alpha, \mathcal{P}_\alpha)$ is one to one and onto map from $C(\bar{\mathbb{F}})$ to the primes of the function field $\bar{K}$.

There is also a natural map from $C(\bar{\mathbb{F}})$ to the primes of K. If $\alpha \in C(\bar{\mathbb{F}})$, let R_α be the set of elements in $f \in K$ which are represented by F/G with $G(\alpha) \neq 0$ where the coefficients of both F and G are in $\mathbb{F}$. Let P_α be the set of elements $f \in R_\alpha$ such that $f(\alpha) = 0$. Then, R_α is a dvr and P_α is its maximal ideal. It is useful to remark that the residue class field, R_α/P_α is generated over $\mathbb{F}$ by adjoining the ratios of the coordinates of α. We call that field $\mathbb{F}(\alpha)$ and note that $\deg_K P_\alpha = [\mathbb{F}(\alpha) : \mathbb{F}]$.

The map $\alpha \to (R_\alpha, P_\alpha)$ from $C(\bar{\mathbb{F}})$ to the set of primes of K is onto, but it is not one to one. In fact, we have $P_\alpha = P_{\alpha'}$ if and only if $\sigma\alpha = \alpha'$ for some $\sigma \in \mathrm{Gal}(\bar{\mathbb{F}}/\mathbb{F})$. An automorphism σ operates on a point $\alpha = [\alpha_0, \alpha_1, \ldots, \alpha_N]$ by $\sigma\alpha = [\sigma\alpha_0, \sigma\alpha_1, \ldots, \sigma\alpha_N]$. It follows from this that the number of points in $C(\bar{\mathbb{F}})$ which correspond to a given prime P of K is equal to $\deg_K P$. An important special case of this remark is that there is a one to one correspondence between rational points on C, i.e. $C(\mathbb{F})$, and primes P of K of degree 1. The number of primes of degree 1 of K was denoted by $N_1(K)$ in Chapter 8. See Proposition 8.18 and also Proposition 5.12.

We define a rational map $\phi : \mathbb{P}^N(\bar{\mathbb{F}}) \to \mathbb{P}^N(\bar{\mathbb{F}})$ by

$$\phi([x_0, x_1, \ldots, x_N]) = [x_0^q, x_1^q, \ldots, x_N^q] \,.$$

This is called the Frobenius morphism. It has the important property that an element is fixed under ϕ if and only if it is in $\mathbb{P}^N(\mathbb{F})$. More generally, an element is fixed under ϕ^n if and only if it is in $\mathbb{P}^N(\mathbb{F}_n)$. Since C is defined over $\mathbb{F}$, $C(\bar{\mathbb{F}})$ is defined by the vanishing of a set of homogeneous polynomials with coefficients in $\mathbb{F}$. It follows easily that ϕ maps $C(\bar{\mathbb{F}})$ to itself and that the fixed points of this action, C^ϕ, is equal to $C(\mathbb{F})$. More generally, $C^{\phi^n} = C(\mathbb{F}_n)$.

As we saw in Chapter 8, the Galois group of $\bar{K}/K$ is isomorphic to the Galois group of $\bar{\mathbb{F}}/\mathbb{F}$. The latter group is generated (topologically) by π, the automorphism that takes $\gamma \in \bar{F}$ to γ^q. We use the same letter π to denote the corresponding automorphism of $\bar{K}$. We think of π as acting on the coefficients of functions. Note that ϕ and π have the same action on points of $C(\bar{\mathbb{F}})$. Using this and the definitions, we find that for $\alpha \in C(\bar{\mathbb{F}})$ we have $\pi \mathcal{P}_\alpha = \mathcal{P}_{\phi(\alpha)}$. Thus, a prime $\mathcal{P}$ of $\bar{K}$ corresponds to a rational point if and only if $\pi \mathcal{P} = \mathcal{P}$.

Our first goal is to establish, under some mild restrictions, an upper bound for $N_1(K)$. We will show

Theorem A1. *Let g be the genus of C and suppose $(g+1)^4 < q$ and that q is an even power of the characteristic p. Then,*

$$N_1(K) \leq q + 1 + (2g+1)\sqrt{q} \,.$$

Before getting to the proof of Theorem A1, we will need a number of preliminary results.

We may assume that $C(\mathbb{F})$ is non-empty since otherwise the Theorem is vacuous. Let o be a rational point, i.e. an element of $C(\mathbb{F})$, and $\mathcal{P}_o$ the corresponding prime of $\bar{K}$. For each positive integer m define $R_m = L(m\mathcal{P}_o) = \{f \in \bar{K} \mid (f) + m\mathcal{P}_o \geq 0\}$. R_m is a finite dimensional vector space over $\bar{\mathbb{F}}$ and we know a lot about this dimension via the Riemann-Roch theorem, Theorem 5.4.

Proposition A2.

1) $\dim R_{m+1} \leq \dim R_m + 1$.

2) $\dim R_m \leq m + 1$.

3) $\dim R_m \geq m - g + 1$ with equality if $m > 2g - 2$.

4) $R_m \circ \phi \subseteq R_{mq}$

5) $f \in R_m$ implies $f \circ \phi$ is a q-th power and $(f \circ \phi) = q\pi^{-1}(f)$.

6) $\dim R_m^{p^e} = \dim R_m$ for $e \geq 0$.

7) $\dim R_m \circ \phi = \dim R_m$.

Proof. To prove 1, note that if f and g both have a pole of order $m + 1$ at $\mathcal{P}_o$ then f/g has order 0 at $\mathcal{P}_o$ and thus is congruent to a constant γ modulo $\mathcal{P}_o$. It follows that $f - \gamma g = g(f/g - \gamma)$ has a pole of order at most m, i.e. $f - \gamma g \in R_m$.

Since R_0 consists precisely of the constants, it has dimension 1. Assertion 2 now follows from 1 and induction.

Assertion 3 is simply Riemann's Theorem, Theorem 6.6, together with $\deg_K \mathcal{P}_o = 1$.

We can deal with 4 and 5 simultaneously. Let f be represented by the quotient of two homogeneous polynomials with coefficients in $\bar{\mathbb{F}}$. Set $\lambda = \pi^{-1} \in \text{Gal}(\bar{\mathbb{F}}/\mathbb{F})$. We find

$$(f \circ \phi)(\alpha) = \frac{F(\phi(\alpha))}{G(\phi(\alpha))} = \frac{\lambda F(\alpha)^q}{\lambda G(\alpha)^q} = \lambda f(\alpha)^q .$$

Thus, $f \circ \phi = \lambda f^q$ and consequently $(f \circ \phi) = q\lambda(f)$ which proves 5. Also, if $f \in R_m$, then so is λf since $\text{ord}_{\mathcal{P}_o}(\lambda f) = \text{ord}_{\pi \mathcal{P}_o}(f) = \text{ord}_{\mathcal{P}_o}(f)$. Thus, $f \circ \phi = \lambda f^q \in R_{mq}$ which proves 4.

The map $f \to f^{p^e}$ is a quasi-linear isomorphism of R_m with $R_m^{p^e}$ which proves 6.

Finally, to show 7 it is enough to check that $f \to f \circ \phi$ is one to one. If $f \circ \phi = g \circ \phi$, then $\lambda f^q = \lambda g^q$ so $\lambda f = \lambda g$ which implies $f = g$ (apply π to both sides).

If A is a subspace of R_m and B is a subspace of R_n we denote by AB the subspace of R_{m+n} generated by all the products fg where $f \in A$ and $g \in B$.

Proposition A3. *If* $lp^e < q$, *then the natural homomorphism from* $R_l^{p^e} \otimes_{\bar{\mathbb{F}}} R_m \circ \phi$ *to* $R_l^{p^e}(R_m \circ \phi)$ *is an isomorphism.*

Proof. By Proposition A2, part 1, we see that R_m has a basis $\{f_1, f_2, \ldots, f_t\}$ such that $\text{ord} f_i < \text{ord} f_{i+1}$ for $i = 1, 2, \ldots, t - 1$. Any element of the tensor product can be written in the form

$$\sum_{i=1}^{t} g_i^{p^e} \otimes f_i \circ \phi ,$$

where the g_i are elements of R_l. If such an element is in the kernel of the natural homomorphism, we would have a relation of the form

$$\sum_{i=1}^{t} g_i^{p^e}(f_i \circ \phi) = 0 .$$

We will show this can't happen unless $g_i = 0$ for all $i = 1, 2, \ldots, t - 1$ and that will establish the Proposition.

Suppose some $g_i \neq 0$ and let r be the smallest such index. Then,

$$g_r^{p^e}(f_r \circ \phi) = -\sum_{i=r+1}^{t} g_i^{p^e}(f_i \circ \phi) \ .$$

Taking the order of both sides at $\mathcal{P}_o$ and using Proposition A2, part 5, we see

$$p^e \mathrm{ord}\ g_r + q\ \mathrm{ord}\ f_r \geq \min_{i > r}\ (p^e \mathrm{ord}\ g_i + q\ \mathrm{ord}\ f_i) \geq -lp^e + q\ \mathrm{ord}\ f_{r+1}\ .$$

Thus,

$$p^e \mathrm{ord}\ g_r \geq -lp^e + q(\mathrm{ord}\ f_{r+1} - \mathrm{ord}\ f_r) \geq q - lp^e > 0\ .$$

It follows that g_r has a zero at $\mathcal{P}_o$. Since $g_r \in R_l$ it has no poles away from $\mathcal{P}_o$ and a zero at $\mathcal{P}_o$. It follows that $g_r = 0$. This contradicts our assumption, and so, completes the proof.

Corollary. If $lp^e < q$, then $\dim R_l^{p^e}(R_m \circ \phi) = (\dim R_l)(\dim R_m)$.

Proof. This follows directly from the Proposition and from Proposition A2, part (7).

We have now completed the preliminaries.

Proof of Theorem A1. The idea is to produce a function with a high order zero at each rational point and a small number of poles. We will see how this works as we go along. We continue to assume that $lp^e < q$.

We begin by defining a $\mathbb{F}$-linear homomorphism δ from $R_l^{p^e}(R_m \circ \phi)$ to $R_l^{p^e} R_m$. Using the notation established in the proof of Proposition A3, this is given by

$$\delta(\sum_{i=1}^{t} g_i^{p^e}(f_i \circ \phi)) = \sum_{i=1}^{t} g_i^{p^e} f_i\ .$$

That δ is well defined follows immediately from Proposition A3. The dimension of the domain of δ is greater than $(l - g + 1)(m - g + 1)$ by Proposition A2, parts 6 and 7 and Riemann's theorem, part 3. Assume that $l, m \geq g$. Then, the image of δ is contained in $L((lp^e + m)\mathcal{P}_o)$ which has dimension $lp^e + m - g + 1$, again using Proposition A2, part 3. Thus,

$$\dim \ker \delta \geq (l - g + 1)(m - g + 1) - (lp^e + m - g + 1)\ .$$

If the quantity on the right is positive, the kernel is not empty. Assume this and let $f = \sum_i g_i^{p^e}(f_i \circ \phi)$ be a non-zero element of $\ker \delta$. If $o \neq \alpha \in C(\mathbb{F})$, we calculate

$$f(\alpha) = \sum_i g_i(\alpha)^{p^e} f_i(\phi(\alpha)) = \sum_i g_i(\alpha)^{p^e} f_i(\alpha) = 0\ .$$

So, f must vanish at every point of $C(\mathbb{F})$ except o. Moreover, by Proposition A2, part 5, and the fact that $p^e < q$ by hypothesis, we see from the expression for f as a sum that f is a p^e-th power. Thus,

$$p^e(N_1(K) - 1) \le \deg_K(f)_0 = \deg_K(f)_\infty \le lp^e + mq .$$

We have used the fact that $R_m \circ \phi \subseteq R_{mq}$ by Proposition A2, part 4. This inequality yields

$$N_1(K) \le 1 + l + mqp^{-e} .$$

Our proof of this inequality is subject to the conditions

a) $lp^e < q$.

(b) $l, m \ge g$.

(c) $(l - g + 1)(m - g + 1) > lp^e + m - g + 1$.

We proceed to make suitable choices for l, m, and e so that these three conditions are fulfilled and makes the above inequality into the one asserted in the statement of the Theorem.

We are assuming that q is an even power of p, so set $q = p^{2b}$ and choose $e = b$. Set $m = p^b + 2g$. We now want to choose l so that condition (c) holds. Simplifying that inequality slightly, we need

$$(l - g)(m + 1 - g) > lp^b$$

or

$$(l - g)(p^b + g + 1) > lp^b$$

or

$$l > \frac{g}{g+1}p^b + g .$$

Let's choose $l = [gp^b/(g+1)] + g + 1$ ($[r]$ denotes the greatest integer less than or equal to r). With these choices for l and m, conditions b and c are fulfilled.

Assuming $(g+1)^4 < q$ we will now show that condition a is also fulfilled. Note that $(g+1)^4 < q$ implies $(g+1)^2 < p^b$ which yields $gp^b + (g+1)^2 < (g+1)p^b$. Thus,

$$\frac{g}{g+1}p^b + g + 1 < p^b \qquad . \tag{1}$$

This inequality implies $l < p^b$, so that $lp^b < p^{2b} = q$ which is condition c.

Let's substitute our choices for l, m, and e into the inequality $N_1(K) < l + 1 + mqp^{-e}$. Since, by Equation 1 we have $l < p^b$ we find

$$N_1(K) < p^b + 1 + (p^b + 2g)p^b = q + 1 + (2g + 1)\sqrt{q} .$$

This completes the proof of Theorem A1.

Having produced a good upper bound for $N_1(K)$, we now take up the task of producing a suitable lower bound. The method will involve consideration of Galois extensions of K.

Let L/K denote a finite, geometric, Galois extension of K with Galois group G. Let $\bar{L} = \bar{\mathbb{F}}L$ and $\bar{K} = \bar{\mathbb{F}}K$. Since we have assumed L/K is geometric, it follows that $\text{Gal}(\bar{L}/\bar{K}) \cong \text{Gal}(L/K) = G$. We will simply identify G with the Galois group of $\bar{L}/\bar{K}$. The Frobenius element π of $\bar{L}/L$ maps to the Frobenius element of $\bar{K}/K$ by restriction. We will denote both of these automorphisms by π. Note that as automorphisms of $\bar{L}/K$, π commutes with G (use the fact that $\bar{L}$ is the composite of L and $\bar{K}$).

Let T denote the set of primes in $\bar{K}$ which lie above rational primes in K, i.e. those primes of K whose degree is equal to 1. As we have seen $|T| = N_1(K)$. Also, we showed earlier that the primes in T are characterized by the condition $\pi\mathcal{P} = \mathcal{P}$. Let $\tilde{T}$ denote the primes in $\bar{L}$ lying above those in T. If $\mathcal{P} \in T$ then the set of primes above $\mathcal{P}$ are acted upon transitively by G. Also, π maps this set into itself since $\pi\mathcal{P} = \mathcal{P}$. Thus, if $\mathfrak{P} \in \tilde{T}$ there is a $\sigma \in G$ such that $\pi\mathfrak{P} = \sigma\mathfrak{P}$. Moreover, the element σ is uniquely determined if $\mathfrak{P}/\mathcal{P}$ is unramified. Let $\tilde{T}' \subseteq \tilde{T}$ be the set of unramified primes in $\tilde{T}$. We have defined a map

$$\eta : \tilde{T}' \to G .$$

Definition. With the above notations, let $\tilde{T}(\sigma)$ denote the set of unramified primes $\mathfrak{P}$ in $\tilde{T}$ such that $\eta(\mathfrak{P}) = \sigma$. Let $N(\bar{L}/\bar{K}, \sigma)$ denote the number of elements in $\tilde{T}'(\sigma)$.

A few observations will be useful. For each prime in T which is unramified in L, there are $|G|$ primes above it in $\tilde{T}$. This follows from from the fundamental relation $efg = |G|$ (see Proposition 9.3) since $f = 1$ because we a working over an algebraically closed constant field $\bar{\mathbb{F}}$. Thus,

$$|\tilde{T}| = |G|N_1(K) + O(1) .$$

The error term depends on the number of ramified primes in $\bar{L}/\bar{K}$ but is independent of q. We will later vary the constant field, i.e consider the fields $K_n = \mathbb{F}_n K$, and so the $O(1)$ term will not matter much.

Since $\tilde{T}' = \cup_{\sigma \in G} \tilde{T}(\sigma)$, disjoint union, we find

$$\sum_{\sigma \in G} N(\sigma, \bar{L}/\bar{K}) = |G|N_1(K) + O(1) . \tag{2}$$

We will need this relation shortly.

For those readers who prefer more geometric language, $\bar{L}$ corresponds to a curve $\tilde{C}$ defined over $\bar{\mathbb{F}}$ and covering C, i.e. there is an epimorphism $\psi : \tilde{C} \to C$. The set T corresponds to the set of rational points $C(\bar{\mathbb{F}})$ on C and the set $\tilde{T}$ corresponds to $\psi^{-1}(C(\bar{\mathbb{F}}))$. The fibers above the points in $C(\bar{\mathbb{F}})$ are mapped into themselves by both ϕ (or π) and G, etc. The whole argument can be given in either language.

Proposition A4. *With the above notations and definitions let $\tilde{g}$ be the genus of $\bar{L}$ (which is the same as that of L) and let σ be an element of G. Suppose $(\tilde{g}+1)^4 < q$ and that q is an even power of p. Then,*

$$N_1(\sigma, \bar{L}/\bar{K}) \leq q + 1 + (2\tilde{g} + 1)\sqrt{q} .$$

Proof. The proof is almost identical with the proof of Theorem A1. One supposes there is at least on rational prime $\mathfrak{P}_o$ in $\bar{L}$ and defines the vector spaces $R_n = L(n\mathfrak{P}_o)$. In the proof of that Theorem we begin with defining a homomorphism δ from $R_l^{p^e}(R_m \circ \phi) \to R_l^{p^e} R_m$. We modify that to get a map δ_σ from $R_l^{p^e}(R_m \circ \phi) \to R_l^{p^e}(R_m \circ \sigma)$ as follows

$$\delta_\sigma\left(\sum_{i=1}^t g_i^{p^e}(f_i \circ \phi)\right) = \sum_{l=1}^t g_i^{p^e}(f_i \circ \sigma) .$$

Just as before, one invokes Proposition A3 to insure this map is well defined. The rest of the proof goes exactly as before with the one exception. After assuming $l \geq \tilde{g}$ and $m \geq \tilde{g}$ one has to show that the image of δ_σ has dimension less than or equal to $lp^e + m - g + 1$.

If $f \in R_m$, then $f \circ \sigma \in L(m\sigma^{-1}\mathfrak{P}_o)$. It follows that the image of δ_σ is contained in $L(lp^e\mathfrak{P}_o + m\sigma^{-1}\mathfrak{P}_o)$ which has dimension $lp^e + m - g + 1$ by Riemann-Roch.

We leave it to the reader to check the remaining details.

Proposition A5. *With the same notations as above, for all $\sigma \in G$,*

$$q + 1 + (N_1(K) - q - 1)|G| + O(\sqrt{q}) \leq N_1(\sigma, \bar{L}/\bar{K}) .$$

Proof. By Proposition A4, we have for each $\sigma \in G$,

$$0 \leq [q + 1 + (2\tilde{g} + 1)\sqrt{q} - N_1(\sigma, \bar{L}/\bar{K})] .$$

Sum over σ and one finds

$$0 \leq \sum_{\sigma \in G}[\ ] \leq (q + 1 + (2\tilde{g} + 1)\sqrt{q})|G| - |G|N_1(K) + O(1) .$$

We have used Equation 2. Since each term in brackets is positive, we deduce

$$q + 1 + (2\tilde{g} + 1)\sqrt{q} - N_1(\sigma, \bar{L}/\bar{K}) \leq (q + 1 + (2\tilde{g} + 1)\sqrt{q})|G| - |G|N_1(K) + O(1) .$$

From this inequality, the Proposition follows immediately.

We are aiming to prove that $N_1(K) = q + O(\sqrt{q})$. Theorem A1 assures us that with mild restrictions $N_1(K) \leq q + O(\sqrt{q})$ so we must derive the inequality in the other direction. Proposition A5 allows us to do this quickly in a special case. Suppose we can find an element $x \in K$ such that $K/\mathbb{F}(x)$ is a finite, geometric, Galois extension. Let's apply Proposition A5 to the pair of field $\bar{K}$ and $\bar{\mathbb{F}}(x)$ instead of $\bar{L}$ and $\bar{K}$. Since the number of rational primes in $\mathbb{F}(x)$ is exactly $q + 1$, we find $q + O(\sqrt{q}) \leq N_1(\sigma, \bar{K}/\bar{\mathbb{F}}(x))$ for each $\sigma \in \mathrm{Gal}(\bar{K}/\bar{\mathbb{F}}(x)$. Summing over σ and using Equation 2 again we find $[\bar{K} : \bar{\mathbb{F}}(x)]q + O(\sqrt{q}) \leq [\bar{K} : \bar{\mathbb{F}}(x)]N_1(K)$. Cancel $[\bar{K} : \bar{\mathbb{F}}(x)]$ and we have our proof that $N_1(K) = q + O(\sqrt{q})$.

In general, we cannot find an element $x \in K$ with all these nice properties. However, since $\mathbb{F}$ is a perfect field, one can find an element $x \in K$ such that $K/\mathbb{F}(x)$ is separable (see Lang [4], Chapter VIII, Proposition 4.1). Let L be the Galois closure of $K/\mathbb{F}(x)$, i.e., the smallest algebraic extension of K that is Galois over $\mathbb{F}(x)$. It can happen that the constant field, $\mathbb{E}$, of L is larger than $\mathbb{F}$. If so replace $\mathbb{F}(x)$ by $\mathbb{E}(x)$ and K by $\mathbb{E}K$. Then, all three extensions are geometric and $L/\mathbb{E}(x)$ is Galois. So, at the expense of making a small constant field extension, we can assume $\mathbb{F}$ is the constant field of L to begin with. We shall see that small constant field extensions will not affect the overall proof.

Theorem A6. *Let $K/\mathbb{F}$ be a function field of genus g over a finite field $\mathbb{F}$ with q elements. Suppose q is an even power of p. Suppose further that there is an element $x \in K$ such that the Galois closure, L, of $K/\mathbb{F}(x)$ is a geometric extension of $\mathbb{F}(x)$. Finally, assume $(g + 1)^4 < q$. Then, $N_1(K) = q + O(\sqrt{q})$.*

Proof. By Theorem A1, $N_1(K) \leq q + O(\sqrt{q})$, so it remains to prove the opposite inequality.

Let $G = \mathrm{Gal}(\bar{L}/\bar{\mathbb{F}}(x))$ and $H = \mathrm{Gal}(\bar{L}/\bar{K})$. Let $\sigma \in G$. Applying Proposition A5 to the extension $\bar{L}/\bar{\mathbb{F}}(x)$ we find

$$q + O(\sqrt{q}) \leq N_1(\sigma, \bar{L}/\bar{\mathbb{F}}(x)) .$$

Sum these inequalities over all elements in $\tau \in H$. We obtain

$$|H|q + O(\sqrt{q}) \leq \sum_{\tau \in H} N_1(\tau, \bar{L}/\bar{\mathbb{F}}(x)) .$$

We will show in a moment that if $\tau \in H$, $N_1(\tau, \bar{L}/\bar{\mathbb{F}}(x)) = N_1(\tau, \bar{L}/\bar{K})$. Assuming this is correct, the sum in the last inequality is

$$\sum_{\tau \in H} N_1(\tau, \bar{L}/\bar{\mathbb{F}}(x)) = \sum_{\tau \in H} N_1(\tau, \bar{L}/\bar{K}) = |H|N_1(K) + O(1) ,$$

using Equation 2 one more time.

Putting the last two relations together, we have $q + O(\sqrt{q}) \leq N_1(K)$ which is the result we are looking for.

It remains to prove $N_1(\tau, \bar{L}/\bar{\mathbb{F}}(x)) = N_1(\tau, \bar{L}/\bar{K})$ if $\tau \in H$. Let $\mathfrak{P}$ be a prime of $\bar{L}$ lying over a rational prime $\mathcal{P}$ of $\bar{\mathbb{F}}(x)$. If $\pi\mathfrak{P} = \tau\mathfrak{P}$ for $\tau \in H$, we have to show that $\mathfrak{P}$ lies over a rational prime of $\bar{K}$. Let $\mathfrak{p}$ lie below $\mathfrak{P}$ in $\bar{K}$. Then, $\pi\mathfrak{P} = \tau\mathfrak{P}$ implies $\pi\mathfrak{p} = \tau\mathfrak{p} = \mathfrak{p}$. However, we have seen that the relation $\pi\mathfrak{p} = \mathfrak{p}$ characterizes the rational primes of $\bar{K}$. This completes the proof.

Our final task is to show that Theorem A6 is equivalent to the Riemann Hypothesis for function fields over finite fields. This is relatively simple.

Theorem A.7 (the Riemann Hypothesis for Function Fields) *Let $K/\mathbb{F}$ be a function field with finite constant field, $\mathbb{F}$, having q elements. Let $\zeta_K(s)$ be the zeta function of K. All the zeros of $\zeta_K(s)$ lie on the line $\Re(s) = \frac{1}{2}$.*

Proof. If we make the substitution $u = q^{-s}$ we have

$$\zeta_K(s) = Z_K(u) = \frac{L_K(u)}{(1-u)(1-qu)} ,$$

where

$$L_K(u) = \prod_{i=1}^{2g}(1 - \pi_i u) .$$

Here, g denotes the genus of K. As we have pointed out in many places, the assertion that all the zeros of $\zeta_K(s)$ lie on the line $\Re(s) = \frac{1}{2}$ is equivalent to the assertion that all the inverse roots, π_i, of $L_K(u)$ have absolute value $\sqrt{q}$.

We first remark that to prove the Theorem, it suffices to prove it for any constant field extension, $K_n = \mathbb{F}_n K$, of K. This follows from Proposition 8.16 which asserts that

$$L_{K_n}(u) = \prod_{i=1}^{2g}(1 - \pi_i^n u) .$$

Thus, if the Riemann Hypothesis is true for K_n we would have $|\pi_i^n| = q^{\frac{n}{2}}$ for all $i = 1, \ldots, 2g$ which implies, obviously, that $|\pi_i| = q^{\frac{1}{2}}$ for all $i = 1, \ldots, 2g$. This is the Riemann Hypothesis for K.

Let's choose n so large that $(g + 1)^4 < q^n$. If q^n is not an even power of p, replace n by $2n$. Next, as we have seen, we may, by taking a finite extension $\mathbb{F}_m$ of $\mathbb{F}_{2n}$, assume there is an $x \in K_m$ such that $K_m/\mathbb{F}_m(x)$ is separable and that the Galois closure L of $K_m/\mathbb{F}_m(x)$ is a geometric extension of $\mathbb{F}_m(x)$. Thus, we have shown that we can find an $m \geq 1$ so that all the conditions of Theorem A6 are fulfilled for K_m. By the last paragraph, to prove the Riemann Hypothesis for K we may as well assume all these properties hold already for $K/\mathbb{F}$. If these conditions hold for $K/\mathbb{F}$, they hold in any constant field extension. Theorem A6 then implies that $N_1(K_n) = q^n + O(q^{\frac{n}{2}})$ for all $n \geq 1$.

We recall some facts proved in the text, namely at the end of Chapters 5 and 8. First, $N_1(K_n) = N_n(K)$ where $N_n(K)$ is defined by $N_n(K) =$

$\sum_{d|n} da_d(K)$. The number $a_d(K)$ denotes the number of prime divisors of K of K-degree equal to d. Moreover,

$$Z_K(u) = \exp\left(\sum_{n=1}^{\infty} \frac{N_n(K)}{n} u^n\right).$$

Taking the logarithmic derivative of $Z_K(u)$ and using the above identity yields

$$u\frac{Z'_K(u)}{Z_K(u)} = \sum_{n=1}^{\infty} N_1(K_n)u^n.$$

Now write $Z_K(u) = \prod_{i=1}^{2g}(1-\pi_i u)/(1-u)(1-qu)$ and calculate $uZ'_K(u)/Z_K(u)$ using this formula. We find

$$u\frac{Z'_K(u)}{Z_K(u)} = \sum_{n=1}^{\infty} (q^n + 1 - \pi_1^n - \pi_2^n - \cdots - \pi_{2g}^n)u^n.$$

Combining these formulas produces the following identity.

$$\sum_{n=1}^{\infty}(N_1(K_n) - q^n - 1)u^n = -\sum_{i=1}^{2g}\sum_{n=1}^{\infty}(\pi_i u)^n.$$

Since $N_1(K_n) = q^n + O(q^{\frac{n}{2}})$, the sum on the left has radius of convergence at least $q^{-\frac{1}{2}}$. The radius of convergence of the sum on the right is exactly the minimum over i of the quantities $|\pi_i|^{-1}$. Thus, $|\pi_i| \leq q^{\frac{1}{2}}$ for all i. We know that $\pi_i \to q/\pi_i$ is a permutation of the set of inverse roots of $L_K(u)$ (see the remarks following Theorem 5.9 where this fact is shown to equivalent to the functional equation for $\zeta_K(s)$). It follows that $|\pi_i| = q^{\frac{1}{2}}$ for $1 = 1, 2, \ldots, 2g$. This is what we wanted to prove!!

Bibliography

Apostol, T.

[1] Introduction to Analytic Number Theory, Springer-Verlag, New York-Heidelberg-Berlin, 1976.

Artin, E.

[1] Quadratische Körper im Gebiete der höheren Kongruenzen I, II, *Math. Zeit.* **19** (1924), 153-246.

[2] Über eine neue Art von *L*-Reihen, in: Collected Papers (S. Lang and J. Tate, eds.), Addison-Wesley, Reading, MA, 1965.

Bilharz, H.

[1] Primdivisoren mit vorgegebener Primitivwurzel, *Math. Ann.* **114** (1937), 476-492.

Bombieri, E.

[1] Counting Points on Curves over Finite Fields, *Séminaire Bourbaki*, No. 430, 1972/3.

Carlitz, L.

[1] The arithmetic of polynomials in a finite field, *Amer. J. of Math.* **54** (1932), 476-492.

342   Bibliography

[2] On certain functions connected with polynomials in a Galois field, *Duke Math. J.* **1** (1935), 137-168.

[3] A class of polynomials, *Trans. Amer. Math. Soc.* **43** (1938), 167-182.

Carlitz, L. and Olsen, F.R.

[1] Maillet's Determinant, *Proc. Amer. Math. Soc.* **6** (1955), 265-269.

Chevalley, C.

[1] Introduction to the Theory of Algebraic Function Fields of One Variable, *AMS Math. Surveys* **6**, New York, 1951.

Coates, J.

[1] p-adic *L*-functions and Iwasawa's Theory, in: Algebraic Number Fields (A. Fröhlich, ed.), Academic Press, London-New York-San Francisco, 1977.

Davenport, H.

[1] Multiplicative Number Theory, Springer-Verlag, New York, 1980.

Delange, H.

[1] Généralisation du théorèm de Ikehara, *Ann. Sci. École Norm. Sup.* **71** (1954), 213–242.

Deligne, P. and Ribet, K.

[1] Values of Abelian *L*-functions at negative integers over totally real fields, *Invent. Math.* **59** (1980), 227-286.

Deuring, M.

[1] Lectures on the theory of Algebraic Functions of One Variable, LNM 314, Springer-Verlag, Berlin-Heidelberg-New York, 1973.

Drinfeld, V. G.

[1] Elliptic Modules (Russian), *Math. Sbornik* **94** (1974), 594-627. English translation: *Math. USSR, Sbornik* **23** (1977), 159-170.

[2] Elliptic Modules II (Russian), *Math. Sbornik* **102** (1977), 182-194. English translation: *Math. USSR, Sbornik* **31** (1977, 159-170.

Eichler, M.

[1] Introduction to the Algebraic Theory of Algebraic Numbers and Algebraic Functions, Academic Press, New York-London, 1966.

Fried, M. and Jarden, M.

[1] Field Arithmetic, Springer-Verlag, New York, 1986.

Fulton, W.

[1] Algebraic Curves, Benjamin, New York, 1969.

Galovich, S. and M. Rosen

[1] The class number of cyclotomic function fields, *J. Number Theory* **13** (1981), 363-375.

[2] Units and class groups in cyclotomic function fields, *J. Number Theory* **14** (1982), 156-184.

[3] Distributions on Rational Function Fields, *Math. Ann.* **256** (1981), 549-560.

Gauss, C. F.

[1] Arithmetische Untersuchungen, Chelsea Publishing Co., New York, 1965.

Gekeler, E.

[1] Zur Arithmetic von Drinfeld-Moduln, *Math. Ann.* *262* (1983), 167-182.

[2] Drinfeld Modular Curves, LNM 1231, Springer-Verlag, Berlin-Heidelberg, 1986.

Goss, D.

[1] The algebraist's upper half plane, *Bull. Amer. Math. Soc.* **2** (3) (1980), 391-415.

[2] The arithmetic of function fields 2; the cyclotomic theory, *J. Algebra* *81* (1983), 107-149.

[3] On a new type of *L*-function for algebraic curves over finite fields, *Pac. J. Math* **105** (1983), 143-181.

344   Bibliography

[4] Basic Structures of Function Field Arithmetic, Ergeb. Math. u. i. Grenz. **35**, Springer-Verlag, Berlin-Heidelberg-New York, 1998.

[5] A Riemann hypothesis for characteristic p L-functions, *J. Number Theory* **82** (2000), 299-322.

Goss, D., Hayes, D.R. and Rosen, M. (ed.)

[1] The Arithmetic of Function Fields, Walter de Gruyter, New York-Berlin, 1992.

Gross, B.

[1] On the Annihilation of Divisor Classes in Abelian Extensions of the Rational Function Field, *Séminaire de Théorie des Nombres, Bordeaux*, Exposé 3, 1980.

Gross, B. and Rosen, M.

[1] Fourier series and special values of L-functions, *Adv. Math.* **69** (1988), 1-31.

Hardy, G.H. and Wright, E.M.

[1] An Introduction to the Theory of Numbers, Fourth Edition, Oxford University Press, New York, 1975.

Hayes, D.R.

[1] Explicit class field theory for rational function fields, *Trans. Amer. Math. Soc.* **189** (1974), 77-91.

[2] Explicit class field theory in global function fields, Studies in Algebra and Number Theory, *Adv. Math. Supplementary Studies* **6** (1979), 173-217.

[3] Analytic class number formulas in global function fields, *Invent. Math.* **65** (1981), 49-69.

[4] Elliptic units in function fields, in *Proceeding of a Conference on Modern Developments Related to Fermat's Last Theorem* (D. Goldfeld, ed.), Birkhäuser, Boston, 1982.

[5] Stickelberger elements in function fields, *Compositio Math.* **55** (1985), 209-239.

[6] A brief introduction to Drinfeld modules, in: The Arithmetic of Function Fields (eds. D. Goss et al), de Gruyter, New York-Berlin, 1992.

[7] Distribution of minimal ideals in imaginary quadratic function fields, in: Applications of Curves over Finite Fields (Seattle, WA, 1997), *Contemporary Math.*, 245, Amer. Math. Soc., Providence, RI, 1999.

Hecke, E.

[1] Eine neue Art von Zetafunktionen und ihre Beziehungen zur Verteilung der Primzahlen, Math. Zeit. **1** (1918), 357-376, II ibid. **6** (1920), 11-51.

[2] Lectures on the Theory of Algebraic Numbers (transls. G. Brauer and J. Goldman), Springer-Verlag, GTM 77, Berlin-Heidelberg-New York, 1981.

Heilbronn, H.

[1] On an inequality in the elementary theory of numbers, *Proc. Cambridge Philos. Soc.* **33** (1937), 207-209.

Hoffstein, J.

[1] Eisenstein Series and Theta Functions on the Metaplectic Group, in: *CRM Proceedings and Lecture Notes*, Vol. 1, 1993.

Hoffstein, J. and Rosen, M.

[1] Average values of L-series in function fields, *J. Reine und Angew. Math.* **426** (1992), 117-150.

Hooley, C.

[1] On Artin's conjecture, *J. Reine und Angew. Math.* **225** (1967), 209-220.

Ireland, K. and Rosen, M.

[1] A Classical Introduction to Modern Number Theory, 2nd Edition, GTM 84, Springer-Verlag, Berlin-Heidelberg-New York, 1990.

Iwasawa, K.

[1] A class number formula for cyclotomic fields, *Ann. of Math.* **76** (1962), 171-179.

[2] On the μ-invariants of cyclotomic fields, *Acta Arithm.* **21** (1972), 99-101.

[3] On $\mathbb{Z}_l$-extensions of algebraic number fields, *Ann. of Math.* **98** (1973), 246-326.

Iwasawa, K. and Tamagawa, T.

[1] On the group of automorphisms of a function field, *J. Math. Soc. Japan* **3** (1951), 137-101. Correction, *J. Math. Soc. Japan* **4** (1952), 203-204.

Jacobson, N.

[1] Basic Algebra I (2nd edition), J.W.H. Freeman and Co., New York, 1985.

[2] Basic Algebra II (2nd edition), J.W.H. Freeman and Co., New York, 1989.

Kani, E.

[1] Bounds on the number of non-rational subfields of a function field, *Invent. Math.* **85** (1986), 185-198.

Kisilevsky, H. and Gold, R.

[1] $\mathbb{Z}_l$-extension of function fields, in: Théorie des Nombres (Québec, 1987), de Gruyter, Berlin-New York, 1987, 280-289.

Kornblum, H.

[1] Uber die Primfunktionen in einer Arithmetischen Progression, *Math. Zeit.* **5** (1919), 100-111.

Knopfmacher, J.

[1] Analytic Arithmetic of Algebraic Function Fields, Marcel Dekker Inc., New York-Basel, 1979.

Lang, S.

[1] Elliptic Functions, Addison-Wesley, Reading, MA, 1973.

[2] Units and class groups in number theory and algebraic geometry, *Bull. Amer. Math. Soc.* **6** (1982), 253-316.

[3] Algebraic and Abelian Functions (2nd edition), GTM 89, Springer-Verlag, New York, 1982.

[4] Algebra (3rd edition), Addison-Wesley, Reading, MA, 1993.

[5] Algebraic Number Theory, GTM 110, Springer-Verlag, New York, 1986.

[6] Cyclotomic Fields I and II (with an appendix by K. Rubin), GTM 121, Springer-Verlag, New York, 1990.

Li, C. and Zhao, J.

[1] Iwasawa theory of $\mathbb{Z}_p^d$-extensions over global function fields, *Expo. Math.* **15** (1997), 315-337.

[2] Class number growth of a family of $\mathbb{Z}_p$-extensions over global function fields, *J. Algebra* **200** (1998), 141-154.

Madan, M.

[1] On class numbers in fields of algebraic functions, *Arch. Math.* **21** (1970), 161-171.

Mason, R.C.

[1] The hyperelliptic equation over function fields, *Math. Proc. Cambridge Philos. Soc.* **93** (1983), 219-230.

[2] Diophantine equations over Function Fields, *London Math. Soc. Lecture Notes Series* 96, Cambridge U. Press, 1984.

Mazur, B.

[1] Rational points of abelian varieties with values in towers of number fields, *Invent. Math.* **18** (1972), 183-266.

Merrill, K.D. and Walling, L.

[1] Sums of squares over function fields, *Duke Math. J.* **71**, Nr. 3 (1993) 665-684.

Milne, J.S.

[1] Jacobian varieties, in arithmetic geometry, (G. Cornell and J. Silverman, eds.), Springer-Verlag, New York-Berlin-Heidelberg, 1986.

Moreno, C.

[1] Algebraic Curves over Finite Fields, Cambridge U. Press, Cambridge-New York-Melbourne, 1991.

Murty, M. R.

[1] Artin's conjecture for primitive roots, *Math. Intelligencer* **10** , No. 4, (1988), 59-67.

Murty, M.R., Rosen, M. and Silverman, J.

[1] Variations on a theme of Romanoff, *Int. J. Math.* **7**, No. 3 (1996), 373-391.

Murty, V.K. and Scherk, J.

[1] Effective versions of the Chebatarov density theorem for function fields, *Compte Rendu* **319** (1994), 523-528.

Mumford, D.

[1] Abelian Varieties, Oxford University Press, Oxford, 1970.

Narkiewicz, W.

[1] Elementary and Analytic Theory of Algebraic Numbers (2nd edition), Springer-Verlag and PWN, New York and Warsaw, 1990.

Neukirch, J.

[1] Algebraic Number Theory, Grundl. Math. Wissens. 322, Springer-Verlag, Berlin-New York, 1999.

Niven, I., Zuckerman, H.S. and Montgomery, H.L.

[1] An Introduction to the Theory of Numbers (3rd edition), John Wiley and Sons, Chichester-Toronto-Brisbane-Singapore, 1991.

Oukhaba, H.

[1] Elliptic units in global function fields, in: The Arithmetic of Function Fields, D. Goss et al, eds., de Gruyter, Berlin-New York, 1992, 87-102.

Pink, R.

[1] The Mumford-Tate conjecture for Drinfeld modules, *Publ. Res. Inst. Math. Sci.* **33** , No. 3, (1997), 393-425.

Pretzel, O.

[1] Codes and Algebraic Curves, Oxford Lecture Series in Math. and Appl., Clarendon Press, Oxford, 1998.

Romanoff, N.P.

[1] Über einiger Sätze der Additive Zahlentheorie, *Math. Ann* **109** (1934), 668-678.

Rosen, M.

[1] The asymptotic behavior of the class group of a function field over a finite field, *Archiv Math. (Basel)* **24** (1973), 287-296.

[2] The Hilbert class field in function fields, *Expos. Math.* **5** (1987), 365-378.

[3] The average value of class numbers in cyclic extensions of the rational function field, Canadian Math. Soc. Conf Proceedings, Vol. 15, 1995.

[4] A note on the relative class number in function fields, *Proc. Amer. Math. Soc.* **125**, Num. 5, (1997), 1299-1303.

[5] A generalization of Merten's theorem, *J. Ramanujan Math. Soc.* **14**, No. 1 (1999), 1-19.

Samual, P. and Zariski, O.

[1] Commutative Algebra, Volumes I and II, Springer-Verlag, New York, 1975-1976.

Schmid, H.L.

[1] *Über die Automorphismen eines algebraischen Funktionenkörper von Primzahlcharakteristik*, J. Reine und Angew. Math. **179** (1938), 5-15.

Schmidt, F.K.

[1] Analytischen Zahlentheorie in Körpern der Characteristik p, *Math. Zeit.* **33** (1931), 668-678.

Serre, J.-P.

[1] Zeta and L Functions, in Arithmetic Algebraic Geometry (O. F. G. Schilling, ed.), Harper and Row, New York, 1965, 82-92.

[2] Local Fields, transl. by M. Greenberg, Springer-Verlag, New York, 1979.

[3] Linear Representations of Finite Groups, transl. by L. L. Scott, Springer-Verlag, New York (third printing) 1986.

350   Bibliography

Siegel, C.L.

[1] Berechnung von Zetafunctionen an ganzzahligen Stellen, *Nachr. Akad. Wissen. Göttingen* (1969) 87-102.

Shu, L.

[1] Class number formulas over global function fields, *J. Number Theory* **48** (1994), 133-161.

Silverman, J.

[1] The S-unit equation over function fields, *Math. Proc. Camb. Phil. Soc.* **95** (1984), 3-4.

[2] Wieferich's criterion and the abc conjecture, *J. of Number Theory* **30** (1988), 226-237.

[3] The Arithmetic of Elliptic Curves, GTM 106, Springer Verlag, New York-Berlin-Heifelberg, 1986.

Sinnott, W.

[1] On the Stickelberger ideal and the circular units of a cyclotomic field, *Ann. of Math.* **108** (1978), 107-134.

[2] On the Stickelberger ideal and the circular units of an abelian field, *Invent. Math.* **62** (1980), 181-234.

Stichtenoth, H.

[1] Algebraic Function Fields and Codes, Universitext, Springer Verlag, New York, 1993.

Tate, J.

[1] Fourier analysis in number fields and Hecke's zeta functions, in Algebraic Number Theory, (J.W.S. Cassels and A. Fröhlich, eds.), Academic Press, London etc., 1967.

[2] Les Conjectures de Stark sur les Fonctions L d'Artin en $s = 0$, Birkhäuser, Boston-Basel-Stuttgart, 1984.

[3] Brumer-Stark-Stickelberger, *Séminaire de Théorie des Nombres, Bordeaux*, exposé *no*. 24, 1980-81.

Thakur, D.

[1] Iwasawa theory and cyclotomic function fields, *Contemporary Math.* **174** (1994), 157-165.

[2] On characteristic p zeta functions, *Compositio Math.* **99** (1995), 231-247.

Washington, L.

[1] Introduction to Cyclotomic Fields (2nd edition), Springer-Verlag, New York, 1997.

Weil, A.

[1] Sur les Courbes Algébriques et les Variétés qui s'en Déduisent, Hermann, Paris, 1948.

[2] Variétés Abéliennes et Courbes Algébriques, Hermann, Paris, 1948.

Yin, L.

[1] Index-class number formulas over global function fields, *Compositio Math* **109** (1997), 49-66.

[2] Distributions on a global function field, *J. Number Theory* **80** (2000), 154-167.

[3] Stickelberger ideals and relative class numbers in function fields, *J. Number Theory* **81** (2000), 162-169.

[4] Stickelberger ideals and divisor class numbers, to appear.

Yu, J.

[1] Transcendence in finite characteristic, in: The Arithmetic of Function Fields, (D. Goss et al., eds.), de Gruyter, New York, Berlin, 1992.

[2] On a theorem of Bilharz, to appear.

Author Index

Subject Index

A-character, 221
A-order, 314
ABC-conjecture, 92
adele ring, 66
additive polynomial, 197
arithmetic function, 15-19, 305-313
Artin
    automorphism, 135
    conjecture on primitive roots,
      149, 150
    conjecture on L-series,116, 130
    conductor, 140, 253
    L-series, 126-131
    map, 135
average value,16-19, 305 ff

Bilharz's theorem, 157,158
Brumer element, 267, 270, 272
Brumer-Stark conjecture, 267

Carlitz
    exponential, 236
    module, 200, 202 ff
canonical class, 49, 73

character
    Artin, 127, 128
    Dirichlet, 34
    even, 291
    Hecke, 140
    imaginary, 291
    odd, 291
    real, 291
Chinese remainder theorem, 3
class
    canonical, 49, 73
    group, 48, 50, 242
    number, 50
class number formulas, 295, 299,
      301
    analytic, 254, 291
    Carlitz-Olsen, 288
    Dirichlet, 284
    Kummer, 285
conorm, 82
constant field, 46
constant field extension, 101 ff
cyclotomic
    number field, 194
    function field, 202

Graduate Texts in Mathematics

(continued from page ii)